Topics in Applied Physics   Volume 24

Topics in Applied Physics Founded by Helmut K.V. Lotsch

Acoustic Surface Waves

Edited by A. A. Oliner

With Contributions by
E. A. Ash   G. W. Farnell   H. M. Gerard
A. A. Oliner   A. J. Slobodnik, Jr.   H. I. Smith

With 198 Figures

Springer-Verlag Berlin Heidelberg GmbH 1978

Arthur A. Oliner, Ph. D.

Polytechnic Institute of New York, 333 Jay Street
Brooklyn, NY 11201, USA

ISBN 978-3-662-30917-9   ISBN 978-3-540-35956-2 (eBook)
DOI 10.1007/978-3-540-35956-2

Library of Congress Cataloging in Publication Data. Main entry under title: Acoustic surface waves. (Topics in applied physics; v. 24). 1. Acoustic surface waves. I. Oliner, Arthur A. II. Ash, E. A. QC176.8.A3A3   530.4'1   77-17957

© by Springer-Verlag Berlin Heidelberg 1978

Originally published by Springer-Verlag Berlin Heidelberg New York in 1978.
Softcover reprint of the hardcover 1st edition 1978

Monophoto typesetting, offset printing and bookbinding: Brühlsche Universitätsdruckerei, Lahn-Giessen
2153/3130-543210

Preface

Acoustic surface waves form the basis of an exciting new field of applied physics and engineering, extending to several disciplines as diverse as nondestructive evaluation (NDE), seismology, and signal processing in electronic systems. This field, often referred to as the SAW (Surface Acoustic Wave) field, has developed enormously during the last decade, particulary within the past half-dozen years, with its principal impact on signal processing, with important applications to radar, communications and electronic warfare.

The extremely low velocity, and therefore extremely small wavelength, of these (ultrasonic) acoustic waves permits them to perform functions in a very simple fashion that would be very difficult or cumbersome to accomplish using any other technology. Devices from the VHF to the low microwave frequency range employing these waves are therefore very practical, in addition to offering dramatically small size and weight, combined with ruggedness and reliability. The vitality and strength of this field derive from its interdisciplinary nature, combining the talents of solid mechanicians, solid-state physicists, and microwave engineers. Great strides were therefore made quickly in both the understanding of these acoustic wave types and in the ingenious engineering developments that have followed from this understanding.

This book is concerned with the *fundamentals* of the acoustic surface wave field, with stress on implications for signal processing. The book includes in one place the following four most important basic aspects of this field: the properties of the basic wave types, the principles of operation of the most important devices and structures, the properties of materials which affect device performance, and the ways by which the devices are fabricated. The attempt throughout has been to stress the the fundamentals so that this book is not likely to be outdated soon. Although a variety of books and journal publications have appeared which present certain basic material or contain broad reviews, there is no single published source, to our knowledge, that duplicates the intent or the contents of this book.

I wish to thank my contributors, each an acknowledged expert in his own specialty in this field, for their fine cooperation in the preparation of the manuscript. I am also very grateful to Mrs. Jean B. Maher for her superb typing skill, her perceptiveness in suggesting improvements, and her creative assistance in the preparation of the index.

Brooklyn, NY   *A. A. Oliner*
January, 1978

Contents

Contributors

Ash, Eric A.
   Department of Electronic and Electrical Engineering,
   University College London, Torrington Place,
   London WCIE 7JE, Great Britain

Farnell, Gerald W.
   Department of Electrical Engineering, McGill University,
   Montreal, Quebec H3A, 2K6, Canada

Gerard, Henry M.
   Building 600, M/S C-241, Hughes Aircraft Company,
   Fullerton, CA 92634, USA

Oliner, Arthur A.
   Microwave Research Institute,
   Polytechnic Institute of New York,
   333 Jay Street, Brooklyn, NY 11201, USA

Slobodnik, Andrew J., Jr.
   Electromagnetic Sciences Division, Deputy for Electronic Technology,
   Rome Air Development Center (AFSC),
   Hanscom AFB, MA 01731, USA

Smith, Henry I.
   MIT Lincoln Laboratory, P.O. Box 73,
   Lexington, MA 02173, USA

1. Introduction

A. A. Oliner

With 2 Figures

The field of acoustic surface waves is concerned primarily with the understanding and exploitation of the properties of elastic waves of very high frequency that can be guided along the interface between two media, at least one of them being a solid. The solid medium is usually piezoelectric, so that interactions with electromagnetic fields become possible, and the second medium is usually air or vacuum.

Although much had been known for many years about some of the fundamental properties of acoustic surface waves, this field has developed enormously during the last decade, particularly within the past half-dozen years. Current concerns with this field extend to several disciplines, as diverse as nondestructive evaluation (NDE), seismology, and signal processing in electronic systems. Although the early interest in acoustic surface waves related almost solely to seismological applications, and although such waves are becoming increasingly important for NDE considerations, the principal impact of the acoustic surface wave field today and over the past few years has been on *signal processing*, with important applications to radar, communications, and electronic warfare. It was the recognition that acoustic surface waves could furnish a new approach to signal processing that gave the acoustic wave field its enormous impetus during the past decade.

The acoustic surface wave field is an interdisciplinary one, and it has derived its great drive and strength from the combination of talents which contributed to its recent development. In addition to the solid mechanics people who furnished the foundations for the field and who continue to contribute to it, the recent rapid development was made possible by the influx of solid state physicists and microwave engineers, who interwove their backgrounds and capabilities with those of the solid mechanicians. As a result, great strides have been made in both the understanding of these wave types and in the ingenious engineering developments that have followed from this understanding.

This book is concerned with the *fundamentals* of the acoustic surface wave field. The chapters which follow present these fundamentals in a way that highlights their relation to signal processing applications. The applications themselves are not treated, but the basic principles underlying the most important devices which are essential to those applications are presented in some detail. In addition to these device principles, the fundamental aspects considered in this book include the wave types themselves, basic guiding structures, the

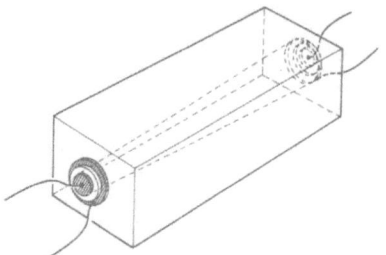

Fig. 1.1. A typical simple bulk-wave delay line. A bulk wave beam propagates within the solid medium between the transducers

properties of materials, and methods for fabricating the required surface wave structures.

The field of acoustic surface waves often uses the designation SAW, standing for "surface acoustic waves". This designation is employed frequently in the remainder of the book.

To tie together the various independent fundamental aspects of the acoustic surface wave field, and to place them in a suitable perspective, this introductory chapter presents first an overview of some features of this field, and then outlines the specific topics treated in each of the following chapters.

1.1 An Overview of the Field

Let us first review why surface acoustic waves (SAW) are of such great interest for signal processing applications. What properties of these waves permit them to be exploited in such a novel (and practical) fashion?

The first, and most important, property is their *extremely low velocity*, about 10^{-5} times that of electromagnetic waves. This property makes acoustic wave structures ideal for long delay lines, a feature which has been recognized for many years in connection with bulk acoustic waves. Because of the low velocity, acoustic waves also possess *extremely small wavelengths*, when compared with electromagnetic waves of the same frequency. The reduction in size is again of the order of 10^{-5}, the precise value depending on the materials used. Acoustic wave devices, when compared with electromagnetic devices, therefore offer *dramatic reductions in size and weight*. In addition, acoustic surface wave devices are fabricated on the surface of a crystal, so that they are also generally more *rugged* and *reliable*.

Early acoustic wave devices employed bulk acoustic waves, as sketched in Fig. 1.1, which represents a typical simple delay line. An incoming electromagnetic wave is first converted into a bulk acoustic wave by a transducer, the acoustic wave traverses the length of the crystal (and the signal is delayed), and the acoustic wave is then transduced back into an outgoing electromagnetic wave. Because the acoustic wave is present in the interior of the crystal, it is difficult to obtain access to the wave in order to modify it or tap into it. This difficulty is overcome by the use of the surface wave structure shown in Fig. 1.2,

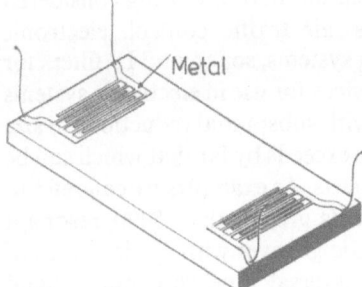

Fig. 1.2. A simple surface wave delay line, which employs interdigital transducers on the surface of a piezoelectric crystal. A Rayleigh surface wave propagates between the transducers

which employs interdigital transducers to excite a Rayleigh surface wave that travels along the surface of the solid and is confined to its vicinity. The clearly *accessible* nature of the surface wave now permits a new order of flexibility which has encouraged the creation of a large variety of novel and effective devices.

The use of surface waves also permits these acoustic wave devices to be compatible with integrated circuit technology and to allow their fabrication by lithographic techniques. Devices using these waves can therefore be mass produced at relatively low cost with precise and reproducible characteristics.

The extremely slow nature of these acoustic surface waves therefore permits a time-varying signal to be completely displayed in space on a crystal surface at a given instant of time. In addition, the lithographic fabrication capability easily permits a complex circuit to be present on the crystal surface. Thus, while the signal progresses from the input end to the output end, one can readily sample the wave or modify it in various ways. As a result, one can perform functions in a very simple fashion that would be very difficult or cumbersome to accomplish using any other technology.

Acoustic surface wave devices can be designed with a center frequency of operation which may lie from the low MHz values up to approximately a GHz, that is, in the VHF or UHF range. Since the size of a circuit element is proportional to the wavelength, the lower frequency limit is governed by the size of available substrates, and the upper limit occurs because of fabrication difficulties.

Despite the short development period, certain aspects of the acoustic surface wave field already correspond to a mature technology. Devices exhibiting exceptional performance have been used to retrofit existing systems (such as pulse compression filters for FM signals used in radar systems), and they are being incorporated into new systems (such as matched filters for phase-coded applications in spread spectrum communications systems). These and other devices such as delay lines, bandpass filters, UHF oscillator control elements, programmable devices for frequency and time domain filtering, frequency synthesizers, correlators, etc., are finding application in or are being considered for radar, spread spectrum communications, air traffic control, electronic warfare, microwave radio relays, data handling systems, sonar, and IF filters for TV use, just to name the major areas. These devices for use in electronic systems not only offer improved reliability, combined with substantial reductions in size and weight, but in many cases their *performance* exceeds by far that which can be achieved by their best electromagnetic counterparts. As examples we can refer to pulse compressors with time-bandwidth products greater than 5000, resonant cavities in the UHF range comprised of periodic grooves with Q's in excess of 50000, and UHF bandpass filters of the transversal type with out-of-band suppression of about 70 dB over the frequency range from DC to 1 GHz.

Despite these impressive accomplishments, many feel that the systematic exploitation of acoustic surface waves is still in its infancy. For one thing, the devices developed so far are not customarily integrated with one another, or with electronic integrated circuit components, on the same substrate surface. Many writers have proposed schemes which could readily benefit from such integration, but the systems needs are not yet apparent. A more widespread utilization of these devices would also occur if the costs of materials and fabrication technology would be significantly reduced. The range of applications for acoustic surface wave devices has so far been limited largely to high-technology needs, where price is secondary to performance; the IF filters for TV use is a notable exception. If mundane commercial applications like the latter become more widespread, costs for substrates should go down, and the popularity of such devices should increase.

In addition, there are other areas of application or potential application which are only embryonic. Examples of such areas, which could well become important, are imaging and nondestructive evaluation (NDE). With respect to imaging, it has been shown that acoustic surface waves can be used to scan an optical image, and to scan and focus an acoustic image. Much more needs to be done before these techniques become practical or competitive with other imaging techniques, but there are implications for medical electronics and for the nondestructive evaluation of materials. With respect to NDE, until relatively recently, simple techniques of testing and interpretation were deemed adequate for most requirements. The recent demands for more quantitative characterizations, which have resulted in the change in terminology from NDT (nondestructive testing) to NDE, have imposed the need for more sophisticated approaches, which the acoustic surface wave field can contribute in the form of better transducers, better theoretical descriptions of wave scattering from defects, novel imaging approaches, and new adaptations of signal processing techniques.

The discussion above has indicated that the acoustic surface wave field has already achieved significant, even spectacular, success in the area of signal processing devices, particularly in connection with high-technology applications, and that there is still much more room for growth, both for application to electronic systems and to new areas of application.

1.2 The Organization of This Book

Underlying all of the applicational developments discussed above are a number of *fundamental* features, ranging from the theoretical understanding of the basic wave types to the essential materials and fabrication technology. This book attempts to include in one place the most significant of these fundamental features, together with the principles of operation of the most important devices and structures in the surface acoustic wave signal-processing field.

Basic to the understanding and design of devices is a knowledge of the properties of the *basic waves* which can propagate in solid volumes and be guided along surfaces and by plates. The fundamental features of such waves are presented in Chapter 2, entitled "Types and Properties of Surface Waves". Although the emphasis is on surface waves, the bulk waves which can propagate in solid media are treated first, before the surface boundary conditions are introduced. These bulk waves are important in their own right, and they come under consideration in later chapters with respect to surface wave device performance, usually because they introduce undesired second-order effects.

Even though most devices employ anisotropic substrates, the properties of the wave types on isotropic substrates are considered in some detail in Chapter 2 to describe the principal behavioral features. Anisotropic substrates, with their added complexity, are treated next, because only with anisotropic substrates can the device designer obtain high piezoelectric coupling and low attenuation at the higher frequencies. The complications relating to the choice of crystal cut and the direction of the propagation of the wave are considered in detail. Power flow in anisotropic media, where the energy-flow and phase-progression directions are usually different, is also examined; this effect leads to beam steering, a property which is also considered in Chapter 6 in connection with specific materials.

Piezoelectric effects are also analyzed in Chapter 2, since the electric fields accompanying the waves represent the way in which the mechanical circuits couple to external electrical circuits or to semiconducting materials. The presence of piezoelectricity leads to stiffened Rayleigh waves, and to an interesting surface wave type, the Bleustein-Gulyaev wave, which can be viewed as a modification of an SH bulk shear wave and which could not exist as a surface wave in the absence of piezoelectricity.

Many devices employ a thin layer of one material placed on a substrate of another material. The presence of the added thin layer not only modifies the properties of the Rayleigh wave which can exist in its absence, but also permits the propagation of another wave type, the Love wave, with polarization different from that of the Rayleigh wave (where polarization is defined by the direction of particle displacement produced by the wave). Chapter 2 also contains discussions on reflections from discontinuities such as corners and steps, and on diffraction effects for a beam of finite width, which results in beam spreading.

The most fundamental component in acoustic surface wave technology is the *interdigital* (ID) *transducer*. Its key importance lies in its dual role as the most common method for transducing electrical input signals into elastic surface waves and as a building block for many surface wave devices in view of its inherent versatility. The first portion of Chapter 3 treats the ID transducer in substantial detail. The discussion there indicates how the ID transducer works, and presents an equivalent circuit model which yields physical insight and which has been found very useful for design purposes. The approach takes into account the three independent elements of the transducer and the interactions among them: the metal electrodes, the properties of the substrate, and the electromagnetic tuning circuits. Later in Chapter 3 it is shown that certain important second-order effects can degrade the performance of devices employing ID transducers, and that modifications in the transducer geometry, such as double electrodes (sometimes called split fingers) and dummy electrodes associated with apodization, can overcome these deleterious second-order effects.

The major objective of Chapter 3 is announced in its title: "Principles of Surface Wave Filter Design". Early in the chapter it is pointed out that the *surface wave filter* is basically an idealized δ-function implementation of a transversal filter. It is shown clearly, however, that the actual filter, which uses the ID transducer as a basic element, is far more complicated, involving elastic, piezoelectric, and electromagnetic variables, since it is incorporated into an electronic system. The chapter reviews the basic principles of transversal filter theory, which are implemented by the repeated delaying and sampling of an input signal. In this context, the discussion explains the importance of amplitude weighting and phase weighting of the required taps, and presents ways of achieving these weightings when ID transducers are employed.

Surface wave filters constitute one of the most important classes of surface wave devices. The technology of surface wave filters has become practical, indeed outstanding, because it has been possible to characterize and control the various second-order effects which cause the actual device to deviate from the ideal one. In practice, however, it is not possible to compensate for all second-order effects simultaneously and, as a result, trade-offs relating to various performance characteristics become necessary. The discussion in Chapter 3 not only presents the principles of operation of these surface wave filters and describes the most important effects which cause deviations from the ideal, but it also shows the engineering steps required to overcome the main difficulties and indicates in some detail what trade-offs must be involved and how they influence the various parameters in filter performance. At the end of the chapter, the process of designing a surface wave filter is reviewed, two examples of filter performance are presented, and two mathematical appendices are included which elaborate on the basic filter principles.

As mentioned earlier, this book is designed to present the fundamentals of surface acoustic waves in a way that highlights their relation to *signal processing* applications, even though much of the material is also of interest to other aspects of the acoustic wave field. Chapter 4, entitled "Fundamentals of Signal Processing Devices", is concerned directly with signal processing, of course, but it stresses the *principles* underlying the most important devices, rather than signal-processing technology or device design details. The basic ID transducer and the surface wave filter analyzed in Chapter 3 are also devices used in signal processing, but the treatment of those devices is presented in a special way to illustrate the engineering designs and trade-offs which arise there. Chapter 4 considers various additional, but also basic, devices and, although the treatment is more descriptive, it nevertheless exposes the key considerations which affect the design and range of applicability of the devices.

The first class of devices considered in Chapter 4 is that of *delay lines*, possibly performing the simplest function and the one recognized earliest as one for which acoustic waves are eminently qualified. The discussion examines the major problems involving the performance of such delay lines, including insertion loss, the suppression of spurious signals (such as "triple transit echo"), bandwidth, and temperature stability. In addition, some special structures are described for use as long delay lines. The next device class explored is that of surface wave *resonators*, which form the surface wave counterpart of the well-known bulk wave resonators (the "quartz crystals") which are used as frequency control elements and as circuit elements for narrow band filters. The surface wave resonators employ periodic arrays of grooves or strips, operated in their stop bands, as reflecting elements, and are found to yield extremely high values of Q, in excess of 50000, with careful design. These grating resonators can also be used as the frequency control element to form a stable signal source. If we arrange a recirculating path for the acoustic energy (or effect the feedback electrically), and include an amplifier in the feedback path to overcome the loop loss, we obtain a surface wave *oscillator*, which is the device treated next in Chapter 4. Such a stable signal source for the UHF range offers much promise; the discussion covers such considerations as stability, mode control, tuning, etc.

The next group of devices described in Chapter 4 is classified as coded "time domain" structures, and they include such highly important devices as the *pulse compression filter* for chirp radar (perhaps the most important single application so far) and *phase-coded* devices for spread-spectrum communications. These devices are composed largely of appropriate modifications in the basic ID transducer, as is the frequency filter treated in Chapter 3. Indeed, much of the discussion there, and the synthesis procedures, are also applicable to these devices. However, in contrast to the use of transfer functions, which express behavior in the frequency domain, these devices are more naturally described in terms of the impulse response, i.e., in the time domain, and this feature forms the basis for the name "time domain" filters chosen in Chaper 4. Considerations relating to pulse compression filters, such as time-bandwidth product, are explored in detail, and alternative geometries, such as the RAC and the RDA, are discussed. Of phase-coded transducers, the binary phase shift keyed (PSK) modulation is stressed; its advantages, and various problems relating to optimal design, are considered, including possibilities for programmable devices, which would permit rapid changing of the code used. In addition, generalizations of linear chirp filters are explored, with the Fourier transformer and variable time delay elements as examples.

Nonlinear signal processing devices are considered next in Chapter 4; nonlinear interactions have led to the development of new classes of device because the nonlinearity of the acoustic medium permits analog signals to be multiplied together instead of only added, as in the linear devices considered up to now. The important convolver and correlator devices, and their implementations using either piezoelectric nonlinearity or the nonlinearity in contiguous semiconductors, are examined in detail. The correlator can be a highly valuable device in a radar system since it permits the radar system to use as a reference the return received from a nearby large target instead of a perfect sample of the transmitted pulse. This return signal can then be correlated with that from the target of interest, much further away; as a result, the performance of the radar is no longer degraded by distortions introduced in the transmitter. A new and exciting development, the storage correlator, which uses an array of discrete diodes on a silicon layer, is also described at some length. The last section in Chapter 4 treats multiport acoustic devices, and considers primarily the *multistrip coupler* (MSC) and its various modifications. The MSC is a versatile device whose incorporation in Chapter 3 in a band-pass filter permits the use of two apodized transducers. Other functions performed by the MSC include bulk wave suppression, track changing, power division, beam compression, multiplexing, etc. The discussion in Chapter 4 indicates why the MSC works and how these functions are accomplished.

Almost all of the acoustic surface wave devices described in Chapters 3 and 4 utilize wide-beam surface waves, despite certain limitations possessed by these beams: beam spreading, inefficient use of the substrate area, and awkwardness in bending their paths. These limitations are all overcome by the use of *waveguides* for surface waves, where the term "waveguide" implies a geometrical structure which confines the lateral extent of the surface wave and binds the wave to itself. The types and the properties of such waveguides are described in Chapter 5, entitled "Waveguides for Surface Waves".

Waveguides are being considered seriously for certain applications, such as long delay lines, as described in Section 4.2; for most other device needs, waveguides are not essential but, with ingenuity, they could be useful in improving device performance. The most intriguing potential application for waveguides is that of a highly-compact sophisticated circuit technology, sometimes referred to as "microsound". So far, the acoustic wave device field has not moved in the direction of circuitry which performs several functions simultaneously; instead, SAW devices generally (but not always) perform a single function, and consist of a simple circuit placed between input and output transducers. When the philosophy changes, and when the performance of *multiple functions on the same substrate* using wide surface-wave beams is seen to result in cross talk between neighboring beams and to require excessive substrate area, waveguides will be logically called upon. Thus, although the applications for waveguides at present are few, the potential for future use could be very great.

Four types of waveguide are described in Chapter 5: 1) overlay waveguides, in which a strip of one material is placed on the substrate of another material; 2) topographic waveguides, which consist of a local deformation of the substrate surface itself; 3) waveguides in which a local change has been produced in the properties of the substrate material; and 4) circular fiber waveguides. Each of the waveguides possesses its own characteristic properties and therefore has a different range of potential applications. Of these properties, the two most important are the acoustic field confinement and the dispersion behavior.

The three types of overlay waveguide are the strip, the slot, and the shorting strip, or $\Delta v/v$, guide. These structures are easy to fabricate and possess mild field confinement with some dispersion. The most important topographic waveguides are the rectangular ridge and the wedge structures. The symmetric mode on the rectangular ridge is essentially dispersionless over a wide frequency range and, for tall ridges, possesses strong field confinement; it should therefore be of substantial practical interest. The wedge waveguide is also essentially dispersionless above a certain minimum frequency, but the fragility of the wedge tip restricts its usefulness to low frequencies. The in-diffused waveguides of the strip type have the potential of low loss, and the circular fiber structures show promise for long delay line applications. Chapter 5 describes the properties of all of these waveguides in detail, and at the end presents a comparative summary of their properties and discusses their actual and potential applications.

As summarized above, Chapters 2 through 5 consider the basic wave types and the most important devices and structures for SAW applications. The remaining two chapters of this book are devoted to the materials and fabrication technology basic to the devices and structures described in Chapters 3–5. Chapter 6, entitled "Materials and their Influence on Performance", examines in detail the principal *properties of materials* which must be taken into consideration and shows how they influence the properties of devices. As was indicated in all of the previous chapters, a judicious choice of the substrate material can result in a significant improvement in device performance. The principal properties of the substrate material which must influence that choice are the wave velocity, the electromechanical coupling constant, the temperature coefficient of delay, the propagation loss, beam steering due to anisotropy, and beam spreading due to diffraction.

Each of these properties is discussed quantitatively in Chapter 6 in a thorough fashion. Some of the results presented were obtained with the help of a laser probe, which is also described in the chapter. Special note should be made of some of the material included. In the discussion on material losses, the effects of surface condition and surface quality are considered, and the contribution to the loss of air or gas loading is incorporated, including its frequency dependence. The diffraction and beam steering treatments are quite extensive, including a review of the various theories and a discussion of the limitations of the parabolic theory, detailed examination of the beam steering vs. diffraction trade-offs, and a description of the minimal diffraction orientations found for $Bi_{12}GeO_{20}$ (which is also discussed briefly in Sect. 2.9). A description of nonlinear effects, in part obtained by the use of the laser probe, is also presented; included are considerations affecting the fundamental frequency, the extent of harmonic generation, and mixing effects.

Although data are presented for the most important current materials, the stress is on the *general principles* underlying the dependence of these performance properties on basic material parameters, since, as time passes, improved materials may well be found. The materials discussed in detail are mostly $LiNbO_3$, which is the most popular material employed in SAW applications, quartz, and $Bi_{12}GeO_{20}$; comments are also included relative to Berlinite, a new material which offers low temperature sensitivity combined with a high coupling constant. Many design curves are presented and practical considerations are emphasized, but always related to basic principles.

The last chapter, Chapter 7, entitled "Fabrication Techniques for Surface Wave Devices", describes the *fabrication procedures* which make the SAW devices possible. These procedures are all variants of the planar fabrication technique which begins with coating the substrate surface with a radiation-sensitive polymer film. The basic procedure involves two stages: a) a specified pattern is formed in the polymer film placed on the substrate surface, and b) this pattern is transferred on to the substrate itself.

The first stage consists itself of two steps: exposure of the polymer film to radiation by the use of a mask or other technique, and development to remove either the exposed or unexposed polymer material to form the pattern. The exposure of the pattern is the most critical step with respect to the quality of the ultimate circuit. The radiation employed may be optical, electron-beam, or x-ray, and the techniques which are described in Chapter 7 for exposing the patterns include several photolithographic methods, electron-beam lithography, and x-ray lithography. Stress is placed on optical conformable-photomask contact printing because it offers a good compromise between high resolution and economy, and on x-ray lithography because of its simplicity and ultimate resolution.

After the polymer film has been exposed and a relief pattern developed, the next step (the second stage) is to transfer this pattern onto the substrate itself. The techniques described in Chapter 7 for this purpose include chemical etching, plasma etching, rf sputter etching, ion bombardment etching, chemical doping by diffusion or ion implantation, and the liftoff technique, which involves first depositing a material on the polymer pattern and then dissolving the polymer. The advantages and disadvantages of the various techniques are summarized, sometimes in tabular form.

When Chapters 2 through 7 are taken as a group, we see that they include the four most important *basic aspects* of the acoustic surface wave signal-processing field. These chapters cover the basic wave types, the most important devices and structures from the signal-processing standpoint, the properties of materials which affect device performance, and the ways by which the devices are fabricated. The attempt throughout has been to stress the fundamentals so that this book will not likely be outdated soon.

Although to our knowledge there is no single published source that duplicates the intent or the contents of this book, a variety of books and journal publications have appeared which present certain fundamental material or contain broad reviews. Selected journal publications which should be noted include special issues devoted to acoustic surface wave properties and devices [1.1–3], and broad review articles [1.4–6]. Certain proceedings volumes are of particular value [1.7, 8]. Many books have been written over the years on topics which relate in one way or another to acoustic surface waves; of these, some are of more direct interest [1.9–12], including pertinent chapters of a series of volumes [1.12], and some offer general relevant background [1.13–15].

References

1.1   R.M.White: Proc. IEEE **58**, 1238—1276 (1970)
1.2   J.D.Maines, E.G.S.Paige: IEE Rev. **120**, 1078—1110 (1973)
1.3   M.G.Holland, L.T.Claiborne: Proc. IEEE **62**, No. 5, 582—611 (1974)
1.4   *Special Issue on Microwave Acoustics*, issued jointly by IEEE Trans. MTT-**27**, No. 11,(1969); IEEE Trans. SU-**16** (1969)
1.5   *Special Issue on Microwave Acoustic Signal Processing*, issued jointly by IEEE Trans. MTT-**21**, No. 4, (1973); IEEE Trans SU-**20** (1973)
1.6   *Special Issue on Surface Acoustic Wave Devices and Applications*, Proc. IEEE **64**, No. 5 (1976)
1.7   *Proceedings of Symposium on Optical and Acoustical Micro-Electronics*(Polytechnic Institute of New York, N. Y. 1974)
1.8   Annual Proceedings of the IEEE Ultrasonics Symposia
1.9   E.Dieulesaint, D.Royer: *Ondes Elastiques dans les Solides: Applications au Traitement du Signal* (Masson et Cie, Paris 1973)
1.10 B.A.Auld: *Acoustic Fields and Waves in Solids*, Vols. I and II (Wiley-Interscience, New York 1973)
1.11 H.Matthews (ed.): *Surface Wave Filters* (Wiley-Interscience, New York 1977)
1.12 W.P.Mason, R.N.Thurston (eds.): *Physical Acoustics*, a series of volumes (Academic Press, New York, London)
1.13 W.M.Ewing, W.S.Jardetsky, F.Press: *Elastic Waves in Layered Media* (McGraw-Hill, New York 1957)
1.14 I.A.Viktorov: *Rayleigh and Lamb Waves*, Translated from the Russian (Plenum Press, New York 1967)
1.15 M.Redwood: *Mechanical Waveguides* (Pergamon Press, New York, London 1960)

2. Types and Properties of Surface Waves

G. W. Farnell

With 26 Figures

The purpose of this chapter is to describe the most important types of elastic waves from the standpoint of the surface acoustic wave (SAW) field, and to present their principal features in appropriate mathematical and graphical form. The chapter begins with basic definitions relating to elastic waves in the context of the wave equation, and discusses the properties of bulk waves as a preliminary to the various surface waves which are of primary interest. The Rayleigh wave on an isotropic substrate and plate modes on an isotropic plate are examined next in order to present the main features of such waves before taking into account the added complexities due to anisotropy and piezoelectrivity.

Since most SAW devices require piezoelectric substrates, which possess anisotropy, the new aspects introduced by anisotropy and piezoelectricity are treated in detail. However, to emphasize the mechanical characteristics, the discussion is restricted first to nonpiezoelectric anisotropic substrates; later, the piezoelectric coupling is reintroduced and the electric boundary conditions considered. The material on anisotropy includes the influence of different wave directions and of different crystal cuts for the substrate face, and the phenomenon of beam steering which arises when one examines the power flow and sees that the directions of phase progression and energy propagation are not the same. The effects of piezoelectricity, which must be considered since it is the mechanism by which the mechanical waves couple to the external electrical circuit, produce stiffened Rayleigh waves and the Bleustein-Gulyaev wave, a shear surface wave which cannot exist in the absence of piezoelectricity.

Propagation in thin layers located on a substrate surface is treated next; it is shown that the properties of the Rayleigh wave are modified and that Love waves, a wave type resembling SH bulk waves but confined to the vicinity of the layer, now become possible. In the last two sections of the chapter certain finiteness effects are examined. One of these involves the reflections produced when certain types of surface discontinuities are present, such as right-angle corners and steps. The other relates to diffraction effects, such as beam spreading, which occur when a surface wave beam is excited by a transducer of finite width.

2.1 The Wave Equations

The displacement components of a point, or more rigorously an infinitesimal volume, of the homogeneous solid medium under consideration are taken to be u_i along the corresponding Cartesian axes x_i. The forces exerted on an infinitesimal volume by contiguous volumes are expressed as a symmetrical stress tensor [2.1, 2] with components T_{ij} where T_{ij} is the component in the x_i direction of the force per unit area acting on a surface normal to the x_j direction. Thus by Newton's third law, if there are no body forces exerted, the equation of motion of each infinitesimal volume is given by [2.3]

$$\varrho \frac{\partial^2 u_i}{\partial t^2} = \frac{\partial T_{ij}}{\partial x_j} \tag{2.1}$$

wherein ϱ is the mass density and the summation convention [2.2] has been introduced in that a repeated subscript in a term implies that the term is summed over the values of the repeated index, for example,

$$\partial T_{ij}/\partial x_j = \frac{\partial T_{i1}}{\partial x_1} + \frac{\partial T_{i2}}{\partial x_2} + \frac{\partial T_{i3}}{\partial x_3}. \tag{2.2}$$

Throughout this chapter the solids are assumed to be elastic with the stress as a linear function of the strain, proportional in the absence of piezoelectricity. However, since piezoelectricity plays a role in most surface wave components, it is useful to introduce it into the constitutive equations at the beginning. Because the disturbances of interest in these devices propagate with sonic rather than electromagnetic velocities, the electric fields involved can be represented as the gradient of a time-varying scalar potential φ [2.4]. Thus the properties of the material are introduced in the coupled constitutive equations, expressed here in terms of the particle displacements [2.2, 3]

$$T_{ij} = c_{ijkl} \partial u_k/\partial x_l + e_{kij} \partial \varphi/\partial x_k$$
$$D_i = -\varepsilon_{ij} \partial \varphi/\partial x_j + e_{ijk} \partial u_j/\partial x_k \quad (i, j, k, l = 1, 2, 3). \tag{2.3}$$

Note that the first terms of the stress equations are of the form of Hooke's law, while the first terms of the electric displacement equations are of the form of the familiar relation between electric displacement and electric field, $D = \varepsilon E$. But the two sets take into account the vector or tensor nature of the variables, the anisotropic nature of the material properties, and the piezoelectric coupling between the electrical and mechanical variables. The latter coupling can often be considered as a perturbation, but it is convenient here to include it explicitly. In (2.3) the elastic behavior is dominated by the elastic constant tensor c_{ijkl} measured at constant electric field in which not all of the components are independent [2.2] because $c_{ijkl} = c_{jikl} = c_{ijlk} = c_{klij}$. Moreover, crystal symmetry further restricts the number of independent elements; for example, an isotropic material has two, a cubic material has three, and a trigonal material such as quartz has six independent elements. The electrical behavior of the solid is dominated by the symmetric second-rank tensor ε_{ij} representing the permittivity measured at constant strain. While there are six possible independent elements, crystal symmetry again restricts the number in specific cases, to one for example in isotropic or cubic materials and to two in a trigonal crystal. The other material constants in (2.3) are the elements of the piezoelectric tensor e_{kij} which intercouple the individual equations of the set [2.3]. Here again the requirement $e_{ijk} = e_{ikj}$ and the crystalline symmetry reduce the number of independent elements.

Substituting (2.3) into (2.1) and into the $\nabla \cdot \boldsymbol{D} = 0$ of a charge-free dielectric gives [2.5]

$$\varrho \frac{\partial^2 u_j}{\partial t^2} - c_{ijkl} \frac{\partial^2 u_k}{\partial x_i \partial x_l} = e_{kij} \frac{\partial^2 \varphi}{\partial x_k \partial x_i} \tag{2.4a}$$

$$\varepsilon_{ik} \frac{\partial^2 \varphi}{\partial x_i \partial x_k} = e_{ikl} \frac{\partial^2 u_k}{\partial x_i \partial x_l}. \tag{2.4b}$$

The left-hand sides of the first three equations ($j = 1, 2, 3$) are basically wave equations for the components of the particle displacement with the individual components intercoupled by the elements of the elastic constant tensor. The fourth equation is basically of the form of an anisotropic Laplace equation, for the potential; in addition, for a piezoelectric solid these four equations are intercoupled in general by the elements of the piezoelectric tensor as indicated by the terms on the right-hand side of (2.4).

While surface waves are of principal concern here, it is useful to digress at this point before the surface boundary conditions are introduced, to consider briefly elastic wave propagation in an unbounded solid. These waves will appear under the name of bulk waves in later chapters, there usually as undesired second-order effects. Consider a plane wave represented by the real part of

$$u_i = \alpha_i \exp[ik(b_i x_i - vt)]$$
$$\varphi = \alpha_4 \exp[ik(b_i x_i - vt)] \tag{2.5}$$

which propagates with a phase velocity v and wave number $k = \omega/v$ along a direction with direction cosines b_i with respect to the Cartesian axes x_i. Since the wave is "plane", there are no variations of particle displacement or potential in directions perpendicular to the direction of propagation. Note that the potential and the associated electric field are not electromagnetic in nature but are a component part of the predominantly mechanical wave propagating with velocity v.

If the direction of the propagation vector $\boldsymbol{k}$ is known, fixed by a transducer, say, then (2.4b) allows the potential component α_4 of (2.5) to be expressed in terms of the mechanical displacements [2.6]

$$\alpha_4 = \alpha_j (e_{ijk} b_i b_k)/(\varepsilon_{pq} b_p b_q) \tag{2.6}$$

and thus it may be eliminated from (2.4a), which becomes

$$\varrho v^2 \alpha_j = c'_{ijkl} b_i b_l \alpha_k \tag{2.7}$$

where

$$c'_{ijkl} = c_{ijkl}(1 + K^2_{ijkl})$$

and

$$K^2_{ijkl} \equiv \frac{e_{mij} e_{nkl} b_m b_n}{c_{ijkl}(\varepsilon_{pq} b_p b_q)}.$$

From these equations it is seen that the effect of piezoelectricity is to increase the effective elastic constants by the factors $(1 + K^2_{ijkl})$, and thus the piezoelectricity is said to "stiffen" the solid. For even the strongest piezoelectrics, the factors K^2_{ijkl} are less than 5% and therefore the stiffened effective elastic constants differ at most by a few percent from the actual values. K^2_{ijkl} is called the electromechanical coupling constant tensor [2.7] and, for a plane wave of given polarization in a given direction, there is a single coupling constant, a constant of great importance in discussing the efficiency of transducers for exciting the given wave.

Equation (2.7) for the mechanical displacements can be written

$$(\Gamma_{jk} - \delta_{jk} \varrho v^2) \alpha_j = 0 \tag{2.8}$$

with $\Gamma_{jk} = b_i b_l c'_{ijkl}$. The three eigenvalues of this homogeneous set give the three phase velocities $v^{(n)}$ and, with the three corresponding eigenvectors $\alpha_j^{(n)}$, represent the three plane wave solutions for the assumed direction of propagation. The unit polarization vectors of these three plane waves are mutually orthogonal, but in general each propagates with a different frequency-independent velocity [2.7]. One wave has its polarization vector more-or-less parallel to the chosen direction of propagation and thus it is called a quasi-longitudinal wave, while the other two have polarizations almost transverse to the direction of propagation and are called quasi-transverse or quasi-shear waves. The velocities of the quasi-shear waves range from about 1000 to 6000 m s^{-1}, depending predominantly on the solid in question but to some extent on the direction of propagation. The quasi-longitudinal waves usually have phase velocities two or three times greater than the corresponding transverse waves. Since in general the velocity of a given wave depends on its direction of propagation in the crystal, the group velocity is not in general parallel to the phase velocity, an effect discussed further below in connection with surface waves. For the bulk waves there are certain directions in which crystalline symmetry requires the modes to be purely transverse and purely longitudinal and the group velocity to be parallel to the phase velocity for each mode. Such directions can be called "pure-mode" directions. Obviously in an isotropic solid any direction is a pure-mode direction. To aid later discussions it is useful to note here that the phase velocity is of the form $\sqrt{c'/\varrho}$, where c' is a linear combination, appropriate to the mode, of the stiffened elastic constants,

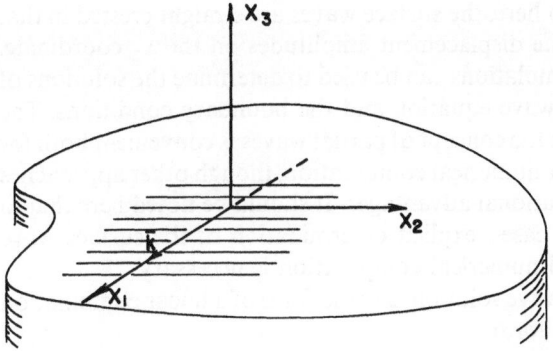

Fig. 2.1. Coordinate system for surface waves showing propagation vector

and thus, from (2.7), the piezoelectricity produces a fractional increase in the phase velocity equal to $K^2/2$ from the value of this velocity in the absence of piezoelectricity, where K^2 is an appropriate electromechanical coupling factor.

2.2 Surface Wave Characteristics

The prototype geometry for a discussion of elastic surface waves shown in Fig. 2.1 has a mechanically traction-free surface of infinite extent separating the infinitely deep solid substrate from the free space above. If the x_3 axis is the outward normal to the surface and the origin of the Cartesian coordinate system is at the surface, the traction-free condition is expressed as [2.5]

$$T_{31} = T_{32} = T_{33} = 0 \quad \text{at} \quad x_3 = 0. \tag{2.9}$$

Thus we are seeking solutions of the coupled wave Eqs. (2.4) which satisfy the mechanical boundary conditions (2.9) and, if the solid is piezoelectric, appropriate electric boundary conditions to be considered below. The fundamental solutions of interest here are surface waves, waves which propagate parallel to the surface with a phase velocity v and whose displacement and potential amplitudes decay rapidly in terms of the wavelength with distance away from the surface. Without loss of generality the direction of propagation can be taken as the x_1 axis, thus defining the (x_1, x_3) plane as the sagittal-plane.

If the propagation geometry is expressed in these Cartesian axes, then the material property tensors must be expressed in the same axes. There are rotation formulae which express the components of the new tensor in terms of the components of the old and the elements of the rotation matrix between the chosen coordinate system and the crystal axes. While it is not necessary to consider the details here, the symmetry properties of the various tensors allow them to be expressed in matrix (two subscript) form [2.2] and thus to be tabulated more conveniently. There are transformation matrices which allow these matrix representations to be rotated also from the crystal axes to the geometrical axes [2.7].

In the prototype problem here, the surface waves are straight crested in that there is no dependence of the displacement amplitudes on the x_2 coordinate. Various approaches and formulations can be used to determine the solutions of this form which satisfy the wave equation and the boundary conditions. The following approach based on the concept of partial waves is convenient both for explanatory purposes and for numerical computation, though other approaches may have particular computational advantages. It should be noted here that in all but the most degenerate cases, explicit determination of the surface wave solutions is not possible and numerical computation is necessary.

Assume that the surface wave solution is in the form of a linear combination of partial waves each of the form

$$u_j = \alpha_j \exp(ikbx_3) \exp[ik(x_1 - vt)]$$
$$\varphi = \alpha_4 \exp(ikbx_3) \exp[ik(x_1 - vt)] \quad x_3 \leqq 0 \tag{2.10}$$

which propagate along x_1 with a phase velocity v and decay in depth with a decay constant kb, where the imaginary part of b must be negative so that the amplitudes vanish for $x_3 \rightarrow -\infty$, a necessary condition for a surface wave. For convenience of expression in the following equations the partial wave is written using the summation convention as

$$u_j = \alpha_j \exp[ik(b_i x_i - vt)]$$
$$\varphi = \alpha_4 \exp[ik(b_i x_i - vt)] \tag{2.11}$$

in which $b_1 \equiv 1$, $b_2 \equiv 0$, and $b_3 \equiv b$.

Now, each partial wave must satisfy the wave equation, and thus substituting (2.11) into (2.4) gives the homogeneous set [2.6]

$$(\Gamma_{ab} - \delta'_{ab}\varrho v^2)\alpha_a = 0 \begin{cases} a, b = 1, 2, 3, 4 \\ \delta'_{44} = 0 \end{cases} \tag{2.12}$$

analogous to (2.8) of the bulk-wave case with $\Gamma_{jk} = \Gamma_{kj} = b_i b_l c_{ijkl}$, $\Gamma_{j4} = \Gamma_{4j} = e_{ikj}b_i b_k$, and $\Gamma_{44} = -\varepsilon_{ik}b_i b_k$. It has been assumed here that each of the material property tensors has been expressed in the axes of Fig. 2.1.

There is appreciable difference between the apparently similar (2.8) and (2.12) because in the latter the value of b_3 is not known a priori. Thus the Γ's are not explicitly specified; moreover, the concept of stiffened elastic constants cannot be applied directly to these equations.

In many practical cases, the crystalline axes are selected in such a manner that (2.12) separates into independent parts, the effects of which are discussed below. However, for the moment we will regard (2.12) as an algebraic equation of eighth order in b with real coefficients for a specified value of v. The roots $b^{(n)}$ of this equation are real or occur in complex pairs, but the roots on the real axis and in the upper half of the complex plane lead to partial waves which do not vanish for

$x_3 \rightarrow -\infty$ and thus are disregarded as parts of a surface wave solution. As a result, there are in general four partial waves satisfying the wave equation and vanishing at large depths, one wave for each of the lower-half plane roots. All four propagate with the same assumed value of phase velocity; a linear combination of the partial waves is formed and an attempt made to satisfy the boundary conditions at the surface by selection of the weighting coefficients of the linear combination. For an arbitrary choice of phase velocity it is not possible to choose a set of weighting coefficients to satisfy the boundary conditions; however, for one particular choice of v the boundary conditions can be satisfied, and this v is the phase velocity of the surface wave. Because of algebraic complexity, numerical techniques, usually involving some form of iterative search procedure, must be used to determine the actual phase velocity and the corresponding particle displacements and potential.

It has been noted previously that piezoelectric effects are little more than perturbations on the predominantly mechanical nature of the waves under consideration. To emphasize the mechanical characteristics, the discussion will be restricted first to nonpiezoelectric substrates; later in Section 2.6, the piezoelectric coupling in the wave equation will be reintroduced and the electric boundary conditions considered.

In the geometry of Fig. 2.1, and with a nonpiezoelectric substrate, the boundary conditions from (2.9) are

$$T_{3j} = c_{3jkl} \frac{\partial u_k}{\partial x_l} = 0 \quad \text{at} \quad x_3 = 0 \tag{2.13}$$

where the particle displacements are linear combinations of the three partial waves

$$u_j = \sum_{n=1}^{3} C_n \alpha_j^{(n)} \exp[ik(b_i^{(n)} x_i - vt)] \tag{2.14}$$

where $b_1^{(n)} = 1$, $b_2^{(n)} = 0$, and $b_3^{(n)} = b^{(n)}$ and the $b^{(n)}$ are the three lower-half plane roots of

$$\det|\Gamma_{ij} - \delta_{ij} \varrho v^2| = 0 \tag{2.15}$$

considered as an algebraic equation in b for a given value of v, and the $\alpha_i^{(n)}$ are the corresponding eigenvectors or partial wave component amplitudes.

Substituting the assumed solution (2.14) into the boundary conditions (2.13) gives [2.6]

$$d_{jn} C_n = 0 \quad (j, n = 1, 2, 3) \tag{2.16}$$

with

$$d_{jn} = c_{3jkl} \alpha_k^{(n)} b_l^{(n)}.$$

Thus the condition for a nontrivial solution is

$$\det |d_{mn}| = 0,$$

which can be considered as an implicit equation in v. Once the value of v satisfying it has been found, the $b^{(n)}$ are then known and the weighting coefficients can be determined from (2.16), with the result that the parameters of the surface wave of (2.14) now satisfying both the wave equation and the boundary conditions are all known.

2.2.1 Isotropic Substrates

In ultrasonic applications the substrates are usually crystalline and thus elastically anisotropic; however, the principal characteristics of surface waves can be illustrated by considering an isotropic substrate. There are two independent elastic constants for an isotropic solid, called c_{11} and c_{44}, and the symmetry restrictions on the elastic constant tensor lead to the following expressions for the Γ in the secular Eq. (2.2), or in (2.15)

$$\begin{aligned}
\Gamma_{11} &= c_{11} + c_{44} b^2 & \Gamma_{12} &= 0 & \Gamma_{13} &= (c_{11} - c_{44}) b \\
\Gamma_{21} &= 0 & \Gamma_{22} &= c_{44}(1 + b^2) & \Gamma_{23} &= 0 \\
\Gamma_{31} &= (c_{11} - c_{44}) b & \Gamma_{32} &= 0 & \Gamma_{33} &= c_{44} + c_{11} b^2.
\end{aligned} \tag{2.17}$$

It should be noted first that, if we return for a moment to the case of bulk waves in an isotropic infinite medium, the same Γ's apply in (2.8) with $b=0$ for any direction of propagation, with the result that the bulk-wave velocities are

$$v_l = \sqrt{c_{11}/\varrho} \quad \text{and} \quad v_t = \sqrt{c_{44}/\varrho} \tag{2.18}$$

for the longitudinal wave and for the two degenerate shear waves, respectively.

Now in the half-space problem again with the Γ's of (2.17) in the secular equation there is a separation into two independent parts because $\Gamma_{12} = \Gamma_{23} = 0$. For one part the eigenvector has only the component perpendicular to the sagittal plane, and this leads directly to the statement that a bulk shear wave, that is, a semi-infinite plane wave, propagating parallel to the free surface and polarized perpendicular to the sagittal plane, satisfies the free-surface boundary conditions. The eigenvectors for the other two partial waves have intercoupling of the two sagittal plane components of displacement and lead to a surface wave solution.

The details of the solution are obtained by determining the two values of $b^{(n)}$ and the corresponding eigenvectors from (2.15) simplified by the separation noted above, and substituting $b^{(n)}$ and $\alpha_i^{(n)}$ as functions of v into (2.16). Here in the

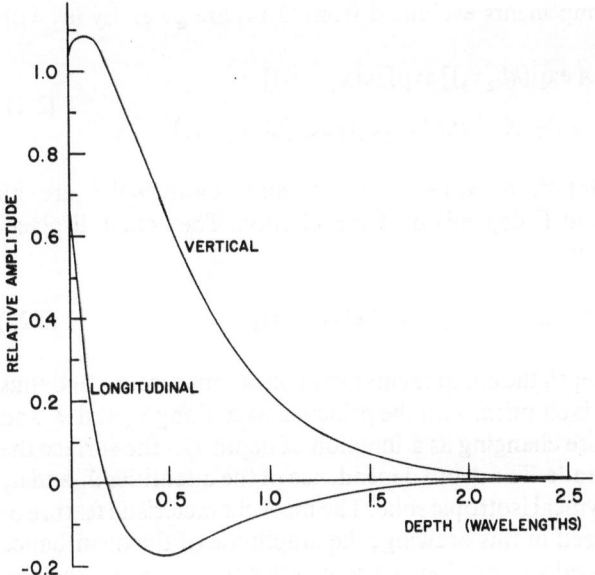

Fig. 2.2. Displacement components as a function of depth for an isotropic material

isotropic case, the functions are explicit and the expression det $|d_{mn}| = 0$ becomes the following implicit equation for the surface wave velocity v [2.8]

$$\left[2 - \left(\frac{v}{v_t}\right)^2\right]^2 = 4\left[1 - \left(\frac{v}{v_l}\right)^2\right]^{1/2}\left[1 - \left(\frac{v}{v_t}\right)^2\right]^{1/2} \tag{2.19}$$

wherein v_t and v_l are given by (2.18). This surface wave is called a Rayleigh wave and its phase velocity contained in (2.19) can be approximated by [2.8]

$$\frac{v}{v_t} \cong \frac{0.72 - (v_t/v_l)^2}{0.75 - (v_t/v_l)^2}. \tag{2.20}$$

Thus since the maximum value of v_t/v_l for a real isotropic solid is 0.5, it is seen that $0.87 < v/v_t < 0.96$, so that the phase velocity of the Rayleigh wave is of the order of 10% less than the bulk shear velocity. It is important to note that the Rayleigh wave propagating along the surface has a phase velocity less than the lowest bulk-wave velocity, and thus its wavelength measured along the surface is greater than the projected wavelength of any bulk wave propagating at any angle to the surface. As a result, the Rayleigh wave on a free isotropic surface cannot phase match to any bulk wave. It will be seen in later sections of this chapter that coupling to bulk waves can occur in certain anisotropic cases, and when there is a discontinuity in the surface boundary conditions.

The displacement components evaluated from (2.14) are given by ($u_2 = 0$)

$$u_1 = C \left[\exp(kb_1 x_3) - A \exp(kb_2 x_3)\right] \exp[ik(x_1 - vt)]$$
$$u_2 = -ikb_1 C \left[\exp(kb_1 x_3) - A^{-1} \exp(kb_2 x_3)\right] \exp[ik(x_1 - vt)] \tag{2.21}$$

in which $b_1 = [1 - (v/v_l)^2]^{1/2}$, $b_2 = [1 - (v/v_t)^2]^{1/2}$, and $A = (b_1 b_2)^{1/2}$ are all positive real quantities and C depends on the excitation. The actual displacements are thus of the form

$$u_1 = \hat{u}_1 \cos[k(x_1 - vt)] \quad \text{and} \quad u_3 = \hat{u}_3 \sin[k(x_1 - vt)] \tag{2.22}$$

and it is seen that at any depth the components are in phase quadrature, and thus the particle displacement is elliptical with the principal axes along x_3 and x_1 and with the shape of the ellipse changing as a function of depth. On the surface the elliptical motion is retrograde. The depth dependence of the quantities $\hat{u}_1$ and $\hat{u}_3$ is shown in Fig. 2.2 for a typical isotropic solid. The most characteristic feature of surface waves is emphasized in this drawing; the amplitude of the disturbance becomes negligible for depths more than a few wavelengths from the surface.

2.3 Plate Modes

Before proceeding with the effects of crystalline anisotropy on surface wave propagation, it is informative to outline the relation between surface waves and waves in a plate. In the remainder of the book, the substrate supporting the surface wave will usually be assumed to be of infinite depth. Obviously, the latter is an approximation valid only if the depth of penetration of the wave is small with respect to the substrate thickness and there is no radiation of energy from the surface of propagation into the substrate. If these conditions are not met, the boundary conditions at the lower surface must also be included. For illustrative purposes consider a substrate of thickness h, a plate with the surface $x_3 = -h$ also traction free. Here the velocity of propagation depends on the plate thickness in wavelengths, that is, there is dispersion, and there are different modes or waves with different distributions of sagittal-plane displacements across the plate [2.5, 7, 9–11]. The modes have either symmetric or anti-symmetric patterns of the particle displacements with respect to the center line of the plate thickness. The symmetric modes with sagittal-plane displacements have the u_1 component symmetric and the u_3 component antisymmetric about the center line. Thus, these modes are dilational in nature, whereas the antisymmetric modes have u_1 antisymmetric and u_3 symmetric so that the plate motion for these modes is flexural. The dispersion curves for the first few of these modes for a typical isotropic plate are shown in Fig. 2.3 [2.9]. Of particular interest here is the behavior with increasing plate thickness of the lowest symmetric and antisymmetric modes. It is seen that the phase velocities of both

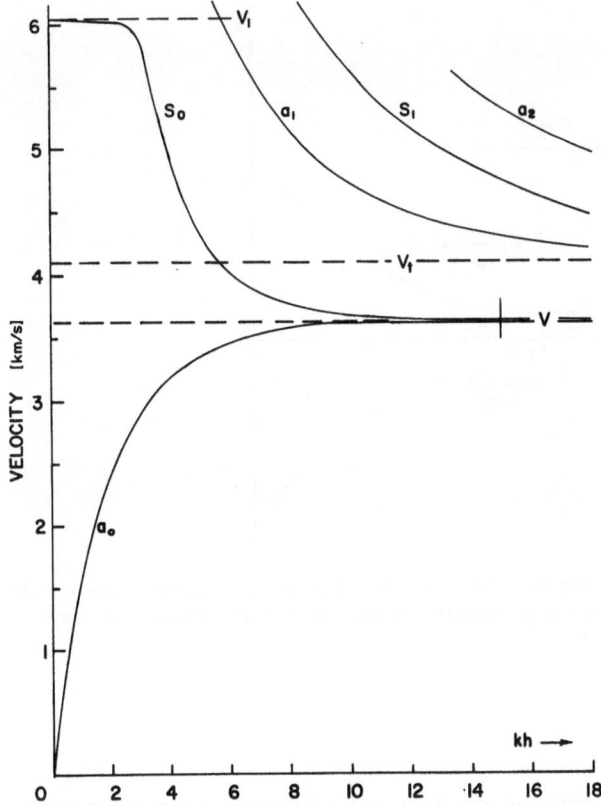

Fig. 2.3. Dispersion curves for symmetric (s) and antisymmetric (a) modes of an isotropic plate of thickness h. Material is fused quartz, and v, v_t, and v_l are the Rayleigh velocity and the bulk transverse and longitudinal velocities, respectively

modes become asymptotic to the same value, and that value is the velocity of a Rayleigh wave on a half-space of the same material. Figure 2.4 shows the u_3 component of particle displacement for the lowest symmetric and antisymmetric modes for $kh = 15$ ($h = 2.4\lambda$), both of which propagate at approximately the Rayleigh velocity for such a thickness. Also on this diagram is the sum of the displacements for the two modes, and it is seen that this sum is approaching the u_3 component of a Rayleigh wave on the surface $x_3 = 0$. A similar statement can be made for the u_1 component. Thus, if it is desirable, the Rayleigh mode on the free surface of a thick but finite substrate can be thought of as the sum of equal amplitude symmetric and antisymmetric modes for a plate of the same thickness [2.7]. Similarly, the difference of the two displacements for the two modes gives a Rayleigh wave on the surface $x_3 = -h$. While these relations are of conceptual importance they are not usually directly applicable to practical surface wave geometries where the substrate is many wavelengths thick and the two surfaces are completely decoupled.

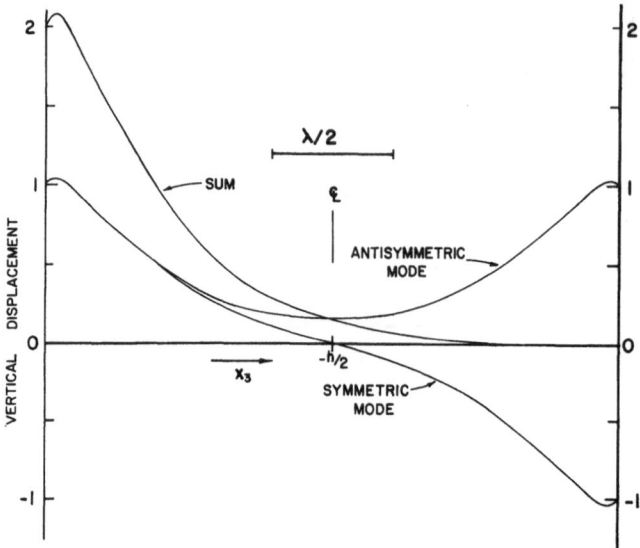

Fig. 2.4. Vertical displacement components for the lowest symmetric and antisymmetric plate modes at $kh = 15$ on Fig. 2.3. Sum of two displacements is almost a surface wave on the surface $x_3 = 0$

2.4 Anisotropy

The previous sections illustrated the first-order features common to elastic surface waves: elliptically polarized displacements decaying to zero a few wavelengths below the free surface and dispersionless phase velocities of the order of the corresponding shear bulk-wave velocities. These properties were derived for isotropic substrates but they are valid in general for anisotropic substrates, and it is important that they be kept continuously in mind as we introduce in turn the effects of anisotropy and piezoelectricity. While the latter effects are of prime concern to the surface wave device designer, they are in a sense second order for the wave propagation itself.

With a given isotropic solid as the substrate, the properties of the surface wave are independent of the choice of plane for the free surface or of the direction of propagation in that plane. Such is not the case for an anisotropic solid where all the properties—the phase velocity, the group velocity, and the form of the displacements—depend on the crystal plane chosen as the free surface and on the direction of propagation of the wave. Yet despite the experimental and computational problems evoked by the use of anisotropic substrates, the surface wave devices are usually constructed on single-crystal, and thus anisotropic, substrates because only among the latter has the designer been able to find such necessary combinations of properties as low attenuation at high frequencies and strong piezoelectric coupling.

Table 2.1 Cubic crystals

Crystal	Anisotropy $\eta = \dfrac{2c_{44}}{c_{11}-c_{12}}$	Shear velocity [100] v_t [m s^{-1}]	Surface wave along [100] on (001)	
			v/v_t	Ellipse eccentricity
Sodium	7.00	2080	0.503	1.03
Copper	3.20	2901	0.693	1.11
Gold	2.85	1526	0.736	1.18
Nickel	2.38	3832	0.759	1.14
InSb	1.99	2284	0.803	1.19
Silicon	1.57	5844	0.841	1.23
Tungsten	0.995	2857	0.926	1.51
PbS	0.508	1818	0.975	2.02
KCl	0.375	1780	0.984	2.31

Almost all of the phenomena to be met using general propagation geometries in anisotropic crystals can be illustrated by consideration of the principal planes of cubic crystals. Thus, for the moment, discussion will be restricted to these planes in nonpiezoelectric cubic crystals.

Cubic crystals have three independent elastic constants [2.2], labelled c_{11}, c_{12}, and c_{44}, and the velocity of bulk longitudinal waves along a crystal axis is $v_l = \sqrt{c_{11}/\varrho}$, while on the same axes there are two degenerate shear waves with velocities $v_t = \sqrt{c_{44}/\varrho}$. In general, the bulk-wave velocities are different for other directions of propagation in the crystal [2.7]. For example, along a [110] axis the two shear velocities are nondegenerate; one wave of velocity $v_{t_1} = \sqrt{c_{44}/\varrho}$ is polarized perpendicular to the basal plane while the other is polarized in the basal plane and has velocity $v_{t_2} = \sqrt{c_{44}/\eta\varrho}$, where

$$\eta = \frac{2c_{44}}{c_{11}-c_{12}}. \tag{2.23}$$

For isotropic crystals $2c_{44} = c_{11} - c_{12}$ and thus $\eta = 1$. This parameter [2.16] is a useful measure of the degree of anisotropy of a cubic crystal, and some typical values [2.9] are shown in Table 2.1 along with the corresponding degenerate shear velocities for propagation along a crystal axis.

The bulk and surface wave behaviors in cubic crystals having η close to unity, e.g., tungsten in Table 2.1, are similar to those of an isotropic solid; thus, for any free surface on such a crystal, the surface wave velocity will be independent of the direction of propagation, its value will be given approximately by (2.20), and the displacements will be of the form shown in Fig. 2.2. Differences in behavior occur when η departs from unity. For illustrative purposes, two different cubic crystals will be considered: nickel with $\eta = 2.38$, appreciably greater than unity, and KCl with $\eta = 0.375$, appreciably less than unity. The behavior for other cubic crystals can be interpolated from the results for these two examples.

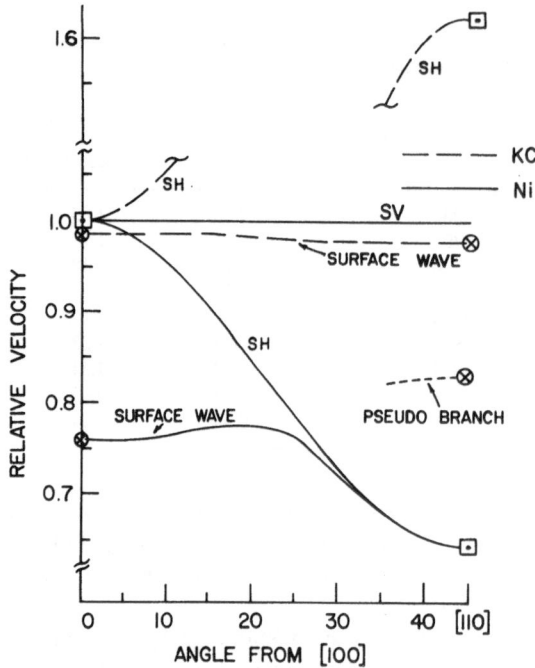

Fig. 2.5. Shear and surface wave velocities on the basal plane of nickel and KCl. ⊗ indicates surface waves with sagittal-plane displacements only. ⊡ indicates SH shear wave satisfying boundary conditions

2.4.1 Basal Plane of Cubic Crystals

Let us consider first that the free surface is the basal or (001) plane of the crystal [2.16–19]. Figure 2.5 indicates the phase velocities of the two shear waves for each of the two crystals normalized to the corresponding degenerate value along a cubic axis. One bulk shear wave labelled SV is polarized vertically, or perpendicular to the free surface, and its velocity is independent of the direction of propagation in this plane. The other wave labelled SH is polarized horizontally, that is, in the basal plane, and its velocity and direction of polarization vary with the angle of propagation. However, on the axis and along the [110] direction the SH wave is polarized perpendicular to the sagittal plane, and this bulk wave in the substrate satisfies the free-surface boundary conditions ($T_{3j} = 0$ at $x_3 = 0$) in these two directions, a characteristic marked by the small squares in Fig. 2.5.

The simplest example of surface wave propagation in an anisotropic medium occurs in these nonpiezoelectric cubic crystals when the propagation vector is along a cubic axis and the free surface is the basal plane [2.6]. Here the sagittal-plane is a plane of mirror symmetry and, as will be seen below, this is a sufficient condition that the wave equation and the boundary conditions separate, leaving the two sagittal-plane components of displacement u_1 and u_3 intercoupled but independent of u_2. The parts of these equations involving u_2 lead to the SH bulk wave satisfying the boundary conditions as discussed in the previous paragraph. The other two components lead to a surface wave solution whose velocity is somewhat less than that of the shear wave and, as shown in Table 2.1, the surface wave velocity in this direction approaches more closely the shear velocity as the anisotropy ratio η decreases. Since only the displacement components u_1 and u_3 are involved, and they are in phase quadrature, the surface displacement is elliptical and the plane of the ellipse is the sagittal-plane, just as for the isotropic case. The ratio of the vertical to the horizontal axis increases as shown in Table 2.1 for decreasing values of η.

It is typical of this propagation geometry on crystals with η appreciably less than unity that the two values of the decay constants b in the appropriate Eqs. (2.12) and (2.14) are purely imaginary and the form of the decay of the displacement components is the same as shown in Fig. 2.2. However, for the same geometry with η appreciably greater than unity there occurs a new phenomenon which is often met in anisotropic cases; the decay constants b have real parts [2.6, 17–19], and thus the depth dependence of the displacement components consists of exponentially damped sinusoids rather than real exponentials. Consequently, the components here, again in phase quadrature at each depth as in (2.22), oscillate as a function of x_3 in the manner shown for nickel in Fig. 2.6. It is perhaps conceptually useful to consider the form of the decay curves for the isotropic case of Fig. 2.6 to be an "envelope" function for the actual decay and then to note that in some cases the "period" of the oscillation is shorter than the envelope decay depth, as in Fig. 2.6, or much larger, giving the form of Fig. 2.2.

We have thus observed in the displacements one possible difference between surface wave propagation on isotropic and on anisotropic substrates. This difference is difficult to measure, but a second difference that is easily observable is the dependence of the phase velocity on the direction of propagation illustrated for the basal planes of KCl and nickel in Fig. 2.5. In this plane for crystals with $\eta < 1$, there is variation of velocity with direction but it is not very dramatic. For propagation along [110], that is, at 45° to the cubic axis on these crystals with $\eta < 1$, the sagittal-plane is again a mirror plane and thus the surface wave contains only two components, and the behavior is much the same as met on the cubic axis. The presence of two sagittal-plane components only is indicated by the symbol $\otimes$ on this and subsequent velocity diagrams. Another difference from isotropic characteristics is introduced for general propagation directions on the basal plane because all three components of displacement are involved, and while the displacement ellipse at any depth again has one major axis perpendicular to the free surface, the other axis is not in the sagittal-plane.

For basal plane propagation on cubic crystals with η appreciably greater than unity, such as nickel in Fig. 2.5, yet another phenomenon is introduced. The behavior for propagation along the cubic axis has been discussed already but here, as the propagation vector is turned away from the cubic axis, the phase

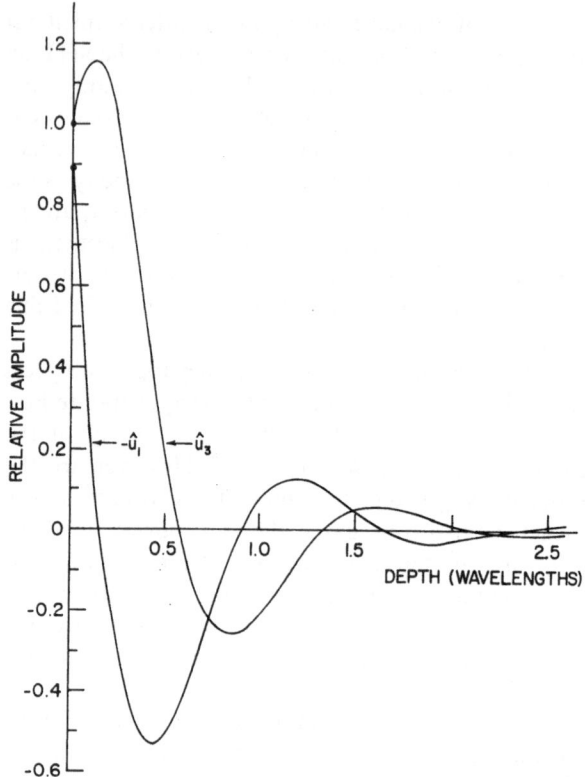

Fig. 2.6. Displacement components as a function of depth for surface wave propagation along a cubic axis on the basal plane of nickel

velocity changes markedly, the effects of which will be considered below; moreover, in this case the character of the surface wave displacements changes continuously. The plane of the ellipse traced out by a particle, say on the surface, turns further and further out of the sagittal plane as the propagation angle from [100] is increased, the ellipse elongates parallel to the surface, and the penetration of the wave into the substrate becomes deeper. As the propagation vector approaches [110], the wave degenerates into the SH bulk shear wave which itself satisfies the free-surface boundary conditions.

For propagation directly along [110] on the basal plane of cubic crystals with η appreciably greater than unity, there is a surface wave solution, having sagittal-plane displacements only and of form much as on the cubic axis. The velocity of this wave marked by $\otimes$ on Fig. 2.5 is intermediate between that of the SV and the SH bulk shear waves. In a sense this wave is a singularity in that it exists only for propagation directly along the [110] direction [2.15, 20, 21]. When the direction of propagation makes a small angle with [110], only the surface wave solution on the lower branch has a real propagation vector; however, there is a second solution marked by the dotted curve of Fig. 2.5 which is very similar to a surface wave except that it contains a small contribution from a bulk wave radiating down into the substrate at an angle to the surface. Thus,

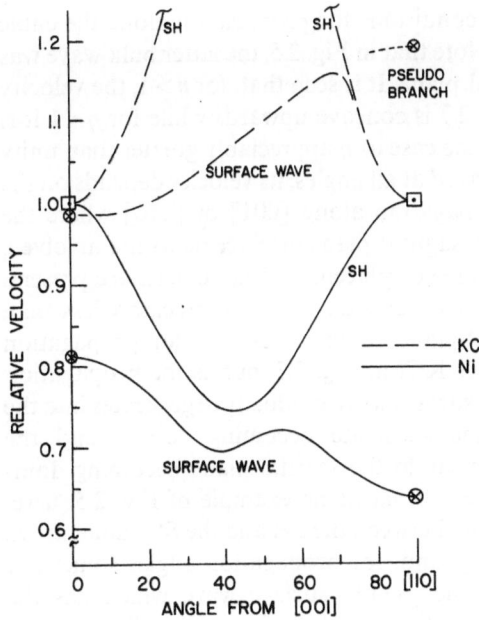

Fig. 2.7. Shear and surface wave velocities on the (110) plane of nickel and KCl

while most of the energy in the wave is transported close to and parallel to the surface, there is a small fraction radiated and thus the particle displacements attenuate in the direction of propagation. The attenuation per wavelength is very small, for example, less than 0.01 dB per wavelength for angles up to 15° away from [110] on nickel, with the result that from an experimental standpoint this wave has all the characteristics of a normal surface wave [2.20, 21]. The wave is similar to the "leaky" waves containing a radiating component met in certain electromagnetic boundary cases and with elastic surface waves if the small acoustic radiation loss into the air above the substrate is included. However, since in the case under discussion the phase match between the tilted radiating wave and the surface wave is permitted only because of the anisotropic nature of the substrate, the term "pseudo surface wave" used in Fig. 2.5 appears to be more appropriate.

2.4.2 (110) Plane of Cubic Crystals

It will be recalled that we are using propagation on principal planes of cubic crystals to illustrate most of the different second-order surface wave phenomena which can be encountered when the substrate is anisotropic. Consider now one of the (110) planes as the free surface [2.17]; thus, the free surface is perpendicular to the basal plane of Fig. 2.5 and contains a cubic axis and the [110] direction. For propagation in the various directions in this plane, there are again SH, SV, and longitudinally polarized bulk waves; the velocities for the SH waves are shown on Fig. 2.7. The small squares indicate that the SH wave satisfies the free surface boundary conditions for propagation along the cubic axis and along the [110] direction. Note that in Fig. 2.5, the latter bulk wave was labelled SV with respect to the basal plane. It is seen that, for $\eta > 1$, the velocity curve for the SH bulk wave in Fig. 2.7 is concave upwards while for $\eta < 1$ it is concave downwards. This time, for the case of η appreciably greater than unity (nickel), the surface wave is well behaved at all angles, its velocity depends on the angle of propagation, and, for propagation along [001] or [110] where the sagittal plane is a mirror plane, only sagittal-plane displacements are involved, but for intermediate directions all three displacement components are present. On the other hand, for this geometry on crystals with η appreciably less than unity, the surface wave has sagittal-plane components only for propagation along the cubic axis, as indicated for KCl in Fig. 2.7, but as the propagation angle approaches the basal plane the surface wave gradually degenerates into the SH wave, the penetration into the substrate becoming deeper, and the displacement component perpendicular to the sagittal plane becoming dominant. However, much as for the $\eta > 1$ basal plane example of Fig. 2.5, here, with $\eta < 1$, a surface wave with velocity between the SH and the SV values arises for propagation directly along [110], and, for propagation angles somewhat away from this direction, there is the pseudo surface wave which has the properties of a surface wave except for the presence of a small radiating term.

2.4.3 (111) Plane of Cubic Crystals

The free surfaces in the previous two principal plane examples considered, Figs. 2.5 and 2.7, were themselves planes of crystalline mirror symmetry. Let us turn now to a case where this is not so and consider propagation on the (111) plane [2.6, 17]. Whereas in the mirror plane cases the bulk shear waves at any angle were polarized either parallel (SH) or perpendicular (SV) to the plane, here, on the (111) plane, the polarization, relative to the plane, of the bulk shear waves is not so simple except along the direction marked 30° in Fig. 2.8, where the polarization of one is SH. Along the [110] direction, the two shear modes are pure shear polarized perpendicular to the propagation direction but are not SV or SH. The velocity normalization in this diagram is the same as in Figs. 2.5 and 2.7.

The bulk shear waves with propagation vector parallel to the free surface have the velocities shown in Fig. 2.8 as a function of the angle from the [110] direction. These shear waves do not satisfy free-surface boundary conditions at any angle. It should, however, be noted in passing that, at the 30° angle in this plane, free-surface boundary conditions are satisfied by SH bulk shear waves which have their propagation vectors not parallel to the plane but tilted upward at a specific angle for nickel and downward for KCl. Despite the tilt of the propagation vector, the direction of transport of energy or of the group velocity of these waves is parallel to the free surface and along the 30° direction. The phase velocity measured along the latter direction is marked by small triangles on Fig. 2.8. This point is the degenerate limit of a pseudo surface wave branch which for nickel has a velocity curve more-or-less parallel to the upper solid bulk-wave curve of Fig. 2.8. The attenuation associated with the wave is small at all angles on the plane; it degenerates into the SH bulk wave at 30°, but not into a pure surface wave at any angle.

The surface wave velocities for propagation on the (111) plane of nickel and KCl are also shown in Fig. 2.8. At all angles except 30° from [110], all three components of displacement are involved. However, for propagation along the 30° direction, only the two sagittal-plane components are present. In contrast to previous cases where the free surface was a mirror plane and thus when only the sagittal-plane components existed, the displacement ellipse was in the sagittal-plane and had one major axis perpendicular to the free surface, here the displacement ellipse is in the sagittal-plane at each depth but the two components are not in phase quadrature and thus the principal axis is not perpendicular to the free surface.

2.5 Power Flow

When an acoustic wave propagates in a piezoelectric solid, the time-average power flow across a unit area perpendicular to the coordinate x_i is, in the quasi-static approximation [2.7, 22],

$$P_i = -\tfrac{1}{2}\operatorname{Re}\{[i\omega(T_{ij}u_j^* - \varphi D_i^*)]\}. \tag{2.24}$$

The vector P is the acoustical equivalent of the time-independent Poynting vector of electromagnetic theory. While this expression is complicated in detail for anisotropic media, comparison with the constitutive relations of the medium, (2.3), shows that there are three types of power-flow terms associated with the wave at each point in space. The first contribution comes from terms of the form $k\omega cu^2$, resulting from the mechanical displacements only in the stress tensor. Besides this purely mechanical contribution, there is a purely electrical one with terms of the form $k\omega\varepsilon\varphi^2$, representing the quasi-electrostatic energy in the electric field propagating along with the wave. Finally, because of the potential terms in the stress and the mechanical displacement terms in the electrical displacement, there is an electromechanical contribution with terms of the form $k\omega e\varphi u$. Note that if we were dealing with a one-component plane wave and the quantities c, ε, and e were scalars, the ratio of the electromechanical energy flow to the geometric mean of the other two is the square root of the electromechanical coupling constant $K^2(K^2 = e^2/c\varepsilon)$ introduced in (2.7). For anisotropic cases and surface waves, this identification of K^2 with the energy ratio is not directly applicable but is sometimes a useful concept for approximation purposes.

It has been noted previously that even for the strongest piezoelectrics useful for ultrasonic applications [2.13] the electromagnetic coupling represented by an appropriate effective value of K^2 is less than 5%, with the result that for an acoustic wave propagating on a source-free region of the solid, the wave is

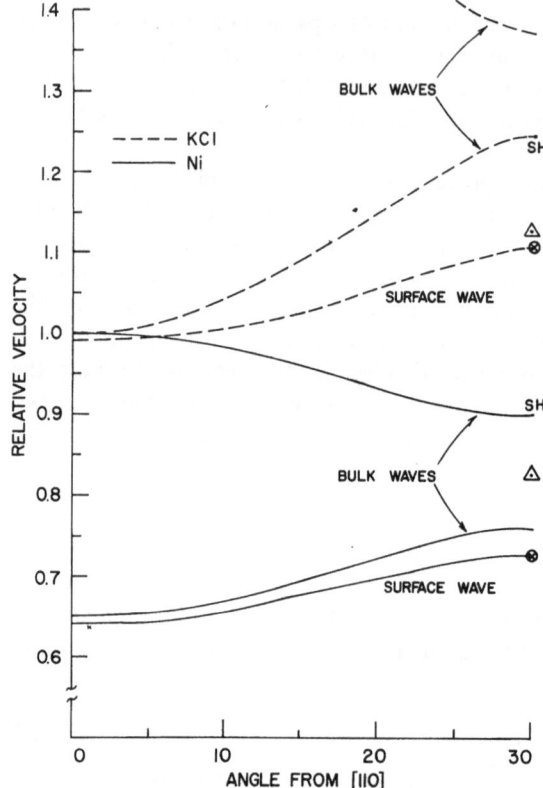

Fig. 2.8. Shear and surface wave velocities for propagation on the (111) plane of Nickel and KCl. △ indicates tilted bulk wave satisfying free-surface boundary conditions

predominantly elastic in that the power flow of (2.24) is dominated by the terms in u^2. Thus, for most cases the electrical terms and the electromechanical terms can be treated as perturbations for energy considerations in freely propagating ultrasonic waves. Of course, in dealing with coupling between electrical circuits and the acoustic wave, such perturbations become of primary importance, and this electromechanical coupling will be illustrated for many surface wave geometries in succeeding chapters. There exists a well-developed perturbation theory [2.7] based on the same concepts as (2.24), which is of great assistance in the discussion of electromechanical coupling with electrodes and of intermode coupling, but here we concentrate on the propagation characteristics of acoustic waves in a source-free region of a homogeneous half-space. Let us thus restrict (2.24) to the case of surface waves propagating on an anisotropic but nonpiezoelectric substrate; therefore, the power flow per unit area at each point in the solid is given by [2.22]

$$P_i = -\tfrac{1}{2}\mathrm{Re}\{i\omega T_{ij}u_j^*\}. \tag{2.25}$$

For surface wave applications, it is not usually the power flow per unit area which is of concern, but rather the total power carried parallel to the surface by the surface wave, per unit width normal to the sagittal-plane; the latter vector is given by

$$W_i = -\tfrac{1}{2} \mathrm{Re} \left\{ \int\limits_{-\infty}^{0} i\omega T_{ij} u_j^* dx_3 \right\} \qquad i = 1, 2. \tag{2.26}$$

There is no power flow normal to the surface for nonleaky surface waves, and thus this vector has only two components. W_1 gives the power flow parallel to the propagation vector per unit width perpendicular to the sagittal-plane. In isotropic media and in specific directions in anisotropic crystals, it is the only component which exists. These cases are called pure-mode directions for surface waves in that the power flow is parallel to the propagation vector, the latter usually being set by the orientation of a transducer. However, for more general anisotropic geometries, W_2 of (2.26) is not zero and there is a component of power flow normal to the sagittal-plane; thus, the power flow vector makes an angle with the propagation vector. In such a case, the "beam" of energy emitted from some transducer which sets up a constant phase front across its aperture propagates at an angle, called the "power-flow angle", to the normal to this phase front, in contrast to the same geometry in an isotropic medium where the beam would propagate along the normal. This so-called "beam steering" is of obvious importance in the optimal location of pairs of transducers to form a transmitter-receiver combination, and numerical examples of the magnitude of the effect for useful crystal orientations are given in Section 6.3.

For computational purposes, the displacement expression for the surface wave can be combined with the constitutive equation for stress and integrated in (2.26) to give [2.6]

$$W_i = -\tfrac{1}{2} \mathrm{Re} \left\{ \sum_{n=1}^{3} \sum_{m=1}^{3} \frac{\omega C_n C_m^* \alpha_k^{(n)} \alpha_j^{(m)*} (c_{ijk1} + c_{ijk3} b^{(n)})}{i(b^{(n)} - b^{(m)*})} \right\}. \tag{2.27}$$

While this expression is appropriate for computation, it is convenient for conceptual purposes to relate the power-flow direction to the phase velocity, in large part because phase velocity data are much more easily measured, calculated, and tabulated than the multitude of quantities in (2.27). If a polar curve is constructed in a plane parallel to the free surface with the radius vector in each direction equal to the inverse of the phase velocity in that direction, the result is a "slowness curve" for surface wave propagation in that plane [2.7]. The normal to this curve at a polar angle corresponding to the direction of the propagation vector gives the direction of the group velocity and the direction of power flow W. Group velocity and energy velocity are the same for the lossless media considered here. The group velocity makes an angle ψ, the power-flow angle, with respect to the propagation vector, and the magnitude of the group velocity v_g is given by

$$v_g = v \cos \psi \tag{2.28}$$

where v is the phase velocity.

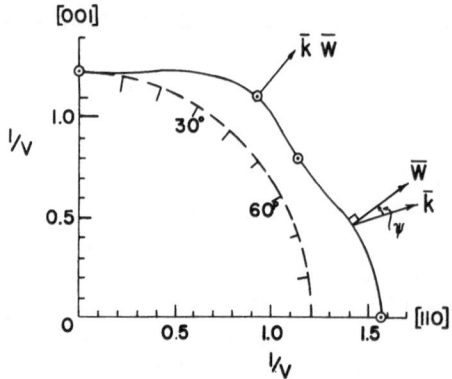

Fig. 2.9. Slowness curve for propagation on the (110) plane of nickel showing typical directions of propagation vector and associated Poynting vector. ⊙ indicates pure-mode directions

It is evident that extrema of the phase velocity curves as a function of propagation angle on the surface, such as those of Figs. 2.5, 2.7, and 2.8 or the corresponding extrema of the slowness curve, are pure-mode axes in that the power-flow angle is zero. Figure 2.9 shows the slowness curve for propagation on the (110) plane of nickel, corresponding to the solid surface wave velocity curve of Fig. 2.7 and using the same normalization. The circled points indicate pure-mode directions; the two on the axes are imposed by cubic crystal symmetry, whereas the locations of the others depend on the particular elastic constants of the material. It should be noted that the power-flow angle away from pure-mode directions can be quite large, values over 10° being common in substrate materials of interest for surface wave applications [2.13].

The common practice in surface wave devices is of course to have the direction of propagation be a pure-mode axis so that the emitted beam from a transducer is perpendicular to the transducer aperture, but because of tolerance limitations on crystal and transducer alignment and because of off-axis effects due to diffraction, it is of interest to have a simple measure of the departure from isotropy near a pure-mode axis. Thus, if v_0 is the phase velocity directly along the pure-mode axis which is an extremum of the velocity curve, the phase velocity at neighboring angles θ from this axis will be [2.23, 24]

$$v(\theta) = v_0(1 - \beta\theta^2). \tag{2.29}$$

A similar measure of the local anisotropy is given by the derivative of the power-flow angle [2.13] as a function of θ at $\theta = 0$ (see related discussions in Sects. 2.9 and 6.3)

$$\beta = -\frac{1}{2}\frac{\partial\psi}{\partial\theta}\bigg|_{\theta=0}. \tag{2.30}$$

The parameter β can be positive or negative, leading to large difference in the diffraction effects from a finite width aperture, as will be seen in Section 2.9.

2.6 Piezoelectricity

As has been noted several times before, piezoelectricity causes only second-order effects on the dominantly mechanical character of acoustic surface waves. Nevertheless, it is by means of the accompanying electric field that a surface wave is usually coupled to external electric circuits, to other surface waves, or to charge carriers in semiconductors, and thus consideration must now be given to some of the piezoelectric effects. The addition of piezoelectricity complicates further the solution of the problem of surface wave propagation in anisotropic media discussed in Section 2.4, first by increasing to four the order of the eigenvectors required in the partial waves in (2.11) [2.12, 14], second by increasing the number of partial waves also to four in general, third by introducing through the constitutive Eq. (2.3) terms containing the potential into the zero-stress boundary condition (2.9), and finally by requiring consideration of the potential in the space above the mechanical surface and the concomitant imposition of some electrical boundary condition at or near the substrate surface.

As far as the first two points are concerned, the formalism for introducing the piezoelectricity into the wave equation has already been included in (2.12), while the last two points have to be included in the homogeneous boundary condition equation [2.6] for the four weighting factors of the partial waves [2.9, 12]

$$d_{mn}C_n = 0 \quad (m, n = 1, 2, 3, 4) \tag{2.31}$$

where here, because of the contribution of the potential to the stress, the first three rows of the d matrix become

$$d_{jn} = c_{3jkl}\alpha_k^{(n)}b_l^{(n)} + e_{k3j}\alpha_4^{(n)}b_k^{(n)} \quad j, k, l = 1, 2, 3 \tag{2.32a}$$

instead of (2.16). As before, the material constants are assumed to be expressed in the axes of Fig. 2.1. The fourth row depends on the electrical boundary conditions imposed by the geometry. Two conditions are of particular interest, first, when the surface of the half-space is electrically free in the sense that there is no conducting layer at the surface and there is effectively free space above it. This condition is frequently called open circuited, but the term "electrically free" may be better because of the penetration of the time-dependent fields into the space above the substrate. A second type of boundary condition of interest will be seen to be the equivalent to covering the surface with a perfectly conducting layer of a material thin enough that the mechanical boundary conditions are unchanged. For this shorted condition, the potential at $x_3 = 0$ in Fig. 2.1 must be zero, with the result that the fourth row of the boundary condition matrix of (2.31) becomes

$$d_{4n} = \alpha_4^{(n)}. \tag{2.32b}$$

For the electrically free case, the potential above the free surface must satisfy Laplace's equation at each instant of time, must vanish for $x_3 \to +\infty$, and must be continuous with the potential within the substrate at each point on the surface. Such a solution is [2.9, 12, 25]

$$\varphi = \Phi(0)\, e^{-kx_3} \qquad x_3 \geqq 0 \tag{2.33}$$

where $\Phi(0)$ is the propagating substrate potential evaluated at $x_3 = 0$. With this form of potential, the condition that the normal component of the electric displacement (2.3) must be continuous at $x_3 = 0$ gives for the fourth row of the boundary condition matrix

$$d_{4n} = e_{3kl}\alpha_k^{(n)} b_l^{(n)} - (\varepsilon_{3k} b_k^{(n)} - i\varepsilon_0)\alpha_4^{(n)}. \tag{2.34}$$

Thus the addition of piezoelectricity leaves the structure of the problem the same but increases the complexity in computation. Little would be added here by further discussion of this general piezoelectric case, but two special examples are important enough in themselves to warrant separate discussion.

Before considering these particular piezoelectric examples, it is worthwhile to note two conditions on the crystal symmetry with respect to the coordinate axes of Fig. 2.1 which guarantee that the axis of propagation is a pure-mode axis and that the various equations involved separate into lower order sets to simplify greatly the computations [2.26].

1) If the sagittal plane is a plane of mirror symmetry of the crystal, x_1 is a pure-mode axis for the surface wave which then involves only the potential and the sagittal-plane components of displacement and may be termed a stiffened Rayleigh wave. The transverse component u_2 is not coupled to the other three either by the wave equation or by boundary conditions of the type discussed.

2) If the sagittal-plane is perpendicular to an axis of twofold rotation, the components u_1 and u_3 are intercoupled, but independent of u_2 and φ, and lead to a surface wave which propagates as though all the piezoelectric constants were zero. The transverse displacement component and the potential are intercoupled, but independent of the sagittal components, and give rise to a type of transversely polarized surface wave, the so-called Bleustein-Gulyaev wave [2.27, 28], which cannot exist on the free surface of a nonpiezoelectric substrate.

The above statements are general as far as substrates are concerned and it should be emphasized that if both are satisfied simultaneously, as for example along (100) and (110) of the basal plane of nickel (Fig. 2.5), the crystal cannot be piezoelectric.

2.6.1 Stiffened Rayleigh Waves

The first of the two piezoelectric examples to be considered is the one most commonly met in current surface wave devices: propagation along the crystalline Z axis on a crystal face perpendicular to the Y axis on the trigonal crystal lithium niobate (LiNbO$_3$ class 3 m). This geometry is referred to as YZ, indicating Y cut, Z propagating. This crystal, and particularly this cut, exhibits

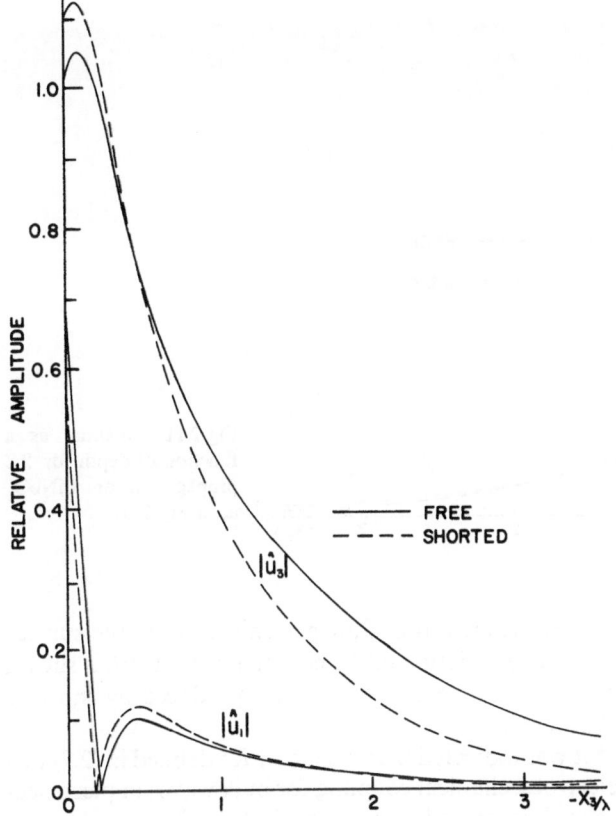

Fig. 2.10. Magnitudes of vertical and longitudinal displacement components for propagation in the Z direction on the Y face (XZ plane) of LiNbO$_3$ for surface free and metallized. Normalization given in (2.39)

high electromechanical coupling, low losses, ease of fabrication, ready availability, and several other attractive attributes for surface waves so that it will be the material most discussed in the later chapters. The YZ geometry for LiNbO$_3$ satisfies the symmetry condition 1) above in that sagittal-plane is then a mirror plane of the substrate crystal. The uncoupled u_2 component leads to a tilted bulk-wave solution independent of the piezoelectricity and satisfying the boundary conditions as discussed for the solution marked with triangles on Fig. 2.8 for the (111) plane of nickel.

The surface wave for YZ LiNbO$_3$ has sagittal-plane mechanical displacement components intercoupled with the potential, and thus the characteristics of the wave are affected somewhat by the choice of electrical boundary conditions. Figure 2.10 shows the magnitude of the two displacement components as a function of depth [2.12, 13], and these curves resemble the corresponding ones for the isotropic case. Here, however, crystal symmetry does not require that a major axis of the ellipse be perpendicular to the surface, and indeed there is a small departure from phase quadrature of the two components. It is seen that the rather extreme change in electrical boundary conditions, from free to shorted, has relatively little effect on the displacements. The curves for the shorted surface

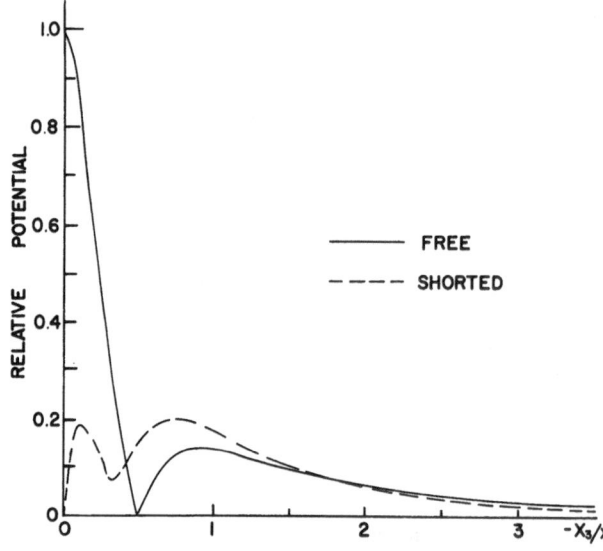

Fig. 2.11. Potential as a function of depth for YZ propagation on LiNbO$_3$ as in Fig. 2.10

are normalized to represent the same power flow per unit width in the surface wave as with the free surface. The variation of the potential with depth is shown in Fig. 2.11, and there is a large change near the surface produced by shorting because, of course, $\varphi = 0$ at $x_3 = 0$.

In Sections 2.1 and 2.5 it was indicated that the factor K^2 defined in (2.7) was a useful measure of the electromechanical coupling for bulk waves in piezoelectric media and that for simple geometries the piezoelectric effect produced a fractional increase in the velocity of a bulk wave of $K^2/2$, where K^2 was an explicitly calculable factor containing terms of the form $e^2/c\varepsilon$. However, for surface waves, because of their inhomogeneity in the x_3 direction, there is no corresponding explicit electromechanical coupling factor. It is nevertheless important to have a single parameter to express this coupling: one that is useful for device analysis and is easily measurable and calculable. Thus, in analogy with the bulk-wave case, an effective electromechanical coupling factor is defined as twice the fractional change in surface wave velocity produced by electrically shorting the mechanically free surface of the piezoelectric substrate [2.12], that is

$$\frac{K^2}{2} = -\Delta v/v = \frac{v - v_s}{v} \tag{2.35}$$

where v is the surface wave velocity with the surface free and v_s with it shorted.

For YZ LiNbO$_3$, K^2 is about 0.045, one of the highest values available in a substrate suitable for high frequency surface wave devices. This concept of an effective electromechanical coupling factor will appear frequently in subsequent chapters and its usefulness will become very apparent.

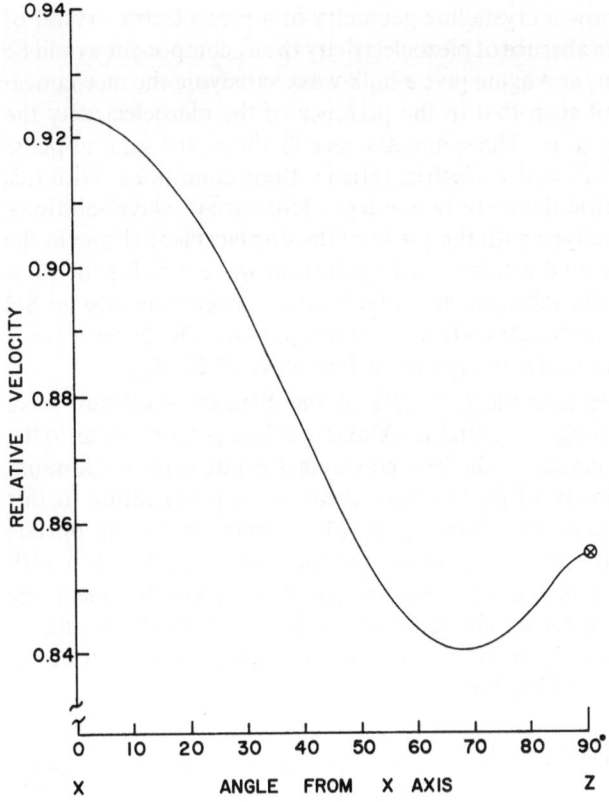

Fig. 2.12. Surface wave velocity for propagation on a free surface of Y cut LiNbO$_3$. Normalized as for Figs. 2.5, etc., here to 4080 m s^{-1}

The surface wave velocity for a free surface of YZ LiNbO$_3$ is marked by the small cross on Fig. 2.12, normalized as in previous cases to the lowest shear velocity along the crystalline X axis (4080 m s^{-1}). Along the Z axis on the Y cut, the velocity is 3488 m s^{-1} and the variation with angle on this XZ surface is shown. For all propagation directions other than along Z on this plane, the surface wave contains the potential and the three components of displacement all intercoupled. The X axis is a pure-mode axis but the wave consists of all four components. The rather odd curvature of this velocity curve near the Z axis will be seen in Section 2.9 to result in rather odd diffraction effects for propagation in the Z direction.

2.6.2 Bleustein-Gulyaev Wave

In the previous example of YZ LiNbO$_3$, where the sagittal-plane was a crystalline mirror plane, the surface wave involved the two sagittal-plane displacement components and the potential, while there was also an independent, transversely polarized bulk wave with its propagation vector tilted to the surface but with its Poynting vector directly along the x_1 axis (Fig. 2.1). It is interesting to consider now a crystalline geometry in a piezoelectric crystal of such symmetry that in the absence of piezoelectricity the u_2 component would be uncoupled from u_1 and u_3 and again give a bulk wave satisfying the mechanical boundary conditions, but such that in the presence of the piezoelectricity the potential is coupled only to u_2. The symmetry case 2) above, the sagittal-plane normal to a twofold axis in a piezoelectric, satisfies these conditions. With this symmetry in a piezoelectric there are two independent surface wave solutions, one of the Rayleigh wave type with the plane of the displacement ellipse in the sagittal-plane and the second a transversely polarized wave which penetrates much more deeply into the substrate and which would degenerate into an SH bulk shear wave if the piezoelectric effect were not present. The latter wave is referred to as a Bleustein-Gulyaev type of surface wave [2.27, 28].

A particularly simple example [2.7, 29] of the Bleustein-Gulyaev wave, occurs for propagation along the crystalline X axis of a face perpendicular to the Y axis of a piezoelectric crystal of the hexagonal class 6 mm, such as cadmium sulphide (CdS). The velocity of an SH bulk shear wave propagating in this orientation in an infinite crystal would be $v_t = (c'/\varrho)^{1/2}$, where c' is the appropriate stiffened elastic constant [see (2.7)], which for this case is $c_{44}(1 + K^2)$ with $K^2 = e_{15}^2/\varepsilon_{11}c_{44}$. It can be shown easily for this particular geometry, using the analysis of Section 2.2, that a simple bulk wave will not satisfy the boundary conditions but a transversely polarized surface wave will. If the surface is shorted, the latter wave is of the form

$$u_2 = C \exp(kbx_3)\exp[ik(x_1 - vt)]$$

$$\varphi = C\frac{e_{15}}{\varepsilon_{11}}[\exp(kx_3) - \exp(kbx_3)]\exp[ik(x_1 - vt)]$$

(2.36)

with

$$b = K^2/(1 + K^2).$$

Thus, since K^2 is small with respect to unity, it is seen that the depth of penetration is larger by a factor of the order of $1/K^2$ with respect to normal Rayleigh waves, where the magnitudes of the factors b are usually about one-half. Also, the decay of the displacement amplitude here is a simple exponential.

This wave may be considered as an SH bulk wave perturbed to meet the electrical contributions to the boundary conditions at the surface, a consideration which is enhanced by noting that the velocity of propagation is

$$v = v_t(1 - b^2)^{1/2}$$

(2.37a)

for a shorted surface and

$$v = v_t(1 - b^2\varepsilon_r^2)^{1/2}$$

(2.37b)

with $\varepsilon_r = \varepsilon_0/(\varepsilon_0 + \varepsilon_{11})$ for a free surface. Since b^2 is much less than unity, the phase velocity is very close to that of the SH bulk wave, even closer for the free surface case because, for most piezoelectrics, $\varepsilon_r \ll 1$. Similarly, the depth of penetration of the wave is much greater for the free than for the shorted surface.

For the case considered, YX propagation in a 6 mm crystal, the algebra is particularly simple because the plane of propagation happens to be a mirror plane. With other crystalline geometries satisfying the necessary condition that the sagittal-plane is perpendicular to a twofold axis, the algebra is somewhat more involved but the general characteristics of the Bleustein-Gulyaev wave are similar to those outlined above for the more restrictive case.

2.7 Propagation in Thin Layers

Surface waves owe much of their importance in devices to the ease with which they can be manipulated by perturbation of the substrate surface. A simple electrical perturbation, namely, the electrical shorting of the free surface, has already been introduced, and it was shown that the effect of this perturbation on the surface wave velocity is a useful measure of the ease with which electromechanical coupling to external circuits can be produced by surface electrodes for the crystal orientation in question. Now we turn to a discussion of the effects of a uniform mechanical perturbation of the substrate surface. This discussion will be restricted to nonpiezoelectric materials and the goal is to illustrate the modifications in the characteristics of surface wave propagation caused by a solid homogeneous layer of uniform thickness h in intimate contact with the substrate. The word "intimate" here implies that the traction stress and the mechanical displacement are continuous across the interface at $x_3 = 0$. Obtaining complete solutions of this problem involves solving the wave equation separately in each of the layer and substrate materials as was done for the substrate alone in Section 2.2, and forming a linear combination of the appropriate partial waves to satisfy the boundary conditions, which are the continuity of the traction stresses and the particle displacements at $x_3 = 0$ and the vanishing of the traction stresses at the free surface, which is now $x_3 = h$. Obviously, the numerical complexity is appreciably increased by the introduction of the layer, but the general methods of solution can be the same as used for the cases without layers [2.9–11, 30, 31].

2.7.1 Rayleigh-Type Waves

From an applications viewpoint, the most dominant effect on surface wave characteristics produced by a thin solid layer on the otherwise free infinite surface of a substrate is the shift of the surface wave velocity and, more particularly, the dependence of this velocity on the frequency of operation. The layer introduces a characteristic dimension into the problem and thus the medium becomes dispersive, with the phase velocity depending on the ratio of the wavelength to this dimension h.

If the layer is very thin with respect to the wavelength ($h \ll \lambda$), then the distribution of the particle displacements as a function of depth and the relative values of the components are not much changed from those appropriate to a free surface of the substrate, e. g., Fig. 2.2 for an isotropic material. The layer material, assumed isotropic, can be characterized by its mass density ϱ_L and by its phase velocities for shear and longitudinal bulk waves, v_{tL} and v_{lL}, respectively. Now, if in the absence of the layer, the surface wave in a given direction propagates with a phase velocity v, and if $\bar{u}_1^2$, $\bar{u}_2^2$, and $\bar{u}_3^2$ are the magnitude squared values of the components of displacement measured at the free surface for the unperturbed wave, the fractional change in phase velocity produced by a layer of thickness h is to first order [2.4, 7]

$$\frac{\Delta v}{v} = \frac{v' - v}{v} = \frac{\omega k h \varrho_L}{4 W_1}$$
$$\cdot \left\{ v_{tL}^2 \left[4 \left(1 - \frac{v_{tL}^2}{v_{lL}^2} \right) \bar{u}_1^2 + \bar{u}_2^2 \right] - v^2 \left[\bar{u}_1^2 + \bar{u}_2^2 + \bar{u}_3^2 \right] \right\} \tag{2.38}$$

where W_1 defined in (2.26) is the power flow for the unperturbed wave in the direction of the propagation vector per unit width normal to the sagittal-plane. It should be noted first that the quantities $\omega \bar{u}_i^2 / 4 W_1$ and v have been tabulated for many crystal cuts and propagation directions, thus facilitating the numerical evaluation.

For example, for YZ LiNbO$_3$

$$\frac{\omega \bar{u}_1^2}{4 W_1} = 3.16 \times 10^{-12} \, \text{m}^3 \, \text{J}^{-1}, \frac{\omega \bar{u}_2^2}{4 W_1} = 0, \frac{\omega \bar{u}_3^2}{4 W_1} = 6.86 \times 10^{-12} \, \text{m}^3 \, \text{J}^{-1}. \tag{2.39}$$

Second, it is seen that in the approximation of (2.38), the change in velocity is proportional to the mass per unit area of the layer, $h \varrho_L$. The extreme limits of the quantity $(1 - v_{tL}^2 / v_{lL}^2)$ are 0.5 and 1. For isotropic substrates [2.8], $1.67 < \bar{u}_3^2 / \bar{u}_1^2 < 3.43$, and for most useful anisotropic cases this ratio lies in the same range. Thus the two expressions in the brackets are usually approximately equal one to the other, with the result that the fractional change of velocity due to a thin layer has a sign which depends predominantly on the relative magnitudes of the shear velocity of the layer material and the unperturbed Rayleigh velocity of the substrate. The perturbed velocity is greater than the unperturbed value if v_{tL} is appreciably greater than v, in which case the layer is said to "stiffen" the substrate, whereas if v_{tL} is appreciably less than v, the phase velocity is decreased by the layer and the latter is said to "load" the substrate. Note that if v_{tL} and v are approximately equal, the exact values of the terms in the brackets determine the sign of the velocity change, but the magnitude of the change will then be small.

Equation (2.38), which gives a linear dependence on kh for a given substrate-layer combination, is based on the assumption that the layer produces only a

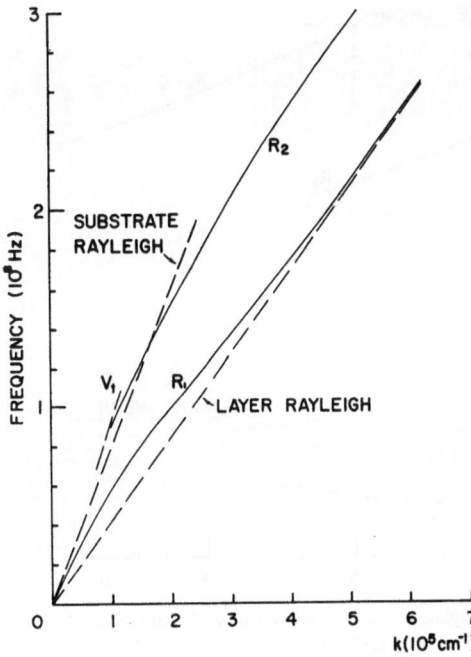

Fig. 2.13. Dispersion curves, frequency as a function of wave number, for the first two Rayleigh modes. 10 μm layer of ZnO on an isotropic silicon substrate

slight perturbation of the displacement distribution and is valid in general only for $kh \ll 1$. It is interesting now to note some of the further effects of the layer on surface wave propagation when the layer thickness becomes an appreciable fraction of a wavelength [2.9, 33]. Here, the substrate as well as the layer will be assumed to be isotropic.

Figure 2.13 illustrates a typical dispersion curve for the surface wave propagating in a layer of fixed thickness when the layer loads the substrate; the parameters chosen correspond to a 10 μm layer of ZnO on silicon, with both materials assumed isotropic. It is seen that with increasing frequency the phase velocity ω/k changes gradually from that of the Rayleigh wave on a free surface of the substrate to the value of the velocity characteristic of a Rayleigh wave on a free surface of the layer material. The phase velocity normalized to the substrate shear velocity v_t is shown more accurately by the solid curve marked R_1 on Fig. 2.14. The initial slope of this curve can be obtained from (2.38), but for larger values of kh the curve deviates appreciably from a straight line and eventually becomes asymptotic to the surface wave velocity of the layer material. The group velocity of this wave is equal to the phase velocity for zero thickness and for large values of kh, but for intervening values there is dispersion in that the group velocity differs from the phase velocity.

Besides the dispersion introduced by the characteristic dimension of the layer there is also the possibility of higher order modes with a layered medium. For the loading case illustrated in Figs. 2.13 and 2.14, these higher modes, such as R_2, always exist for sufficiently large values of kh. Each higher mode has a lower

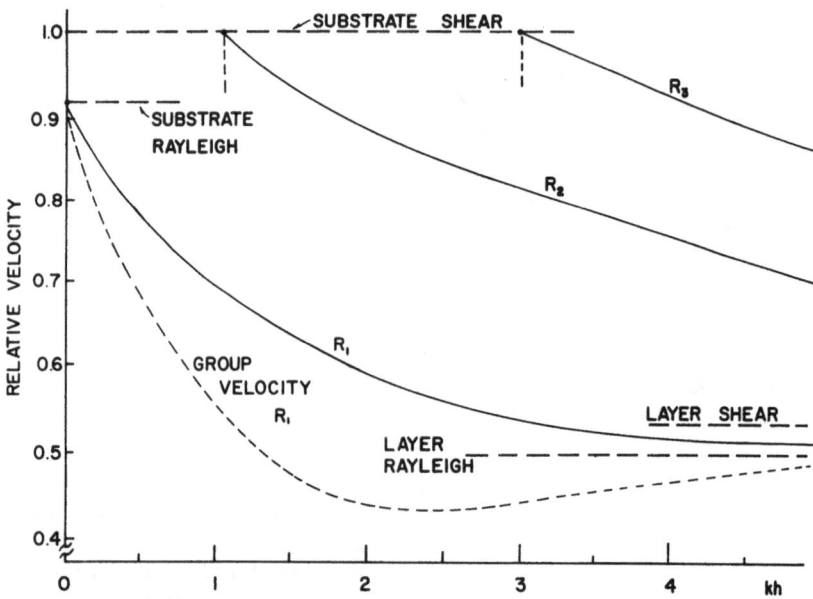

Fig. 2.14. Normalized dispersion curves for Rayleigh modes of a ZnO layer on an isotropic silicon substrate. Broken curve is group velocity of first Rayleigh mode

cutoff value of kh at which it phase matches to a bulk shear wave propagating parallel to the surface; the penetration into the substrate of the coupled wave becomes very deep and the group velocity equals the phase velocity. For values of kh larger than the corresponding cut-off value, the phase velocity of each of these higher modes decreases and at large kh becomes asymptotic to the velocity of a bulk shear wave in the layer material.

The changing displacement distribution of the first mode is indicated in Fig. 2.15, which shows the vertical component of displacement as a function of depth for different values of kh. For $kh \ll 1$, the displacement is characteristic of the wave on the free surface of the substrate, while for large values it is seen to correspond to a surface wave on the free surface of the layer and little of the energy of the wave is carried in the substrate. Similar gradual changes occur in the profile for the longitudinal component of displacement. Note that for $kh = 10$ the layer thickness is 1.6 wavelengths.

The behavior of the displacements of the second Rayleigh mode near cut-off is illustrated in Fig. 2.16, the broken curves show the displacement components for kh some 41 % above the cut-off value here of $kh = 0.2267$, and it is seen that the penetration of both components is only a wavelength or so, typical of surface waves. The profiles with depth of the two displacement components for this particular mode are interchanged from those of the first mode (Figs. 2.2 and 2.15). For a value of kh only 0.09 % above cut-off, the displacement profiles are as shown by the solid curves of Fig. 2.16, and here the vertical component maintains a constant amplitude to large depths so that this wave very close to cut-off might

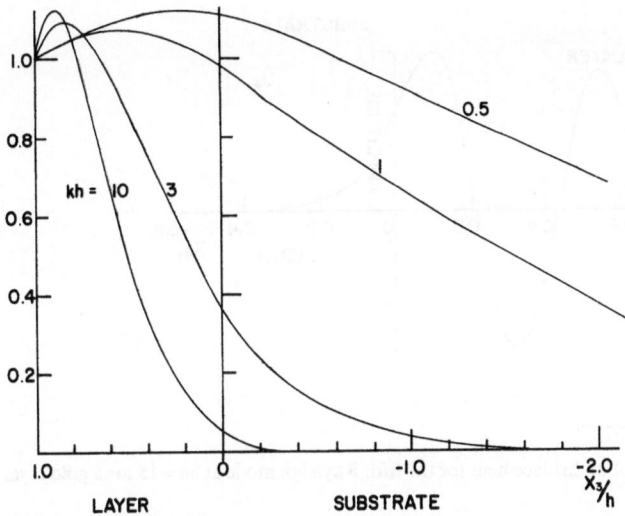

Fig. 2.15. Depth dependence of vertical component of displacement for the first Rayleigh mode of Fig. 2.14 at several values of kh

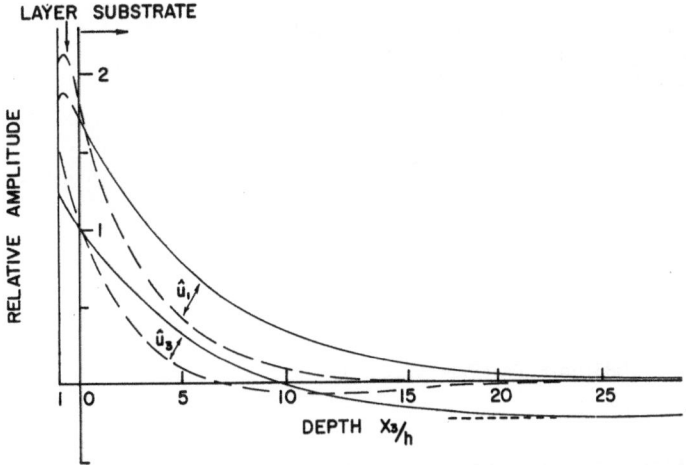

Fig. 2.16. Displacement components for kh only 0.09 % (solid curve), and 41 % (broken curve), above cut-off for the second Rayleigh mode of a gold layer on a fused quartz substrate

be described in terms of a SV bulk wave perturbed by the surface boundary conditions rather than in terms appropriate to a surface wave.

As the value of kh becomes much greater than the corresponding cut-off value, the character of each of the higher Rayleigh modes approaches that of a plate mode appropriate to a plate of thickness h of the layer material, but with perturbations introduced by the existence of the substrate below $x_3 = 0$ and the

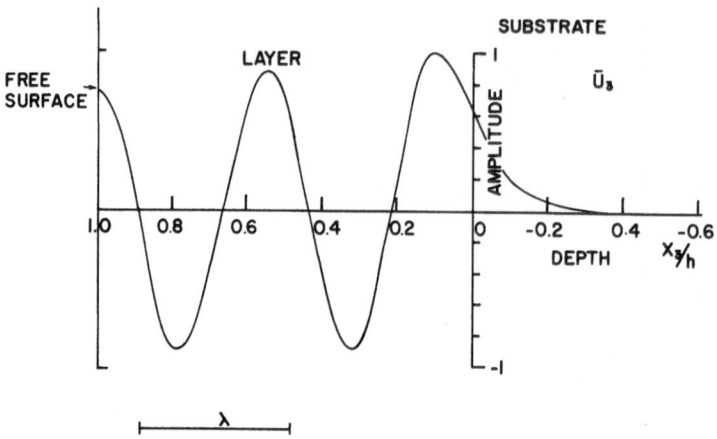

Fig. 2.17. Vertical component of displacement for the fifth Rayleigh mode at $kh = 15$ for a gold layer on a fused quartz substrate

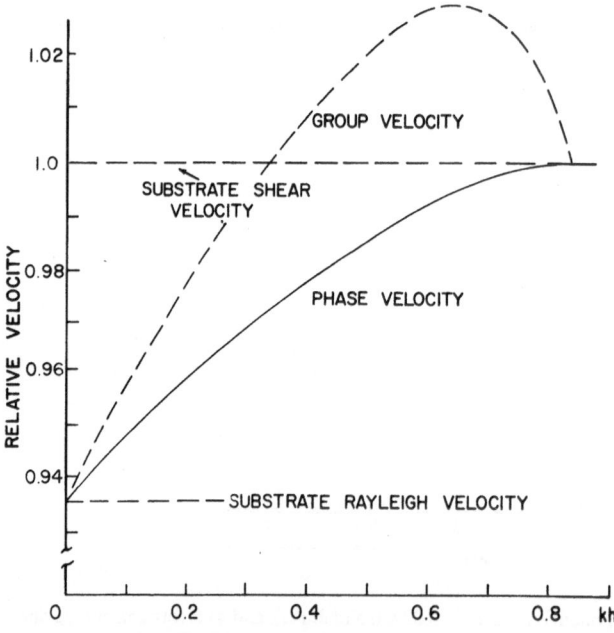

Fig. 2.18. Normalized dispersion curves for an isotropic silicon layer on a ZnO substrate. Broken curve gives the group velocity

small fraction of the energy carried in that substrate. The latter behavior for the fifth Rayleigh mode at $kh = 15$ for a gold on fused quartz system is shown in Fig. 2.17, where only the vertical component is plotted.

We have been discussing the main effects on Rayleigh wave propagation of layers which load the substrate. It was seen in (2.38) that if, on the other hand, the layer shear velocity is much larger than that of the substrate, the phase velocity is increased by the presence of the layer. The maximum increase in this velocity is limited, as illustrated by the solid curve of Fig. 2.18. As kh is increased, the phase velocity of the mode (only one exists) increases from the substrate surface wave velocity towards the bulk shear wave velocity of the substrate, and eventually equals the latter velocity. Very near this cut-off point the group and phase velocities are equal and the penetration of the vertical component of displacement becomes very deep, analogous to Fig. 2.16, so that the wave resembles at this point an SV bulk wave perturbed at the layer but with most of the energy being carried in the substrate.

2.7.2 Love Waves

It was shown in Section 2.4 that, for isotropic conditions and for many high symmetry anisotropic conditions, the wave equation and the boundary conditions separated, the sagittal-plane components leading to a Rayleigh-type surface wave and the transverse component leading to an SH bulk wave satisfying free-surface boundary conditions. In the previous sections, the effects on the propagation of the Rayleigh wave produced by the addition of a thin layer were discussed. If the layer material loads the substrate then, in the presence of the layer, the SH bulk wave noted above becomes a surface wave mode with energy carried within a few wavelengths of the surface. Such modes of propagation in a layered medium, with their single, transverse component of displacement, are referred to as Love waves [2.32]. The dispersion curves for a typical isotropic combination, gold on fused quartz, are shown in Fig. 2.19. The phase velocity is given implicitly by the relation

$$\tan b_L kh = \varrho v_t^2 b/\varrho_L v_{tL}^2 b_L \qquad (2.40)$$

where

$$b_L = \left[\left(\frac{v}{v_{tL}}\right)^2 - 1\right]^{1/2}, b = [1-(v/v_t)^2]^{1/2},$$

and v_t and v_{tL} are the bulk shear velocities of the layer and substrate materials, respectively, with $v_{tL} < v_t$. It is seen in Fig. 2.19 that for each mode the phase velocity given by (2.40) is equal to v_t at a cut-off value of $kh(kh=0$ for $L_1)$ and decreases with increasing kh to become asymptotic to the layer shear velocity v_{tL} at large values of kh. The displacement profile for each mode is sinusoidal within the layer and decays with a simple exponential form into the substrate as given by

$$u_{2L} = C[\cos b_L k(h-x_3)]/\cos b_L kh \qquad 0 \leq x_3 \leq h \qquad (2.41)$$

and

$$u_2 = C \exp(bkx_3) \qquad x_3 \leq 0$$

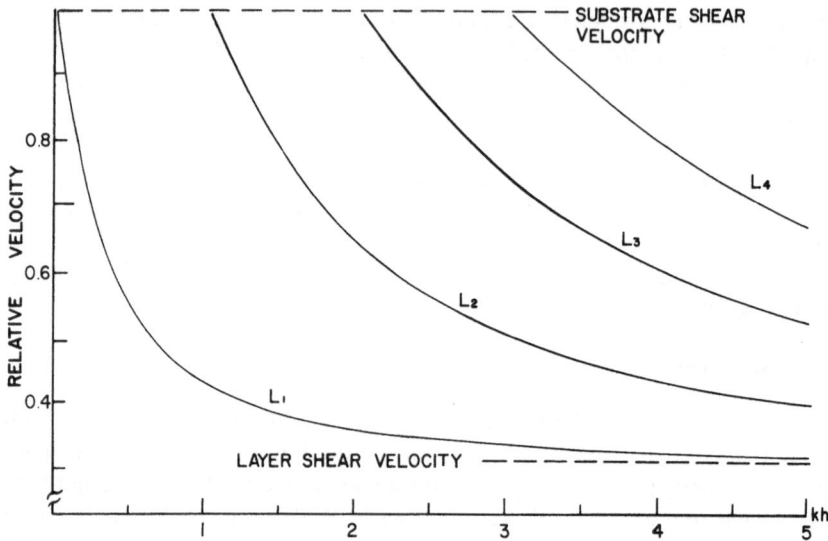

Fig. 2.19. Normalized dispersion curves for the Love modes of a gold layer on a fused quartz substrate

omitting the usual propagation factor. Figure 2.20 illustrates for different values of kh the displacement profiles of the first (solid curves) and the second (broken curves) Love modes corresponding to the dispersion curves of Fig. 2.19. For values of kh near cut-off, the penetration of the wave is very deep into the substrate and a large fraction of the energy of the wave is carried in this substrate. As kh is increased, a larger and larger fraction of the energy of the wave is carried in the layer, and in the limit when the wavelength becomes less than the layer thickness, the displacement profiles approach those of a plate free on one surface and clamped on the other, with but a small perturbation in the interface region.

2.7.3 Anisotropy

If the layer or the substrate is anisotropic or piezoelectric, the solutions for propagation in this layer are much more complex in detail than with the isotropic cases discussed above, and these solutions do not in general separate into Rayleigh and Love modes; that is, the general solutions include the potential and all three components of displacement in both the layer and the substrate [2.9, 33]. The general anisotropic layered geometry will not be considered further here; however, same comments should be made regarding certain high symmetry anisotropic cases. If for both media the sagittal plane is a plane of reflection symmetry, the Love modes are independent of the potential and the corresponding dispersion curves are similar to those of Fig. 2.19 except for a necessary redefinition of the bulk shear wave velocities involved because

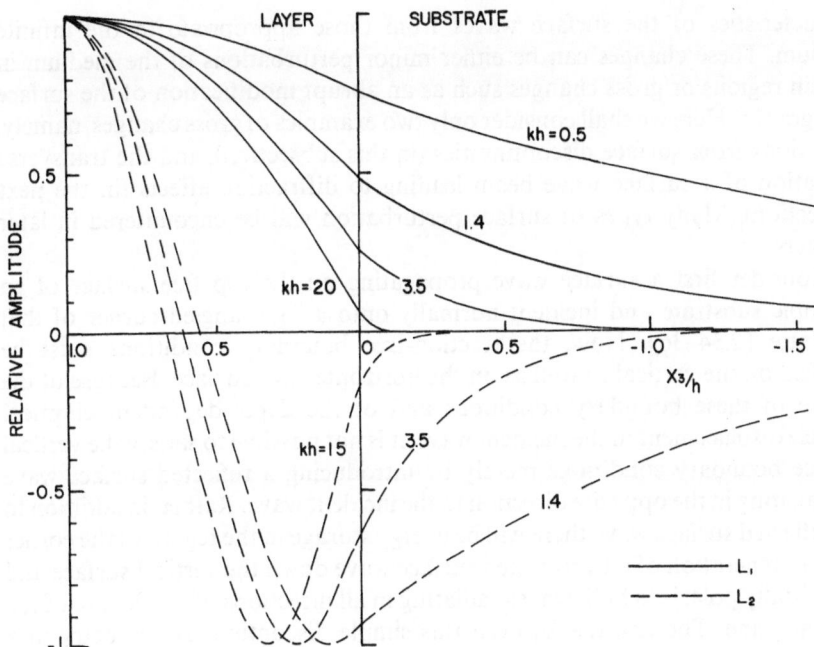

Fig. 2.20. Displacement profiles for the first two Love modes of Fig. 2.19 at several values of kh

such velocities are functions of angle and polarization. The sagittal-plane mechanical displacements for this symmetry are coupled to the potential but the general form of the dispersion curves is similar to that of Figs. 2.14 or 2.18, again with an appropriate choice of bulk shear velocity. On the other hand, if the sagittal-plane is perpendicular to a twofold axis of both the layer and the substrate, the Love modes have the accompanying potential while the Rayleigh-type modes have only sagittal displacements and propagate independently of the piezoelectricity existing in either crystal. For either type of symmetry, the modes are pure in the sense that the Poynting vector is in the same direction as the propagation vector. However, when, for the same layered system, the angle of propagation is changed away from the pure-mode axis, there will be changes in the phase velocity and intercoupling of all the displacement components and the potential.

2.8 Reflections

In all of the discussion so far in this chapter, the propagation medium has been taken to be infinite both in the direction of propagation and perpendicular to this direction. There are many fashions in which the propagation medium can be changed as a function of distance in these two directions thereby altering the characteristics of the surface waves from those appropriate to the infinite medium. These changes can be either minor perturbations of the medium in certain regions or gross changes such as an abrupt modification of the surface topography. Here we shall consider only two examples of gross changes, namely, reflections from surface discontinuities (in this subsection), and the transverse limitation of a surface wave beam leading to diffraction effects (in the next subsection). Many types of surface perturbation will be encountered in later chapters.

Consider first a surface wave propagating on the top free surface of an isotropic substrate and incident normally onto a right-angled corner of that substrate [2.34–36]. Now, the traction-free boundary conditions must be satisfied on the vertical as well as on the horizontal free surface. Because of the nature of these boundary conditions and of the depth-dependent elliptical particle displacement in the incident wave, it is not possible to satisfy the vertical surface boundary conditions merely by introducing a reflected surface wave propagating in the opposite direction to the incident wave. Rather, in addition to the reflected surface wave there will be energy storage in the region of the corner and the generation of a transmitted surface wave down the vertical surface and of sagittally polarized bulk waves radiating in all directions of the 90° arc of the sagittal-plane. The analysis of even this simple discontinuous geometry in a homogeneous medium is very difficult and will not be discussed here, but a few results obtained by numerical iteration techniques will be presented to indicate some of the important features.

In the right-angled or quarter-space geometry under consideration, there is no characteristic dimension and hence no dispersion. Moreover, the behavior of the corner for such parameters as reflection coefficient depends only on the ratio of the velocity of the bulk shear wave to that of the bulk longitudinal wave in the substrate material and rather weakly on this ratio at that.

The reflection and transmission behavior of a right-angled corner for a typical homogeneous solid $v_t/v_l = 0.581$, can be summarized by the following table:

Transmission coefficient

Amplitude	0.64
Phase	$-79°$

Reflection coefficient

Amplitude	0.36
Phase	$38°$

Energy conversion to bulk modes

46%

where the amplitude of the reflection coefficient is the ratio of the amplitude of a displacement component, say the vertical component on the surface, of the reflected wave measured several wavelengths from the corner, to the amplitude

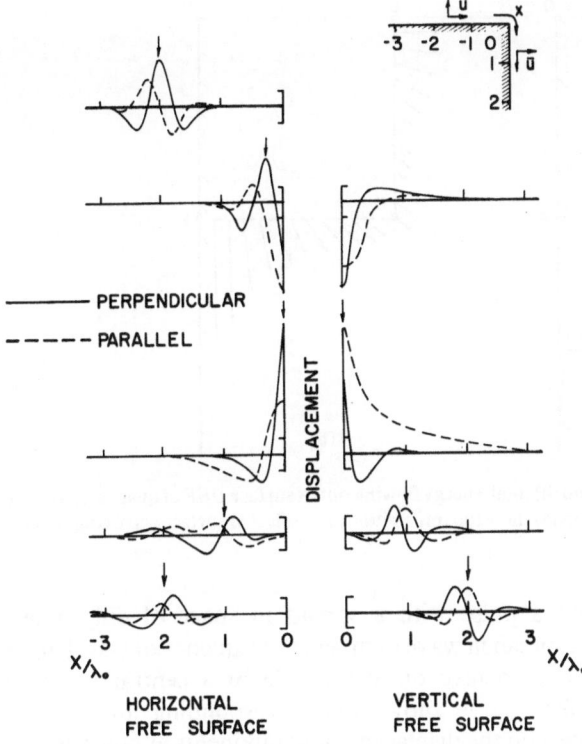

PERPENDICULAR
PARALLEL

HORIZONTAL
FREE SURFACE

VERTICAL
FREE SURFACE

Fig. 2.21. Inset shows geometry and distances normalized to λ_0. Graphs show displacement components perpendicular and parallel to the horizontal surface (left-hand curves) and to the vertical surface (right-hand curves) at successive incidents of time. Arrows "propagate" within the Rayleigh velocity

of the same component in the incident wave. The phase of the reflection coefficient is the additional phase lag between reflected and incident wave over and above that introduced by the geometrical path length at the surface wave velocity from the surface point of observation of the incident wave to the corner and back to the surface point of observation of the reflected wave. Similar definitions apply for the transmission coefficient.

From the above table, it is seen that 41 % of the incident energy is transmitted around the corner and propagates down the vertical surface as a surface wave and 13 % is reflected as a surface wave propagating away from the corner on the top surface, while 46 % of the incident energy is converted into bulk modes radiating away from the corner into the substrate. Because of the phase shift in the transmission coefficient, the surface wave arriving at a point far down the vertical surface appears to have originated from an effective line source some 0.22λ below the corner.

The details of the reflection can be visualized by considering the time sequence of events as a pulse of surface wave energy, rather than a continuous wave, impinges on the corner. Many pulse shapes could be used, but a particularly useful form [2.35] is represented by the curves in the upper left diagram of Fig. 2.21. The solid curve shows the vertical component of the particle displacement on the surface at a particular instant of time. This

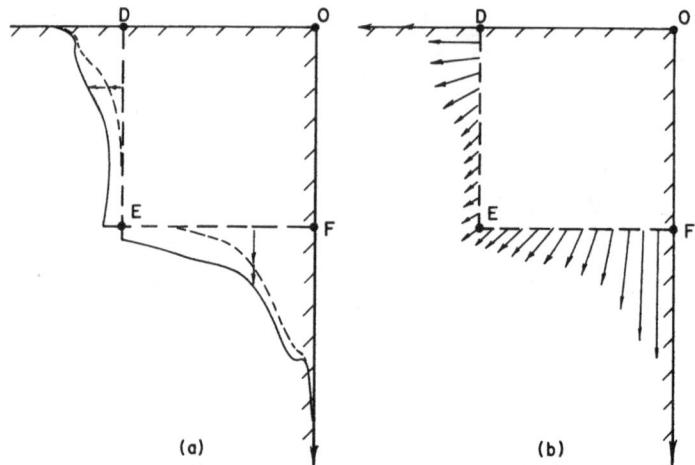

Fig. 2.22. (a) Normally directed, and (b) total, energy flowing out of surface DEF of quarter-space. DE $= EF = 1.9 \, \lambda_0$. The broken curves in (a) show the transmitted and reflected surface wave components of the normal energy flow

particular waveform allows a pulse with a spread in space of only a few wavelengths, yet requiring a spread in wave numbers or frequency or wavelength to create it of only about an octave on either side of a central value of $k_0 = \omega_0/v = 2\pi/\lambda_0$. The broken curve shows the corresponding longitudinal displacements on the surface, and the displacement components at other depths are also chosen to correspond for each wavelength to the depth dependence of Fig. 2.2. This pulse is assumed to be propagating to the right and, since the free surface is dispersionless for surface waves, the pulse maintains its shape as it propagates towards the corner $X = 0$.

The sequence of graphs in Fig. 2.21 shows the evolution of the surface displacements as the pulse is incident onto the corner, the right-hand set giving the components on the verical surface. The small vertical arrows "propagate" at the Rayleigh velocity and thus indicate the successive time instants. At the corner itself, the perpendicular component on the top surface is of course the same as the component parallel to the vertical surface. The reflected and transmitted pulses that have almost completely evolved in the last row propagate unchanged in shape. The reflection and transmission coefficients cited above are given by the magnitude and phase of any one Fourier component, for example at λ_0, of the latter pulses. These pulses differ in shape from the incident one because each component undergoes a fixed phase shift in degrees.

Figure 2.21 was concerned with the surface displacements and hence the surface waves excited at the corner, but it is of interest to consider also the development of the bulk waves. The solid curves constructed on lines DE and EF of Fig. 2.22a show normally directed energy per unit area flowing out of the surface DEF where $OD = OF = 1.9\lambda_0$. The total area under these curves is equal to the energy of the incident surface wave pulse. The broken curve shows the contribution of the two scattered surface waves. The remaining area under the solid curves indicates the fraction of the incident energy which is converted into bulk waves. At this short distance from the corner, it is not possible to resolve the total displacement at, say, point E into longitudinal and shear wave contributions because propagation "direction" cannot be defined at this distance. However, if the instantaneous power-flow vector is integrated over the duration of the pulse at each point on DEF, the total outward flowing energy per unit area is given by the vectors of Fig. 2.22b. As mentioned previously, about 46% of the incident energy is converted to bulk waves, and of this converted energy about 90% is radiated into a sector defined by planes at 15° to the two free surfaces.

If the incident surface wave impinges on a 270° corner rather than on the 90° corner discussed above, that is, the substrate is a three-quarter rather than a quarter-space, the scattering is given by ($v_t/v_l = 0.581$)

Reflection coefficient

    Amplitude   0.09

    Phase   −125°

Transmission coefficient

    Amplitude   0.28

    Phase   140°

Energy conversion to bulk modes

        91%.

Here the reflected wave is small; there is some transmitted wave propagating up the vertical surface, but most of the energy is radiated from the corner as bulk waves into the substrate beyond the corner.

As a final example of gross discontinuities in the surface wave propagation path, consider that the wave is incident onto a downward step, as indicated on the insert sketch of Fig. 2.23. The solid curve gives the amplitude of the transmission coefficient, and, because of the characteristic dimension, the step height, the transmission and reflection now depend on the wavelength of the incident wave. For large values of relative step height, the amplitude transmission coefficient approaches the value appropriate to a 90°-corner followed by a distant 270°-corner, $0.64 \times 0.28 = 0.18$. The phase of the transmission coefficient plotted in Fig. 2.23 indicates the phase lag from the phase of the transmitted wave which would exist to the right of the step location if the steps did not exist. Figure 2.24 shows the percentage of the incident energy which is transmitted as a surface wave, reflected as a surface wave, and converted to bulk waves by a downward step as a function of the relative step height. For large step height, the converted energy curve becomes asymptotic to $46\% + (0.64)^2 \times 91\% = 83\%$, corresponding to the two corners in cascade.

The cases of small discontinuities which can be treated by perturbation *theory neglecting the bulk modes* will not be treated here, but several examples of

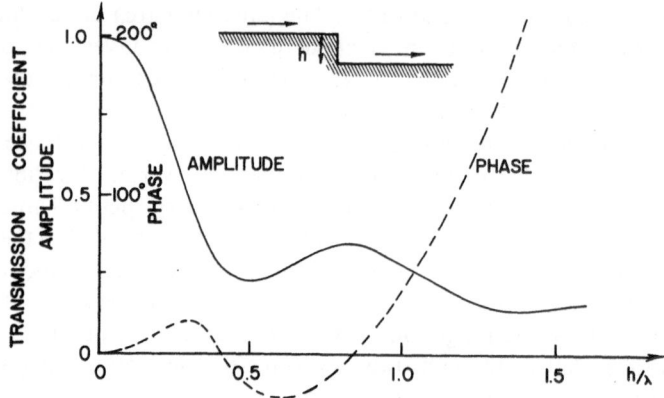

Fig. 2.23. Amplitude and phase of the transmission coefficient for a surface wave normally incident on a downward step

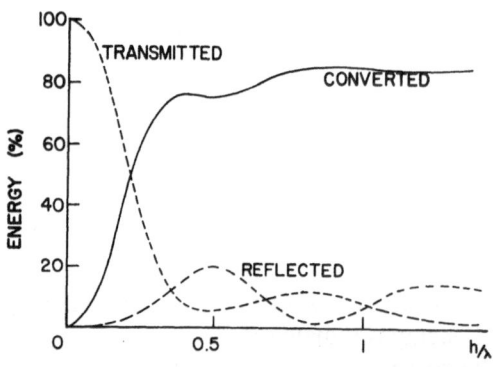

Fig. 2.24. Percentage of incident energy reflected and transmitted as Rayleigh waves (broken curve) and converted to bulk waves (solid curve) for downward step of Fig. 2.23

reflection from such perturbations will arise in later chapters. For example, in Chapter 4, reflections from an array of shallow grooves are considered and it is found that the amplitude reflection coefficient from each discontinuity is proportional to the step height in wavelengths. Similarly, practical methods for minimizing the effects of reflections from the electroded part of an interdigital transducer is discussed in Section 3.2.2.

2.9 Diffraction

The surface waves discussed previously in this chapter, even for the anisotropic substrates, have all been without structure in the direction perpendicular to the sagittal-plane, the (x_1, x_3) plane of Fig. 2.1. Thus it was assumed that both the medium and the wave were without variation in this x_2 direction, with the result that the wave was identical for different sagittal-planes chosen at different values of x_2. In devices and experiments, however, the width of the surface wave beam

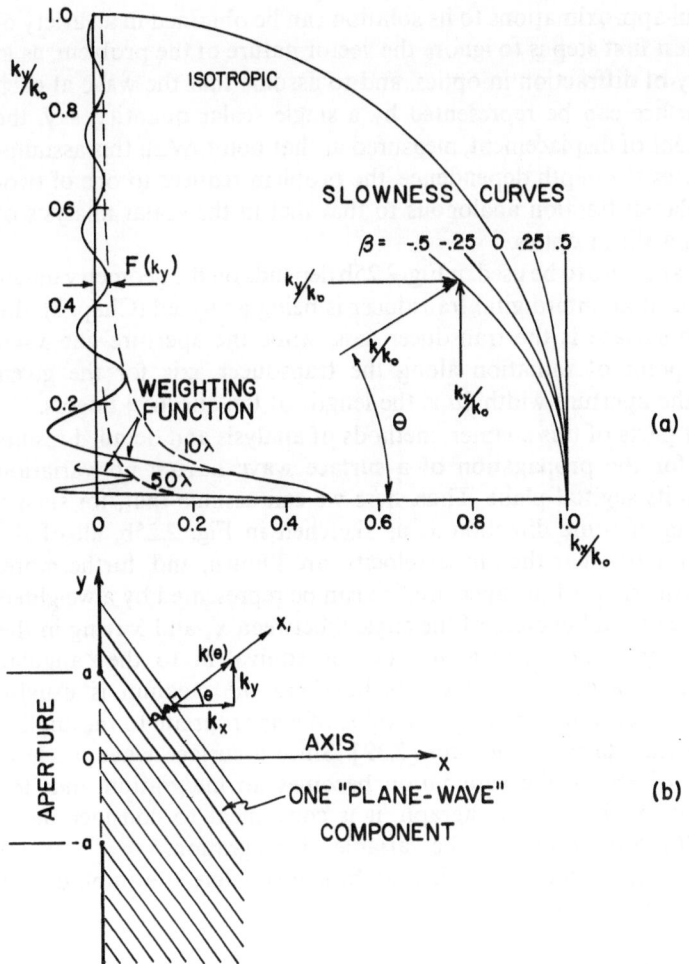

Fig. 2.25. (a) Slowness curve for an isotropic substrate and various parabolic approximations near the axis for anisotropic substrates. Also shown are curves of $F(k_y)$ in (2.45) for $2a=10\lambda_0$ and $2a=50\lambda_0$. (b) Geometry for diffraction calculation using angular spectrum of plane waves

in the x_2 direction will be limited either by the transducer exciting the wave or by changes in the propagation medium in the x_2 direction, as, for example, in the waveguiding structures of Chaper 5. Here we wish to discuss the diffraction effects produced by limiting the width of a surface wave beam in one region of an otherwise semi-infinite substrate [2.23, 24, 37–40].

In Fig. 2.25b, the (x, y) plane is the surface of the substrate which in general is anisotropic. It is assumed that a surface wave of known characteristics is incident onto the plane $x=0$ so that the displacement components are known on this plane and that some form of aperture causes the displacements to be zero for $|y|>a$ on a plane just to the right of $x=0$. This general problem has not been solved, but useful approximations to its solution can be obtained in a variety of ways. The simplest first step is to ignore the vector nature of the problem, as in the scalar theory of diffraction in optics, and to assume that the wave at each point on the surface can be represented by a single scalar quantity, say, the vertical component of displacement, measured at that point. With this assumption, which ignores the depth dependence, the problem reduces to one of two-dimensional scalar diffraction analogous to that met in the scalar analysis of diffraction from a slit in optics.

The effective aperture to be used in Fig. 2.25b depends on the geometry under consideration, but if an interdigital transducer is being analyzed (Chap. 3), the line $y=0$ on the surface is the transducer axis, while the aperture line $x=0$ is the effective point of radiation along the transducer axis for the given frequency, and the aperture width $2a$ is the length of the effective fingers.

In the earlier parts of this chapter, methods of analysis and detailed results were discussed for the propagation of a surface wave having no variation perpendicular to its sagittal-plane. Thus, here we can assume that, for such a wave propagating in some direction x_1 as sketched in Fig. 2.25b, all of the characteristics, in particular the phase velocity, are known, and, furthermore, that the wave to the right of the aperture line can be represented by a weighted sum of such waves for all choices of the angle θ between x_1 and x lying in the range $-\pi/2 \leq \theta \leq \pi/2$. The latter assumption is equivalent to the "angular spectrum of plane waves" used in optics but here the medium is usually anisotropic and, for a given frequency, the value of k appropriate to the angle θ must be used in each term of the sum [2.39]. Since a surface wave mode is possible in each direction, the summation becomes an integration and, for reasons to be met in the next paragraph, it is convenient to consider the y component of $k(\theta)$ rather than θ as the variable of integration.

At any point on the surface to the right of the aperture line, the amplitude of the scalar quantity is thus given by

$$u(x,y) = \int_{-\infty}^{\infty} F(k_y) \exp[i(k_x x + k_y y)] dk_y \quad x \geq 0 \tag{2.42}$$

where the $\exp(-i\omega t)$ time dependence is implied and k_x, treated as a function of k_y, is given by

$$k_x = +[k^2(\theta) - k_y^2]^{1/2}. \tag{2.43}$$

As illustrated by the vectors in Fig. 2.25a, for each value of k_y, here normalized to the value k_0 along the axis, the vector k terminating on the slowness curve appropriate to the crystalline geometry is determined and thereby the corresponding x component k_x is determined. The treatment of the values of k_y larger than k in the complete range of real values of k_y in the integral of (2.42) will be considered later.

In the limit as $x \to 0$, (2.42) becomes

$$u(0,y) = \int_{-\infty}^{\infty} F(k_y) \exp(ik_y y) dk_y \tag{2.44}$$

where y in this expression is measured along the aperture line. Since (2.44) is the Fourier transform of the function $F(k_y)$, the latter function can be expressed as an inverse transform

$$F(k_y) = \frac{1}{2\pi} \int_{-\infty}^{\infty} u(0, y) \exp(-ik_y y) dy \qquad (2.45)$$

but the aperture "illumination" $u(0, y)$ has been taken as a known function; thus, the weighting function of (2.42) is the Fourier transform of the aperture illumination. Knowing the aperture function and the details of (2.43), the amplitude and phase of the acoustic field can be determined at each point by evaluating (2.42) either by numerical integration or suitable approximate analytical methods.

     If the aperture illumination is of uniform amplitude and constant phase for $y < |a|$ and vanishes for $y > |a|$, then the weighting function is of the form $\sin(k_y a)/k_y a$, which has been plotted as a function k_y/k_0 for $2a = 10\lambda_0$ and $2a = 50\lambda_0$ along the vertical axis of Fig. 2.25a, where $\lambda_0 = 2\pi/k_0$. It is seen that even for apertures as small as ten wavelengths across, all the significant contributions to the integral come from small values of k_y/k_0. As a result, only the parts of the slowness curve near the axial value are important in determining the diffraction patterns of the aperture, provided the aperture is many wavelengths in width. This is true even if the illumination departs somewhat from a constant value across the aperture. Returning to (2.43), the values of $|k_y|$ greater than $k(\pm\pi/2)$ do not usually contribute to the diffraction integral and thus can be ignored or perhaps better included approximately by taking $k_x = +i|k_y^2 - k(\pi/2)|^{1/2}$ in the range $k_y > k(\pi/2)$ (evanescent waves).

     A typical complete slowness curve was shown in Fig. 2.19, but it has just been indicated that only a small section of this curve is important for most diffraction calculations, the small section being about the direction perpendicular to the aperture if the phase of the illumination is symmetrical about this axis. The section would be offset by an angle from this axis if the phase varied linearly across the aperture. When only a small section of the slowness curve is significant it can be convenient to approximate it in this region by an algebraic expression. In many practical applications the axis perpendicular to the transducer is chosen to be a pure-mode axis of the crystalline substrate. Under these circumstances, the velocity can be written [2.23, 24] as in (2.29) and (2.30)

$$v = v_0(1 - \beta\theta^2). \qquad (2.46)$$

This choice of v gives for the slowness surface relation of (2.43)

$$k_x \cong k_0[1 - (1 - 2\beta)(k_y/k_0)^2]^{1/2} \qquad (2.47)$$

and several examples of the corresponding slowness curves are illustrated in Fig. 2.25a.

The diffraction integral now becomes

$$u(x, y) = \exp(ik_0 x) \int_{-\infty}^{\infty} F(k_y) \exp\left\{i\left[-\frac{k_y^2}{2k_0}(1-2\beta)x + k_y y\right]\right\} dk_y. \tag{2.48}$$

Since $\beta = 0$ corresponds to an isotropic medium, (2.48) differs from that of an isotropic medium only in that x is replaced by $(1-2\beta)x$. Thus, the amplitude profile calculated as a function of y in this parabolic approximation to an anisotropic case will be the same at a distance x as the profile calculated for the same geometry with the isotropic medium at a distance $(1-2\beta)x$ from the aperture.

It is interesting to note that for the case in which $\beta = 1/2$, k_x is independent of k_y as shown in Fig. 2.25a, and (2.48) reduces to

$$u(x, y) = u(0, y) \exp(ik_0 x) \tag{2.49}$$

which implies that in this special case the beam profile as a function of transverse distance at any value of x is the same as the profile at the aperture.

It is convenient to measure the distance from the aperture in terms of the dimensionless parameter

$$X = \frac{\lambda_0(1-2\beta)}{a^2} x \tag{2.50}$$

so that for a given aperture illumination the diffraction profile as a function of the transverse distance y/a is the same for all physical distances giving the same value of this parameter. The diffraction profiles $u(x, y)$ for uniform-amplitude constant-phase illumination of the aperture are shown in Fig. 2.26 for different values of X. For values of X less than unity, the surface wave energy radiated from the aperture is in the form of a beam in that most of the energy remains within a transverse distance about the axis approximately equal to the aperture width. There are ripples in the amplitude, as seen in Fig. 2.26, but the phase variation over the region $(-a < y < a)$ is much less than 2π and thus, as will be seen in Chapter 6 for a constant-phase type of receiving transducer of width $2a$, such as an interdigital comb, the contributions from each transverse region will add in phase in this Fresnel region, $X < 1$. The fraction of the beam intercepted is discussed quantitatively in Chapter 6.

In the Fraunhofer region $X > 1$, the surface wave energy diverges such that the profile as a function of angle from the aperture axis remains constant for increasing distance. The phase in each lobe of this divergent beam is constant over an arc centered approximately at the aperture.

From the form of the parameter X, it is evident that positive values of the anisotropy parameter β lead to a certain degree of autocollimation in that the larger the value of β the larger the physical distance from the aperture must be before a given diffraction profile is met [2.23, 24, 40]. As noted previously, for $\beta = 1/2$ the autocollimation is complete provided the aperture is large enough for

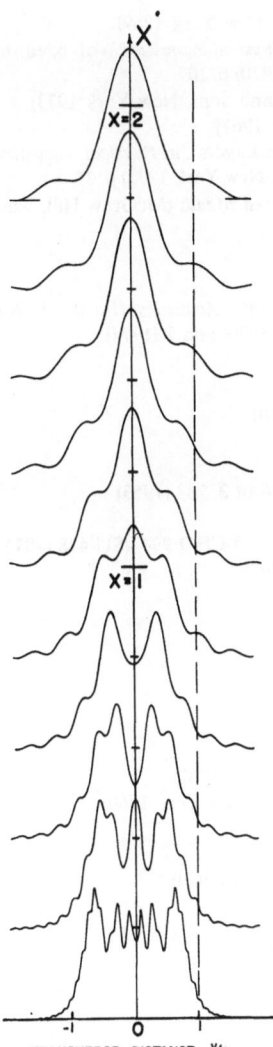

TRANSVERSE DISTANCE y/a

Fig. 2.26. Transverse diffraction profiles at different values of the normalized distance from the aperture

the parabolic approximation to hold. Table 6.6 shows that practical choices of crystal geometries for $Bi_{12}GeO_{20}$ do lead to values of $\beta = -(1/2)(d\Psi/d\theta)$ approaching the autocollimation condition.

References

2.1   G. Nadeau: *Introduction to Elasticity* (Holt, Rinehart and Winston, New York 1964)
2.2   J. F. Nye: *Physical Properties of Crystals* (Oxford University Press, London 1957)
2.3   W. G. Cady: *Piezoelectricity* (McGraw-Hill, New York 1946)
2.4   H. F. Tiersten: J. Appl. Phys. **40**, 770 (1969)

2.5   H.F.Tiersten: *Linear Piezoelectric Plate Vibrations* (Plenum, New York 1969)
2.6   G.W.Farnell: "Properties of Elastic Surface Waves", in *Physical Acoustics*, Vol. 6, ed. by W. P. Mason, R. N. Thurston (Academic Press, New York 1970) p. 109
2.7   B.A.Auld: *Acoustic Fields and Waves in Solids* (John Wiley and Sons, New York 1973)
2.8   I.A.Viktorov: *Raleigh and Lamb Waves* (Plenum, New York 1967)
2.9   G.W.Farnell, E.L.Adler: "Elastic Wave Propagation in Thin Layers", in *Physical Acoustics*, Vol. 6, ed. by W. P. Mason, R. N. Thurston (Academic Press, New York 1972) p. 35
2.10 W.M.Ewing, W.S.Jardetsky, F.Press: *Elastic Waves in Layered Media* (McGraw-Hill, New York 1957)
2.11 J.P.Jones: J. Appl. Mech. **31**, 213 (1964)
2.12 J.J.Campbell, W.R.Jones: IEEE Trans. SU-**15**, 209 (1968)
2.13 A.J.Slobodnick, Jr., R.T.Delmonico, E.D.Conway: *Microwave Acoustics Handbook*, Air Force Cambridge Research Laboratories, Bedford, Mass. **1** (1970) and **2** (1974)
2.14 T.C.Lim, G.W.Farnell: J. Appl. Phys. **39**, 4319 (1968)
2.15 K.A.Ingebrigtsen, A.Tonning: Phys. Rev. **184**, 942 (1969)
2.16 D.C.Gazis, R.Herman, R.F.Wallis: Phys. Rev. **119**, 533 (1960)
2.17 J.J.Campbell, W.R.Jones: J. Appl. Phys. **41**, 2796 (1970)
2.18 R.Stoneley: Proc. Roy. Soc. **232A**, 447 (1955)
2.19 V.T.Buchwald, A.Davis: Quart. J. Mech. and Appl. Math. **16** pt 3, 283 (1963)
2.20 T.C.Lim, G.W.Farnell: J. Acous. Soc. Amer. **45**, 845 (1969)
2.21 D.Penunuri, K.M.Lakin: Proc. Ultrasonics Symp., IEEE Cat. 75 CHO 994-4SU-478 (1975)
2.22 G.A.Coquin, H.F.Tiersten: J. Acous. Soc. Amer. **41**, 921 (1967)
2.23 E.P.Papadakis: J. Acous. Soc. Amer. **40**, 863 (1966)
2.24 M.G.Cohen: J. Appl. Phys. **38**, 3821 (1967)
2.25 C.-C.Tseng, R.M.White: J. Appl. Phys. **38**, 4274 (1967)
2.26 C.Lardat, C.Maerfeld, P.Tournois: Proc. IEEE **59**, 355 (1971)
2.27 J.L.Bleustein: Appl. Phys. Lett. **13**, 412 (1968)
2.28 Y.V.Gulyaev: ZhETF Pis. Red. **9**, 63 (1969)
2.29 C.-C.Tseng: Appl. Phys. Lett. **16**, 253 (1970)
2.30 J.D.Achenbach, H.Epstein: Proc. Amer. Soc. Civil Eng. EM**5**, 27 (1967)
2.31 L.M.Brekhovskikh: *Waves in Layered Media* (Academic Press, New York 1960)
2.32 P.Tournois, C.Lardat: IEEE Trans. SU-**16**, 107 (1969)
2.33 L.P.Solie: Appl. Phys. Lett. **18**, 111 (1971)
2.34 Z.Alterman, D.Loewenthal: Geophys. J. Royal Astr. Soc. **20**, 101 (1970)
2.35 M.Munasinghe, G.W.Farnell: J. Appl. Phys. **44**, 2025 (1973)
2.36 W.H.Haydl: Proc. Ultrasonics Symp. IEEE Cat. 73 CHO 807-8SU-363 (1973)
2.37 T.L.Szabo, A.J.Slobodnik, Jr.: IEEE Trans. SU-**20**, 240 (1973)
2.38 I.M.Mason: J. Acous. Soc. Amer. **53**, 1123 (1973)
2.39 M.S.Kharusi, G.W.Farnell: Proc. IEEE **60**, 945 (1972)
2.40 I.M.Mason, E.A.Ash: J. Appl. Phys. **42**, 5343 (1971)

3. Principles of Surface Wave Filter Design[1]

H. M. Gerard

With 26 Figures

This chapter presents a circuit model characterization of the interdigital (ID) surface acoustic wave transducer that is valuable in understanding the fundamentals of transducer design trade-offs. Since virtually all surface wave components employ ID transducers, the circuit model [3.1] (with refinements [3.2–5]) forms the cornerstone of surface wave technology.

We begin by placing surface wave technology in perspective with other techniques for implementing transversal filters. The principles of transversal filter design are then briefly reviewed in order that they may be used later as a measure of the validity of the surface wave filter design principles.

Next, the circuit model characterization of the electric to acoustic transfer function of a generalized ID transducer is derived. In the weak-coupling limit, this model is shown to become equivalent to a δ-function model which is, itself, a representation of an ideal transversal filter. Therefore, departures in the predictions of the circuit model from those of the δ-function model represent systematic errors in the ID transducer.

The details of transversal filter theory comprise an extensive subject in its own right [3.6, 7], which is not appropriate for discussion here. Instead, circuit-model trade-off relationships are presented for use in achieving transducer performance that is within a specified tolerance of a general ideal transversal filter response. The relationships among amplitude errors, phase errors, and insertion loss are explored. The dependence of ID transducer performance on substrate material, transducer geometry, design bandwidth, and on the details of the electrical tuning circuit is explained, thus completing a unified approach to the design of surface wave filters.

3.1 Historical Background

Since 1965, three major factors have spurred a growing interest in surface wave filters. First is the substantial microminiaturization achievable because an acoustic wavelength is 10^5 times smaller than the electromagnetic wavelength at

---

[1] Motivation for the preparation of much of the material in this chapter was stimulated by Robert Janowiak of the National Electronics Conference. That portion of this material was presented at NEC Professional Growth Seminars in Acoustic Surface Wave Techniques during 1972–1973.

Fig. 3.1. Uniform periodic interdigital (ID) transducer with single electrodes

the same frequency. Second is the particular convenience of the surface wave mode. Being both nondispersive and guided, since acoustic wave propagation follows the stress-free surface, the wave is ideally suited for implementing tapped delay lines. The wave type is discussed in detail in Chapter 2. Furthermore, as there is only one surface mode, which is easily excited, detected, and terminated on a piezoelectric substrate, efficient, spurious-free operation results. Third is the ease of precision fabrication of ID transducer structures, leading to extremely high filter accuracy and reliability. With respect to fabrication, surface wave filter development has prospered greatly from the advances in photofabrication technology, promoted by a burgeoning semiconductor industry. Further details regarding fabrication are presented in Chapter 7.

3.1.1 The Interdigital (ID) Transducer

Interdigital (ID) transducers are the building blocks of surface wave filters. As shown in Fig. 3.1, they generally consist of interleaved combs of metal electrodes, with each set extending from a common contact pad. The ID transducers are photodeposited on the highly polished surface of a precisely oriented piezoelectric crystal. Upon application of a voltage to the contact pads, an electric field distribution having the spatial period of the electrodes is established between the electrodes. By means of piezoelectric coupling, these surface-concentrated fields produce a corresponding elastic strain distribution. With fields localized at the free surface, the coupling to surface acoustic waves can be made quite strong. It also follows that the transducer efficiency is a maximum at the rf excitation frequency for which the surface wave propagates one transducer period in one rf period. This frequency is called the synchronous frequency.

3.1.2 The Surface Wave Filter

As a consequence of the bilateral symmetry of the structure in Fig. 3.1, the elastic strains produced by the electrical input signal leave the transducer symmetrically from both ends. A surface wave filter is created when a second ID transducer is

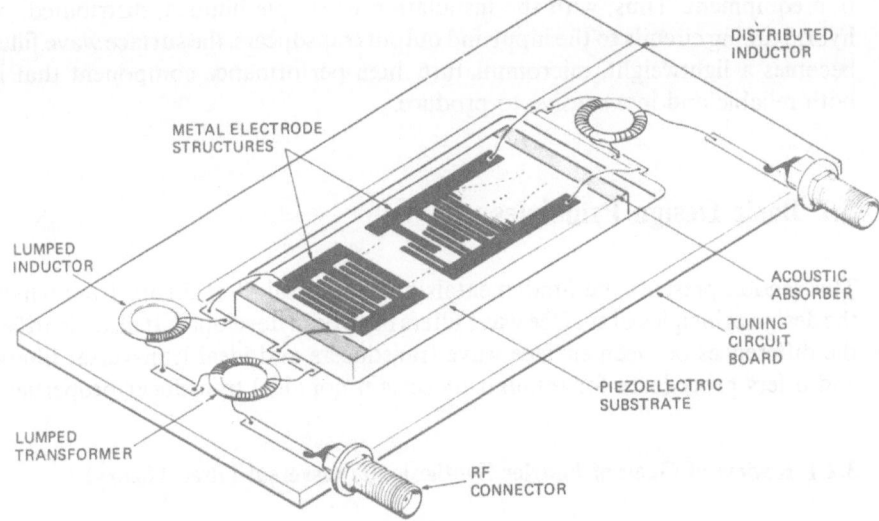

DISTRIBUTED
INDUCTOR

METAL ELECTRODE
STRUCTURES

LUMPED
INDUCTOR

ACOUSTIC
ABSORBER

TUNING
CIRCUIT
BOARD

PIEZOELECTRIC
SUBSTRATE

LUMPED
TRANSFORMER

RF
CONNECTOR

Fig. 3.2. Typical surface wave filter showing substrate, transducers, and tuning circuits

used to detect the piezoelectric waves emanating from one end of the first transducer. Detection occurs by means of inverse piezoelectric coupling. The principal merit of the ID transducer circuit model is that it predicts the transfer function relating the input voltage to the output acoustic signal. This transfer function provides the basic description of the transducer design/trade-off relationships that is required to optimize a surface wave filter design [3.8].

In addition to the two ID transducers, the complete surface filter generally contains electrical tuning circuits for optimizing the interfaces with external electrical components. Figure 3.2 shows the three basic elements of a surface wave filter, namely, the substrate, the metal electrode transducers, and the electromagnetic tuning circuits, combined in a simple rf package.

Dependent as it is on elastic, piezoelectric, and electromagnetic variables, the surface wave filter is far more complicated than its simplest theoretical representation, namely as an idealized δ-function implementation of a transversal filter [3.9]. In practice, this complexity increases the difficulty of developing a high-performance surface wave filter. Nevertheless, the effort is usually well justified, since the volume production of rather sophisticated high-frequency filters becomes a highly repeatable operation.

In addition to inherent producibility, the surface wave filter has the advantage of high reliability deriving from the quality of raw materials and fabrication processes. Substrates are obtained from single-crystal boules which are x-ray oriented and precision sliced and polished to remove pits and scratches. All metal structures are vacuum deposited to carefully controlled thickness or resistivity. Moreover, surface wave filter accuracy is often derived directly from the positioning accuracy of laser-interferometer-controlled photomask generation equipment. Thus, with the installation of simple lumped, distributed, or hybrid tuning circuits to the input and output transducers, the surface wave filter becomes a lightweight, microminiature, high-performance component that is both reliable and inexpensive to produce.

3.2 Basic Design Principles

This section presents the fundamentals of transversal filter theory from which the design principles of surface wave filters have been developed. It also identifies the differences between surface wave transducers and ideal transversal filters, and offers procedures for minimizing certain nonideal transducer properties.

3.2.1 Review of General Fourier Synthesis (Transversal Filter Theory)

Acoustic surface wave filters are implemented via the repeated delaying and sampling of an input signal and are thereby classified as transversal filters. Similarities between surface wave filters and ideal transversal filters lead to design simplicity and versatility. However, the acoustic device is not strictly equivalent to the ideal transversal filter, and it is important to understand how the differences affect performance and how they are controlled by transducer design.

Figure 3.3 is a schematic representation of an ideal transversal filter composed of a bandpass filter in series with a uniformly tapped delay line (TDL). Amplitude and phase weights are applied to the tap outputs, which are then summed to form the transversal filter output.

Let us calculate the impulse response of this composite structure. First, the impulse response of the ideal bandpass filter $s(t)$ is found from the Fourier transform of the bandpass filter transfer function, namely,

$$s_1(t) = \left(\frac{1}{2\pi}\right) \int_{\omega_0 - \pi B}^{\omega_0 + \pi B} \exp(i\omega t)\, d\omega = B\, \frac{\sin \pi B t}{B t} \exp(i\omega_0 t),$$

where B is the bandwidth (in frequency—not ω) of the bandpass filter. The action of the TDL is to delay $s_1(t)$ by an amount $n\tau$, where n is the tap number and τ is the intertap delay. Thus, if the input to the TDL is $s_1(t)$, the output of the nth ideal tap is $s_1(t - n\tau)$. As shown in Fig. 3.3, this output is multiplied by complex weight $A_n \exp(i\phi_n)$, then added to the weighted outputs of the other $N - 1$ taps to form the transversal filter response $s(t)$, given by

$$s(t) = B \sum_{n=1}^{N} A_n \frac{\sin \pi B(t - n\tau)}{\pi B(t - n\tau)} \exp\{i[2\pi f_0(t - n\tau) + \phi_n]\}. \tag{3.1}$$

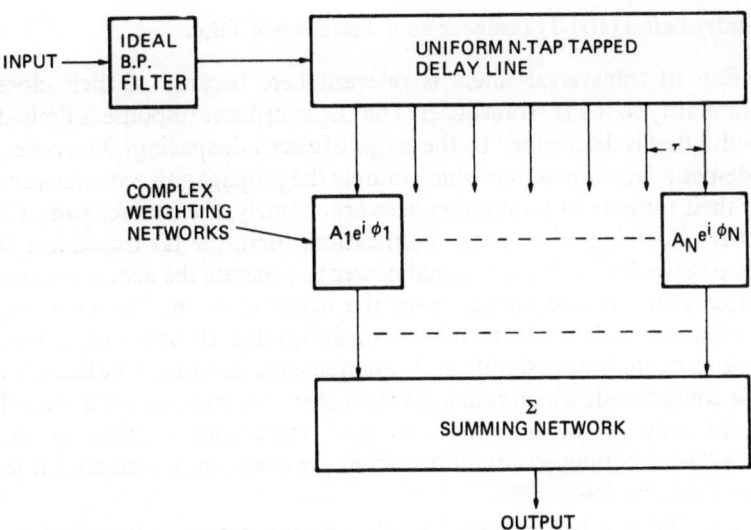

Fig. 3.3. Ideal transversal filter based on uniform-tap delay line

The transfer function of the transversal filter $S(f)$ is obtained by taking the Fourier transform (F.T.) of the above impulse response. We thus have,

$$S(f) = \text{F.T.}|s(t)| \equiv \int_{-\infty}^{\infty} s(t) \exp(-i2\pi f t) dt.$$

The resulting integral can be readily cast in the form of the standard Dirichlet integral (see, for example, [3.10], p. 198, integral number 858.9) which results in:

$$S(f) = \begin{cases} \sum_{n=1}^{N} A_n \exp(i\phi_n) \exp(-i2\pi f n\tau), & |f - f_0| \leq B/2 \\ 0, & \text{elsewhere}. \end{cases} \tag{3.2}$$

This is a most significant expression; it shows how a desired transfer function may be synthesized by using a uniform tapped delay line with tap weights adjusted in correspondence with the impulse response of the desired filter. Furthermore, these N tap weights prove to be the exact expansion coefficients of the desired transfer function, expressed as an N-term Fourier series. Therefore, from the properties of the Fourier series, the filter thus synthesized provides the N-term least-mean-squared fit to the desired transfer function.

Although transversal filter design is a simple and powerful technique, its practical value depends on the availability of components with which to implement the structure in Fig. 3.3. The most critical element is the tapped delay line, which must contain taps that are small, accurate, and noninteractive (in order that the weighted samples be independent).

3.2.2 The Interdigital (ID) Transducer as a Transversal Filter

The discussion of transversal filters is relevant here because of their close structural similarity to the ID transducer. The ID transducer response is limited to a bandwidth that is determined by the range of electrode spacings. Moreover, the electrode-pairs function as taps which sample the propagating acoustic wave and deliver their outputs to a summing network, namely, the contact pads. Of course, this is something of an oversimplification. First, the ID transducer is inherently a passive device; it takes signal power to generate the acoustic wave, and taps necessarily absorb energy from the acoustic beam. The taps are, therefore, not strictly independent, and care must be taken that the output taps do not couple so strongly as to significantly attenuate the main acoustic beam. In addition, the contact pads which sum individual electrode currents act like ideal sum networks only in the limit of zero load impedance terminating the transducer: a "weak coupling" condition which, we shall see, is tantamount to infinite filter insertion loss.

Surface wave filter technology has become practical because it is possible to characterize and control the nonideal transducer properties. The circuit model transfer function specifies limits for the strength of tap-coupling for a given design in order that the filter amplitude and phase response remain within tolerable bounds. That is, it establishes a loss-accuracy trade-off relationship. Although the transfer function accurately predicts the consequences of strong tap coupling and of multiple tap iterations, such as "regeneration", there are several other problems that can arise because of oversimplifications in the basic circuit model representation. Effects such as acoustic propagation attenuation and acoustic diffraction loss cannot always be neglected. Often they must be offset by predistorting the characteristics of the desired filter response [3.11, 12]. The resistance of the metal electrodes, also neglected in the present circuit model treatment, can be included *a posteriori*, in most cases, by adding a lumped-circuit resistor in the equivalent electrical circuit of each transducer [3.13].

There are two other common problems which are best overcome through clever transducer geometry. The cures are discussed in the prefacing remarks, because, once implemented, the problems virtually vanish and do not bear on the general techniques of surface wave filter design. Both effects result from the *mechanical* presence of the tapping electrodes, which is normally neglected in the basic circuit model development. In reality, a metal electrode alters the local boundary condition for piezoelectric surface wave propagation by imposing a (nearly) short circuit at the free surface. The shorted condition leads to a wave discontinuity and a decrease in velocity, both of which are proportional to the strength of piezoelectric coupling [3.14, 15]. Both of these consequences also result, to a small extent, from the finite mass of the electrode, independent of piezoelectricity.

Let us focus on the first of these two problems: electrode reflections produced by the wave discontinuity. These reflections have been well characterized in extended circuit models [3.2, 15, 16]. Analysis shows, however, that unless

Fig. 3.4. Uniform periodic interdigital (ID) transducer with double electrodes

the reflections are eliminated *a priori*, the derivation of a general transducer design procedure becomes hopelessly complex. If we return to the transducer geometry in Fig. 3.1, we see that the excitations from consecutive gaps are spaced by one-half an array period, and that they have opposite electrical polarity. Hence, at the synchronous frequency, the propagation phase shift between consecutive gaps is π radians and there is an equal shift in phase of the electrical drive. Unfortunately, reflections from adjacent electrodes *also* add in phase at the synchronous frequency. Thus, while the geometry in Fig. 3.1 maximizes surface wave generation at the synchronous frequency, it also maximizes the deleterious effects of electrode reflections.

Figure 3.4 shows a simple modification to the transducer geometry which substantially reduces the problem of electrode reflections by essentially doubling the periodicity of the metal electrodes without altering the periodicity of the *excitation* gaps. With this "split-finger", or "double-electrode", geometry, the propagation phase shift for reflections from adjacent electrodes is reduced to π, without changing the electrical excitation periodicity. Thus, excitations continue to add in phase at the synchronous frequency, while the individual reflections now *cancel* in pairs. The double-electrode geometry has been well proven, both by circuit model analysis and laboratory measurement [3.17, 18]; hence it is strongly recommended for *all* surface wave transducers.

The second problem referred to above is caused by the decrease in surface wave velocity resulting from the presence of metal electrodes. While it is straightforward to reposition the electrodes to correct for a uniform change in velocity (i.e., change in delay time) caused by electrodes in the path of the acoustic wave [3.19], a more troublesome problem can occur when the transducer electrodes are *apodized*, i.e., when they have varying overlap length. Let us consider the electrodes indicated by solid bars in the long transducer of Fig. 3.5. A surface wave propagating along the middle of this transducer crosses more electrodes and, hence, is slowed more substantially than a wave propagating closer to the contact pads. For this reason, a straight-crested wavefront incident upon the long transducer does not remain straight-crested as it

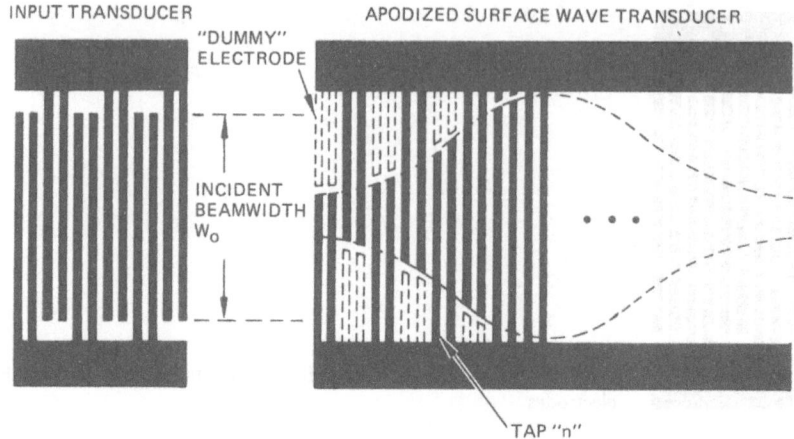

INPUT TRANSDUCER   APODIZED SURFACE WAVE TRANSDUCER

"DUMMY"
ELECTRODE

INCIDENT
BEAMWIDTH
W_0

· · ·

TAP "n"

Fig. 3.5. Typical filter geometry employing one apodized and one uniform transducer

traverses the apodized transducer. Curvature of the wavefront is equivalent to defocusing of the beam, and it produces severe distortion in the transducer response [3.20]. A simple remedy for this problem is to include "dummy" electrodes, as indicated by the dashed bars in Fig. 3.5. This change does not affect the electrical properties of the transducer, but it does make the physical conditions within the transducer more nearly equivalent to those assumed for the circuit model upon which the filter design prescription is based. With dummy electrodes, the surface wave velocity remains uniform over the width of the transducer aperture and defocusing is prevented. For this reason dummy electrodes should always be employed in the geometry of apodized transducers.

3.2.3 Surface Wave Filter Design Prescription

Having digested these few words of caution concerning the physical realities of metal electrodes on piezoelectric substrates, we shall now concentrate on the surface wave filter design procedure. Transversal filter theory (see Sect. 3.2.1) has demonstrated

    1) that a desired transfer function can be implemented using a tapped delay line with complex tap weights, and

    2) that these tap weights correspond to the amplitude and phase of the impulse response of the desired transfer function.

We view the ID transducer as a transversal filter with the ID electrodes providing the taps. The implementation of the required tap weights is our first consideration. Amplitude weights are readily introduced through *apodization*, i.e., the tap efficiency is to be controlled by the fraction of the incident wave that is intercepted by the overlapped electrodes (see Fig. 3.5). Phase weights are more difficult. It is clear, by the manner in which tap outputs are summed, that the samples must have polarity of ± 1 (effective phase weights of 0 or π) determined

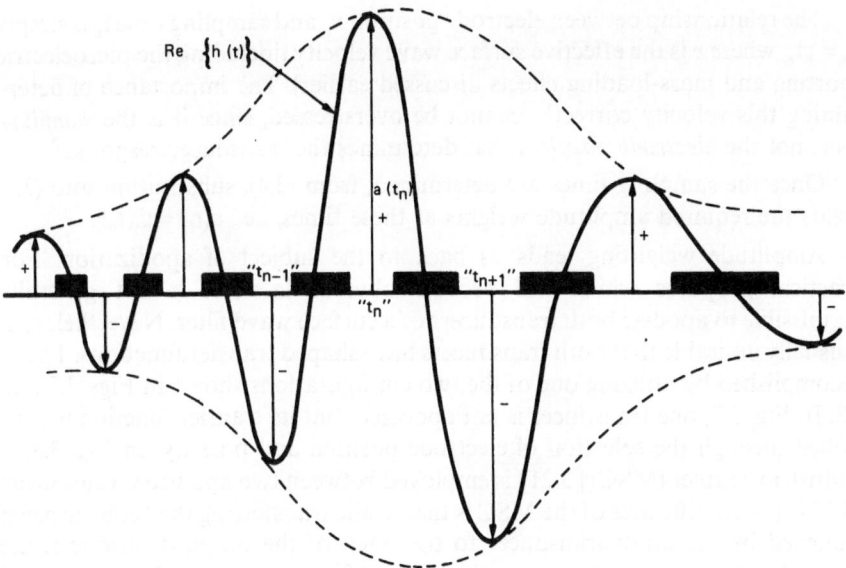

Fig. 3.6. Relationship between sampling times and electrode positions

by the contact pad to which the sampled signal is routed. In general, however, the required phase weights for implementing an arbitrary transfer function with a uniform tapped delay line are continuous between $\pm\pi$. This difficulty could be overcome if a means were available to place a controlled phase-weighting network between each electrode and the contact pad, but this is not at all practical. Thus, the surface wave taps cannot always be spaced uniformly, but must be repositioned to those sampling *times* for which the desired filter impulse response is real. That is, the electrode sampling times are chosen such that the required tap phase weights become 0 or π.

If the desired transducer impulse response (which is perhaps derived from the Fourier transform of a desired transfer function) is represented in complex notation by

$$h(t) = a(t)\exp[i\theta(t)], \tag{3.3}$$

then the phase-weighting constraint leads to the condition

$$\theta(t_n) = 0 \quad \text{or} \quad \pi. \tag{3.4}$$

Assuming that the physical sampling point is in the center of a gap, (3.4) determines the required sampling times, and the contact pad to which each electrode is connected. The significance of this sampling prescription is illustrated in Fig. 3.6, showing the real part of the impulse response superimposed on the electrode pattern.

The relationship between electrode position x_n and sampling time t_n is simply $x_n = vt_n$, where v is the effective surface wave velocity (including the piezoelectric shorting and mass-loading effects discussed earlier). The importance of determining this velocity correctly cannot be overstressed, since it is the *sampling time*, not the *electrode position*, that determines the transducer response[2].

Once the sampling times are determined, from (3.4), substitution into (3.3) yields the required amplitude weights at these times, i.e., $a(t_n) = |h(t_n)|$.

Amplitude weighting leads us back to the subject of apodization. For practical purposes which will become obvious later, it is not generally permissible to apodize both transducers of a surface wave filter. Nevertheless, it is usually desirable that both transducers have shaped transfer functions. This is accomplished by utilizing one of the two configurations shown in Figs. 3.7 and 3.8. In Fig. 3.7, one transducer is not apodized, but its transfer function is controlled through the selection of electrode position and polarity. In Fig. 3.8, a multistrip coupler (MSC) [3.21] is employed between two apodized transducers [3.22]. The significance of the MSC is that, while transferring the acoustic beam launched by the input transducer to the track of the output transducer, the output beam is caused to become *uniform* in profile, i.e., as if it has been generated by an unapodized transducer. The MSC approach is thus attractive because it permits maximum freedom in apodizing both transducers, but it is useful only for a strong coupling piezoelectric substrate, for which the MSC implementation is practical. Further information relating to the MSC is given in Section 4.7.

For either of the configurations in Figs. 3.7 and 3.8, the complete filter transfer function $H(f)$ is the product of the transfer functions of the individual transducers, $H_1(f) \cdot H_2(f)$. Therefore, it is necessary to compute two Fourier transforms, namely $h_1(t)$ and $h_2(t)$, which serve to define the electrode positions and amplitude weights of the respective two transducers. If the MSC configuration is used, both transducers may be apodized, and identical transducers could be used to implement the desired filter. In this case, the transducer transfer function is $H_1(f) = H_2(f) = \sqrt{H(f)}$.

For situations in which the MSC configuration is impractical, such as on weakly piezoelectric quartz substrates[3], an unapodized transducer must be designed. The unapodized transducer in Fig. 3.7 illustrates one technique for amplitude weighting, employing selective reversals of the electrode phases (as indicated by the vertical arrows). If the unapodized transducer is designed such that it has a center of symmetry, then the phase reversals provide amplitude weighting and contribute only a linear phase (constant delay) to the filter response (see Appendix A). When required, nonlinear phase characteristics are implemented through the design of a nonsymmetrical apodized transducer.

---

[2] This point is emphasized because modern photomask techniques can position electrodes with extreme precision; yet, the filter response might well be in error as a result of poor control of the surface wave velocity due to fabrication variations.

[3] Quartz substrates are particularly attractive when high-temperature stability is required.

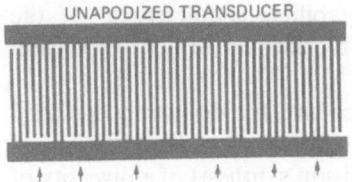

Fig. 3.7. Filter configuration employing an apodized and a phase-reversal unapodized transducer

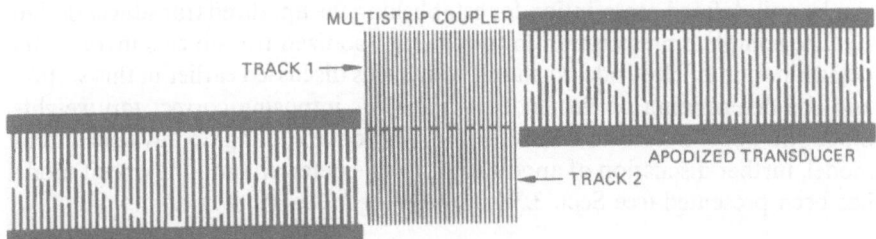

Fig. 3.8. Filter configuration employing two apodized transducers and a multistrip coupler

Symmetrical phase reversing of the electrodes is one method for shaping the unapodized filter response [3.23]; however, a systematic procedure for achieving a "best-fit" to a specified response is preferable. Many diverse techniques exist for amplitude weighting unapodized transducers, and a comprehensive treatment is beyond the scope of this chapter [3.16, 24, 25]. For one example, however, we shall consider the method reported by *Hartmann* [3.24]. He implements amplitude weighting by selectively removing electrodes from the unapodized transducer, thereby weakening the tapping efficiency of the surrounding electrodes in a coarsely controllable manner. Using relative electric field intensities, tabulated for a variety of electrode configurations, he has developed a filter-synthesis computer program for iteratively determining whether to remove a given electrode, based on that desired tap amplitude and on the electrode configurations on both sides of the tap location. His treatment also indicates the amount by which the subsequent electrode positions must be shifted to compensate the sampling time for the removal of electrodes. Recently, *Smith* and *Pedler* [3.16] reported a similar treatment for accurately computing the transfer functions of unapodized transducers with arbitrary electrode polarity sequences. Hence, a satisfactory, although not necessarily unique, approach to the design of an apodized/unapodized surface wave filter is

    1) Utilize a synthesis program, e.g., the technique of *Hartmann*, to develop an optimized unapodized transducer geometry based on the specific desired impulse response;

    2) Use the results of *Smith* and *Pedler* to compute the *exact* transfer function of the unapodized transducer $H_1(f)$, based on the geometry from 1;

    3) Complete the filter design for transfer function $H(f)$ by designing the *apodized* transducer to yield transfer function, $H_2(f) = H(f)/H_1(f)$, (see [3.38]).

Thus, the precision transfer function capability, characteristic of the apodized transducer, is employed to compensate for the relative lack of precision in the synthesis of the unapodized transducer.

The subsequent sections focus on the procedure for designing apodized transducers. The motivations for emphasizing apodized transducer design are

1) Versatility: Apodization permits the precision synthesis of a diversity of sophisticated filters, even when one transducer is constrained to be unapodized;

2) Uniqueness: Given a desired transducer transfer function, there is a single, well-defined prescription for establishing the apodized transducer design.

The complete prescription for designing apodized transducers involves the determination of electrode positions, which was discussed earlier in this section, and the determining of electrode overlaps for imposing correct tap weights. Since the apodization law derives from the details of the equivalent circuit model, further discussion of apodization will be deferred until the circuit model has been presented (see Sect. 3.3.4).

3.3 Equivalent Circuit Model

This section deals with a (frequency domain) circuit model representation of a general ID transducer which has proven remarkably successful in characterizing the excitation and detection of surface waves. When second-order effects, such as electrode resistance, electrode discontinuities, and propagation losses, are neglected, the model can be expressed by an equivalent (time domain) representation employing δ-function (i.e., discrete impulse) acoustic sources, summed with appropriate phasors [3.2, 9, 26]. The δ-function representation, illustrated in Fig. 3.6, is particularly convenient for filter synthesis, while the frequency domain representation is preferable for analyzing filter performance. In addition, the circuit model lends physical insight into the factors which affect tap coupling strength and into the conditions which lead to nonideal transversal filter performance.

3.3.1 The Crossed-Field Model

The crossed-field model was derived to characterize a class of piezoelectric bulk-wave transducers [3.27]. Rigorously, the model relates to one-dimensional, plane wave acoustic propagation in a direction orthogonal to that of the piezoelectric driving field, as shown in Fig. 3.9. The model relates the coupling between an applied uniform electric field and a plane bulk wave to the capacitance between the electrode plates, a piezoelectric coupling constant, and an acoustic wave impedance. It was adapted for surface waves in a phenomenological manner by presuming a correspondence between the bulk-wave parameters and similar ones for surface waves [3.1] (see Fig. 3.10). For the surface wave, the values of effective coupling constant and wave impedance are, within the

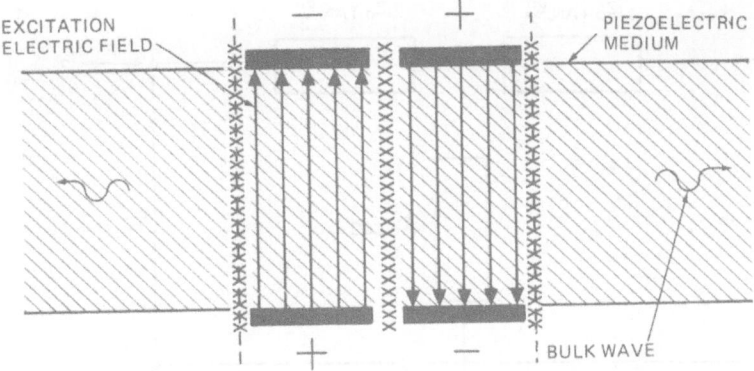

Fig. 3.9. Crossed field bulk-wave representation for one interdigital electrode pair

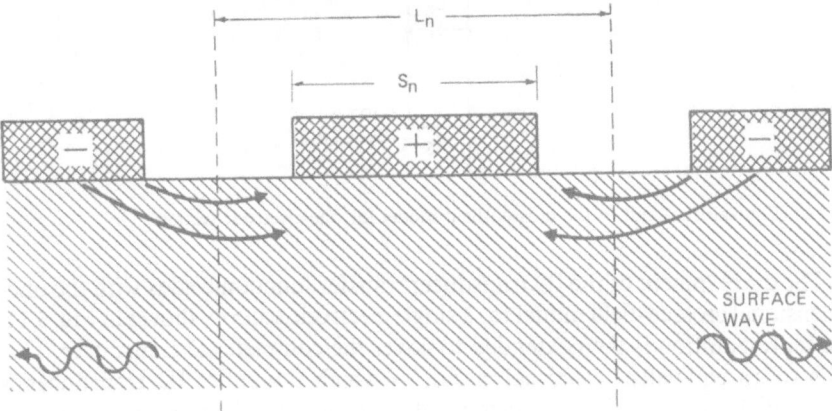

Fig. 3.10. Geometry of a one-electrode section of an interdigital surface wave transducer

circuit model formulation, empirically determined. Extensive theoretical and experimental work has been done to establish that the concepts of effective coupling constant and wave impedance are compatible and numerically consistent with the postulates of the crossed-field model [3.14, 15, 28–30]. The model has been extensively tested and proven for surface waves, although it was derived for bulk waves whose geometrical field distribution is quite different. *Smith* and *Pedler* [3.16] have shown that the crossed-field model may be generalized for arbitrary excitation field distributions (see also [3.37]). Their computer model successfully predicts most aspects of transducer performance, but the computational complexity obscures the insights of the simple crossed-field model. The simpler model is stressed here for two reasons:

1) it resembles a simple δ-function representation;

2) it yields accurate predictions for a wide range of transducer designs, with the improved model representing essentially a higher order refinement.

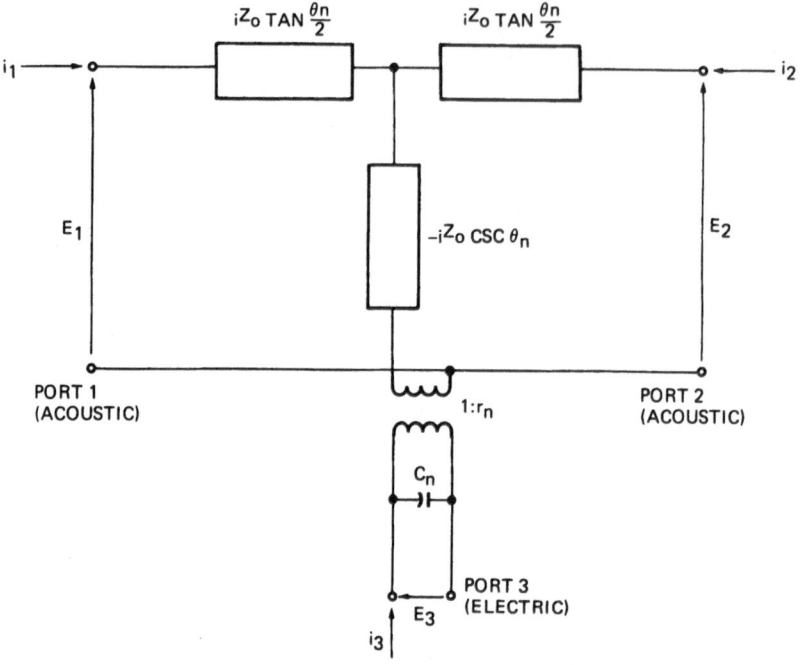

Fig. 3.11. Equivalent circuit of the crossed-field model for one-electrode section

The development of the ID transducer transfer function begins with the one-dimensional crossed-field model for a single electrode section, shown in Fig. 3.11. The representation is then extended to cover the complete ID transducer. This leads to an equivalent electrical circuit in which the entire ID transducer is represented as a section of (acoustic) transmission line of characteristic impedance Z_0 and length L, which is coupled to the transducer electrical terminals through a transformer and a capacitor (see, for example, Fig. 3.14).

The model of a single electrode section has three ports, corresponding to the two symmetric acoustic ports and the electric port. The stresses and particle velocities at the acoustic ports are represented by equivalent voltages and currents, respectively. A convenient method for describing the relationship between these currents and voltages, $i_i = Y_{ij}E_j$, is in terms of the admittance matrix. The Y_{ij}'s for the nth section are the matrix elements in the equation

$$\begin{pmatrix} i_1 \\ i_2 \\ i_3 \end{pmatrix}_n = \begin{bmatrix} -i\cot\theta_n & i\csc\theta_n & -ir_n\tan\theta_n/2 \\ i\csc\theta_n & -i\cos\theta_n & -ir_n\tan\theta_n/2 \\ -ir_n\tan\theta_n/2 & -ir_n\tan\theta_n/2 & i(\omega C_n + 2r_n^2\tan\theta_n/2) \end{bmatrix} \begin{pmatrix} E_1 \\ E_2 \\ E_3 \end{pmatrix}_n , \qquad (3.5)$$

where θ_n is the acoustic transit angle of the nth section and Z_0 has been arbitrarily set equal to 1. The transit angle of the nth transmission line segment is given by

$$\theta_n = 2\pi f L_n/v = \pi f/f_n,$$

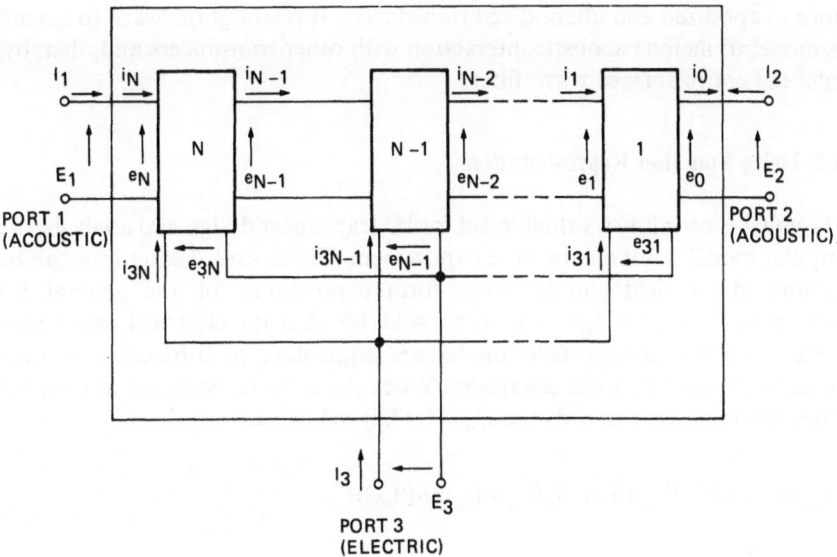

Fig. 3.12. Block diagram illustrating the cascading of one-electrode sections for a general transducer

where the local synchronous frequency is given by $f_n = v/2L_n$ [see also (B.6)] and L_n is the length of the nth section.

The transformer ratio is given by [3.2]

$$r_n = (-1)^n \sqrt{2f_n C_n k^2 Z_0} [K(1/\sqrt{2})/K(q_n)] \tag{3.6}$$

where C_n is the capacitance of the nth section and k^2 is the effective surface wave piezoelectric coupling constant for the particular substrate. The factors of $K(1/\sqrt{2})$ and $K(q_n)$ are Jacobian complete elliptic integrals of the first kind with $q_n = \sin[\pi(S_n/2L_n)]$, where S_n is the width of the nth electrode. These integrals arise from the dependence of capacitance on stripe to gap ratio of the electrodes, namely

$$C_n = (w_n/2) \sqrt{\varepsilon_\parallel \varepsilon_\perp} [K(q_n)/K(\sqrt{1-q_n^2})], \tag{3.7}$$

where w_n is the aperture of the nth section. The dielectric constants $\varepsilon_\parallel$ and $\varepsilon_\perp$ correspond to the substrate dielectric anisotropy between the directions parallel to acoustic propagation and normal to the surface, respectively. When the electrode widths equal the gaps (i.e., $S_n = L_n/2$), the integrals drop out, which for simplicity will be assumed henceforth.

The model for one section is extended to the many-electrode ID transducer by joining the individual sections with their acoustic ports in cascade and their electrical ports in parallel, as shown in Fig. 3.12. A computer program for implementing this combination is valuable for analyzing the three-port performance of apodized and unapodized transducers. It is straightforward to extend this model to include acoustic interaction with other transducers and, thereby, model the entire surface wave filter.

3.3.2 Delta Function Representation

Although it is certainly a valuable aid for ID transducer design and analysis, the computer model is not an absolute requirement. The cascaded equations can be transformed to yield simple closed-form expressions for the general ID transducers. Using the Y_{ij}'s in (3.5) we shall see that the electrical excitations described by the crossed-field model are equivalent to δ-function sources located in the gaps between electrodes. When the acoustic ports are terminated in their characteristic impedance $Z_0 = 1$, (3.5) reduces to

$$i_{1n} = -i \cot \theta_n E_{1n} + i \csc \theta_n E_{2n} - i r_n \tan(\theta_n/2) E_3 .$$

When a voltage is applied to port 3, the stresses induced at the acoustic ports are symmetrical, implying $E_{1n} = E_{2n}$. Noting that for matched acoustic terminations $i_{1n} = -(1 \cdot E_{1n})$, solving for E_{1n}, and shifting the origin of θ from the center of a gap to the center of the adjacent electrode gives the frequency response of stress produced by the applied voltage, namely,

$$E_{1n} = i r_n \sin(\theta_n/2) E_3 , \tag{3.8}$$

where $\theta_n/2 = \pi f / 2 f_n$. The Fourier transform of (3.8) gives the equivalent δ-function time domain sources, namely,

$$e_{1n}(t) = r_n E_3 [\delta(t - 5\tau_n/4) - \delta(t - 3\tau_n/4)] , \tag{3.9}$$

where the effective sample times are given by $\tau_n \equiv 1/f_n$. These sources are proportional to E_3 and weighted by the factor r_n.

Thus, when the circuit model coupling factors r_n are adjusted (by apodization) to correspond to the desired tap weights $a(t_n)$, the frequency domain crossed-field transducer representation becomes equivalent to the δ-function time domain model (see Fig. 3.13). In fact, the apodization law that we are seeking is precisely the expression that establishes the correspondence between $a(t_n)$ and r_n.

3.3.3 Acoustic/Electric Transfer Function

To obtain the transfer function of the complete ID transducer, it is necessary to sum the N individual electrode responses. Based on the form of (3.5), this is easy to do. That is, when an ideal voltage source ($G_L = \infty$) is employed to generate the

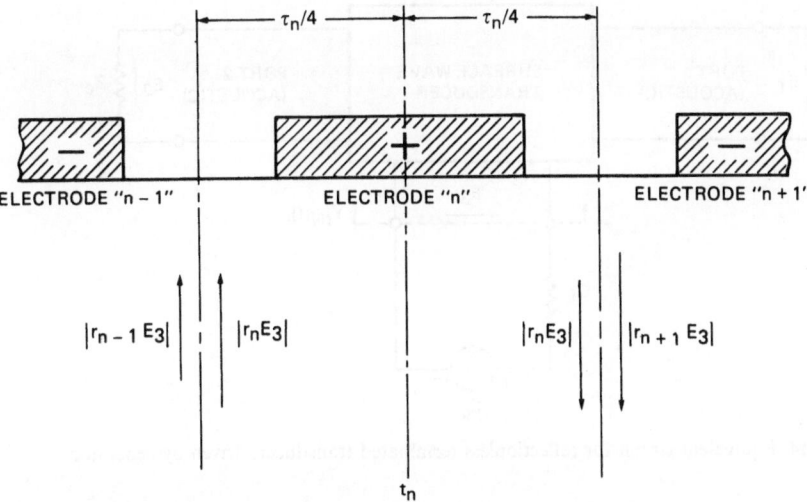

Fig. 3.13. δ-function stress pattern corresponding to applied electrical impulse

voltage impulse, there is a short circuit of the electrical port, and this implies $e_3(t)=0$ for all times after the initial impulse. Returning to the admittance matrix, we see that under this condition the acoustic forces and particle velocities are related by

$$\begin{pmatrix} i_1 \\ i_2 \end{pmatrix} = \begin{pmatrix} -i\cot\theta_n & i\csc\theta_n \\ i\csc\theta_n & -i\cot\theta_n \end{pmatrix} \cdot \begin{pmatrix} E_1 \\ E_2 \end{pmatrix},$$

which is exactly the representation of a reflectionless length of transmission line. Thus, with an *ideal* voltage source, the impulse response of the entire array is given by the sum of the N individual responses, from (3.8) with appropriate phasors, namely

$$E_1 = i \sum_{n=1}^{N} r_n \sin(\pi f/2f_n)\exp(-i2\pi f t_n)E_3. \tag{3.10}$$

For a transducer driven by a *real* generator (see Fig. 3.14) the voltage across the transducer terminals is related to the generator voltage E_g by

$$E_3 = E_g G_L/[G_L + Y_{in}(f)], \tag{3.11}$$

where $Y_{in}(f)$ is the Norton equivalent circuit admittance of the transducer and G_L is the generator conductance.

The electric to acoustic transfer function $T_{13}(f)$ (normalized such that $|T_{13}|^2 \equiv p_{13}$, where p_{13} is the electric to acoustic power transfer efficiency) is given by

$$T_{13}(f) = (2/\sqrt{Z_0 G_L})(E_1/E_g). \tag{3.12}$$

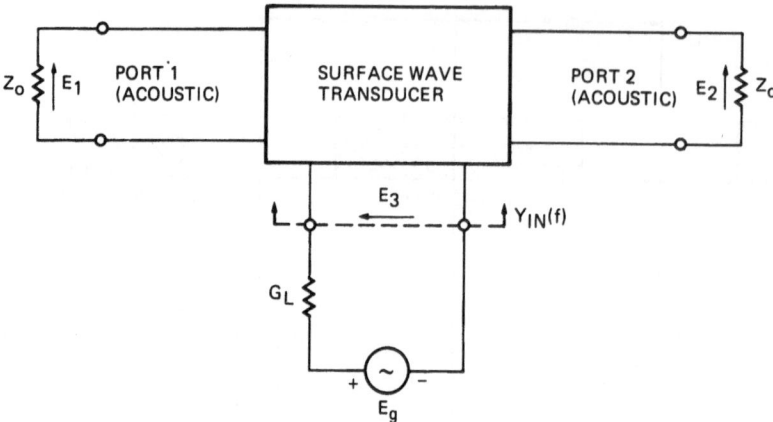

Fig. 3.14. Equivalent circuit for reflectionless terminated transducer, driven by generator

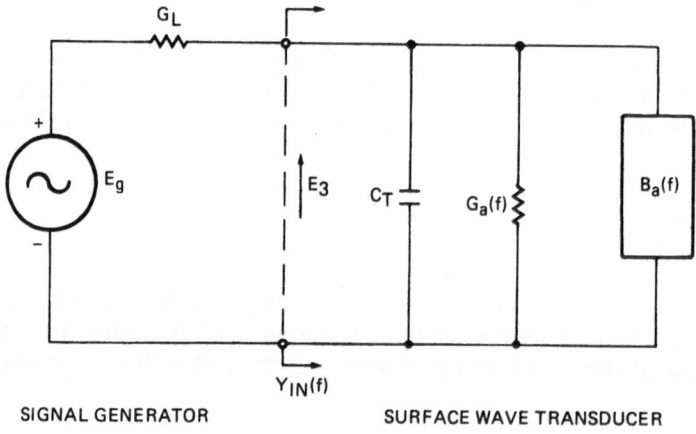

Fig. 3.15. Equivalent electrical circuit for electrical excitation of untuned transducer in a reflectionless delay line

Substituting the ratio of E_1/E_3 from (3.10), and the value of E_g/E_3 from (3.11), gives

$$T_{13}(f) = i(2G_L/\sqrt{Z_0 G_L}) \frac{\sum_{n=1}^{N} r_n \sin(\pi f/2f_n) \exp(-i2\pi f t_n)}{G_L + Y_{in}(f)} \qquad (3.13)$$

where

$$Y_{in}(f) = G_a(f) + i[\omega C_T + B_a(f)], \qquad (3.14)$$

$$r_n = (-1)^n \sqrt{2w_n f_n C_0 k^2 Z_0}, \qquad (3.15)$$

Table 3.1. Surface wave transducer material constants for YZ lithium niobate and ST quartz

Material	k^2	C_0 [pF/in.]
YZ LiNbO$_3$	4.4×10^{-2}	6.9
ST quartz	1.7×10^{-3}	0.65

and C_0 is the capacitance per unit length of an electrode. $\dot{Y}_{in}(f)$ is the admittance of the transducer equivalent circuit, which includes $G_a(f)$, the acoustic radiation conductance, and $B_a(f)$, the radiation susceptance. The transducer equivalent circuit is shown in Fig. 3.15. For this circuit, the power delivered to the acoustic conductance, p_{13}, is equivalent to the total surface wave power generated by the transducer. Therefore, from simple circuit analysis,

$$p_{13} = \frac{2G_L G_a(f)}{[G_L + G_a(f)]^2 + [2\pi f \, C_T + B_a(f)]^2},$$ (3.16)

where an extra factor of 1/2 has been included to account for the fact that only half the generated acoustic power radiates from one end of the transducer. Comparing (3.13) and (3.16) leads to the identification

$$G_a(f) \equiv 4k^2 C_0^2 \left| \sum_{n=1}^{N} (-1)^n \sqrt{w_n f_n} \sin(\pi f / 2f_n) \exp(-i2\pi f t_n) \right|^2,$$ (3.17)

which expresses the equivalent acoustic radiation conductance in terms of the transducer design parameters. The remaining equivalent circuit component values are found by summing the individual electrode capacitances, namely,

$$c_T = \sum_{n=1}^{N} C_n = C_0 W_0 \sum_{n=1}^{N} (w_n/W_0),$$ (3.18)

and by using the Hilbert transform,

$$B_a(f) = (1/\pi) \int_{-\infty}^{\infty} du [G_a(u)/(u-f)].$$

Some useful constants for determining numerical values of the circuit components are given for quartz and lithium niobate in Table 3.1.

At this point, we must recall our goal is to derive the appropriate transducer transfer function for describing the two transducers that compose a surface wave filter. This distinction is made because T_{13} in (3.13) yields the *total* acoustic power radiated from one end of a generalized transducer. When we consider the *MSC-coupled* apodized transducer configuration in Fig. 3.8, for example, we recognize that only the fraction w_n/W_0 of the incident beam is coupled across the tracks [3.21]. Hence, if the value of $T_{13}(f)$ were reduced by the factor $\sqrt{w_n/W_0}$, then the effect of the MSC would be included. Similarly, since the beam leaving the MSC illuminates the full aperture of the second transducer, only a fraction, w_n/W_0, of the energy intercepts the transducer aperture. Therefore, the transfer function of the second transducer must also include the extra factor of $\sqrt{w_n/W_0}$. When we consider the details of the apodized-unapodized surface wave filter configuration in Fig. 3.7, we see that while the unapodized transducer launches a beam of width W_0, only the fraction w_n/W_0 intercepts the apodized aperture of width w_n. Hence, it is once again appropriate to include the factor $\sqrt{w_n/W_0}$ in the general transducer transfer function. We therefore make the generalization that in a surface wave filter configuration the ID transducer has an *effective* transfer function $\hat{T}_{13}(f)$ given by

$$\hat{T}_{13}(f)=i2\frac{\sqrt{2W_0C_0k^2G_L}}{[G_L+Y_{in}(f)]}\sum_{n-1}^{N}(-1)^n(w_n/W_0)\sqrt{f_n}\sin(\pi f/2f_n)\exp(-i2\pi f t_n),$$

(3.19)

where the extra factor of $\sqrt{w_n/W_0}$ is included in $\hat{T}_{13}(f)$.

Equation (3.19) thus provides an expression for the transfer function of each transducer in the surface wave filter, based upon a set of tap locations t_n and tap weights w_n. Furthermore, the product of the $\hat{T}_{13}(f)$ of the two ID transducers in a surface wave filter yields the correct transfer function for either configuration described in Section 3.2.3.

3.3.4 The Apodization Law

Starting with $\hat{T}_{13}(f)$ in (3.19), it is now a simple matter to establish the appropriate relationship between the electrode overlaps w_n's and the required tap weight amplitudes $a(t_n)$'s. It is shown in Appendix B that, for a time-sampled replica $h_s(t)$ of the desired transducer impulse response $h(t)$, the Fourier transform of $h_s(t)$ is given by

$$H(f)=\text{F.T.}\{h_s(t)\}=(1/\pi)\sum_{n=1}^{N}(-1)^n[a(t_n)/f_n]\sin(\pi f/2f_n)\exp(-i2\pi f t_n). \quad \text{(B.7)}$$

We observe that there is a strong similarity between the summation terms in (3.19) and (B.7). If we specify the apodization law as

$$(w_n/W_0)\equiv(f_n/F)^{-3/2}a(t_n)/A,$$

(3.20)

then (3.19) can be rewritten as

$$\hat{T}_{13}(f)=i2\pi\frac{\sqrt{2W_0C_0k^2G_L}}{G_L+Y_{in}(f)}[\text{F.T.}\{h_s(t)\}].$$

(3.21)

Thus, the apodization law in (3.20) is justified because the transfer function it produces [i.e., (3.21)] is proportional to the Fourier transform of the desired impulse response. The constants A and F in (3.20) are for normalization; A is the maximum value of $a(t_n)$, and F is determined such that the maximum value of w_n is W_0.

The significance of (3.20) is that it demonstrates that the overlap lengths in apodized transducers must be proportional to the amplitude of the desired impulse response. It also shows that the overlaps must be compensated for changes in the instantaneous tap frequencies f_n whenever the electrode spacings are not constant. The significance of (3.21) is that it proves the validity of the filter synthesis prescription based on the selected apodization law, (3.20). Equation (3.21) is important for another reason: it clearly demonstrates how the transducer circuit elements impact the accuracy of the filter design. The $Y_{in}(f)$ term in the denominator of (3.21) is not a constant; therefore, $\hat{T}_{13}(f)$ is not *strictly* proportional to the Fourier transform of $h_s(t)$ as it would be for an ideal transversal filter. Instead, this term leads to an error whose magnitude must be controlled by transducer design trade-offs.

3.4 Transducer Performance Characteristics

At this point, the apodization law, (3.20), has been derived, which, in combination with the electrode positioning rule, (3.4), constitutes a design prescription for producing a transducer transfer function, (3.21), that approximates the desired ideal response. Since the transducer response is inherently nonideal, it is necessary to understand the trade-off relationships among insertion loss, bandwidth, and filter accuracy. It is also important to understand the significance of substrate material constants and of the electrical tuning circuit parameters so that the surface wave filter design can be optimized.

3.4.1 Design Trade-Off Relationships

Surface wave filter insertion loss is considered first. For purposes of simplicity, the only transducer loss considered here is that which results from mismatch between the signal source (or load) and the transducer. Since each ID transducer is equivalent to a reactive electrical circuit, there is a constraint upon the minimum insertion loss that can be maintained over a specified bandwidth.

Bode and *Fano* have established that the reflection loss $|\Gamma(f)|$ for a resonant RC network obeys the following integral equation [3.31, 32]:

$$\int_0^\infty [\ln|\Gamma(f)|]^{-1} df = \pi f_0/Q,$$

where f_0 is the resonant frequency and Q is the quality factor of the network. Therefore, the *ideal* transducer matching conditions are $|\Gamma(f)| = \exp(-\pi/\beta Q)$

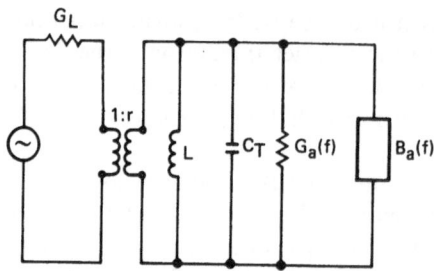

Fig. 3.16. Equivalent electrical circuit for transducer with tuning inductor and transformer

over the passband, and $|\Gamma(f)|=1$ elsewhere, where β is the fractional bandwidth. Thus, the minimum in-band insertion loss L_{min} is

$$L_{min} = -10\log_{10}[1-\exp(-2\pi/\beta Q)]/2\,^4.\tag{3.22}$$

It is significant that L_{min} is dependent only on β and Q, and that the minimum loss increases with each factor. In practice, no attempt is made to implement the ideal transducer tuning network. Nevertheless, (3.22) serves as a valuable yardstick for evaluating more practical tuning networks. The tuning circuit, shown in Fig. 3.16, has the practical advantage of requiring only one transformer and one resonant coil [3.33]. Moreover, for most designs, it also proves to be remarkable close to ideal.

If electrode resistance and other parasitic losses are neglected, the transducer insertion loss is given by $|\hat{T}_{13}|^2$. When describing the general insertion loss properties of apodization transducers it is, however, much simpler to deal with p_{13}, the power delivered to the acoustic beam. As described in Section 3.3.3, these two terms differ by the extra factor[5] of $\sqrt{w_n/W_0}$ in $\hat{T}_{13}$. The actual insertion loss level computed from (3.19) is therefore generally slightly greater than that estimated from p_{13}, using (3.16). Since this difference in loss depends on the details of the apodization function, there is no difference in the case of an unapodized transducer.

When the shunt resonant inductor and transformer are included, (3.16) becomes

$$p_{13}(f)$$

$$= \frac{2(Q_L/Q_R)[G_a(f)/G_a(f_0)]}{\{1+(Q_L/Q_R)[G_a(f)/G_a(f_0)]\}^2+\{Q_L(f/f_0-f_0/f)+(Q_L/Q_R)[B_a(f)/G_a(f_0)]\}^2}\tag{3.23}$$

where

$$Q_L \equiv 2\pi f_0 C_T/(G_L/r^2), \qquad Q_R \equiv 2\pi f_0 C_T/G_a(f_0),\tag{3.24}$$

and

$$f_0 = 1/2\pi\sqrt{LC_T}.$$

---

[4] Note the factor of 1/2 included to account for bidirectional loss in a transducer.
[5] This loss is sometimes referred to as apodization loss.

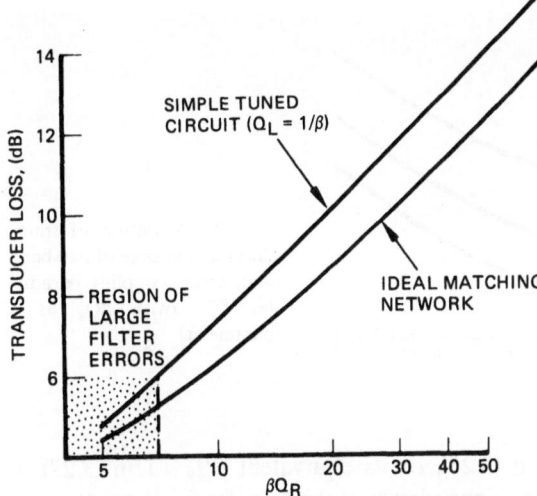

Fig. 3.17. Comparison of mid-band transducer insertion loss for ideal and simple-tuned matching circuits

Equation (3.23) shows that the factor Q_L/Q_R multiplies the frequency-dependent $G_a(f)$ and $B_a(f)$ terms in the denominator of the transfer function. When this factor is large, the frequency dependence of G_a or B_a becomes significant, thereby distorting the filter characteristics from the desired response, which is contained in the numerator (often called the *array factor*). For frequencies near f_0, and with $Q_L/Q_R \ll 1$, (3.23) reduces to

$$p_{13}(f) \cong \frac{2(Q_L/Q_R)[G_a(f)/G_a(f_0)]}{[1+4Q_L^2(f-f_0)^2/f_0^2]}, \qquad Q_L/Q_R \ll 1. \tag{3.25}$$

As it is impossible to set Q_L equal to zero without also reducing the transducer efficiency to zero, (3.25) retains some frequency distortion in the denominator. Thus, we are left with no alternative but to specify a Q_L that is consistent with a favorable trade-off between filter errors and efficiency.

The transducer efficiency at the passband edges, from (3.25), is

$$p_{13}(f_0 \pm \Delta f/2) \cong \frac{2(Q_L/Q_R)[G_a(f)/G_a(f_0)]}{1+\beta^2 Q_L^2}, \qquad Q_L/Q_R \ll 1 \tag{3.26}$$

where $\beta \equiv \Delta f/f_0$ defines the transducer fractional bandwidth. Maximizing (3.26) with respect to Q_L gives the value of Q_L that corresponds to maximum transducer efficiency at the $(-3\,\mathrm{dB})$ passband, edges, namely, $Q_L = 1/\beta$. For this value of Q_L, the midband transducer efficiency is

$$p_{13}(f_0) \cong 2/\beta Q_R, \qquad \beta Q_R \gg 1. \tag{3.27}$$

In Fig. 3.17, the midband insertion loss for the simple-tuned network is compared with the ideal network loss, from (3.22). It is important to remember

p13 (dB)

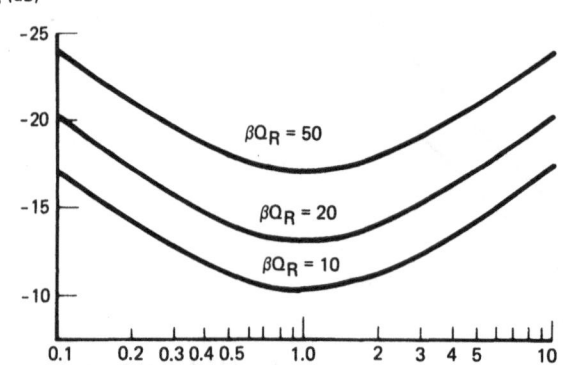

Fig. 3.18. Variation of transducer loss at edge of passband with circuit coupling parameter βQ_L (with βQ_R as a parameter)

that the restriction $Q_L/Q_R \ll 1$ in (3.25), or its equivalent $\beta Q_R \gg 1$ in (3.27), is imposed to prevent the frequency dependence of the transfer function denominator (often called the *circuit factor*) from distorting the transducer response excessively.

The condition $Q_L = 1/\beta$ defines a relationship between the source conductance, array capacitance, and transformer ratio, namely

$$2\pi f_0 C_T/(G_L/r^2) = 1/\beta. \tag{3.28}$$

Thus, (3.28) may be used together with (3.18) to determine W_0, once the relative apodization is specified. If this value of W_0 proves to be impractical, it can be scaled appropriately by adjusting r, the transformer turns ratio.

Perhaps the most important factor controlling the design trade-offs is Q_R, defined in (3.24). Since Q_R depends on $G_a(f_0)$, it is necessary to specify the electrode positions and overlaps and compute $G_a(f_0)$ in order to arrive at the *exact* value of Q_R. However, for most cases Q_R has been found to be closely approximated by

$$Q_R \cong (\pi\beta/4k^2). \tag{3.29}$$

Thus, from (3.27) we see that, in general, transducer efficiency increases in proportion to k^2 and decreases as $1/\beta^2$. For example, with $Q_L = 1/\beta$ and $\beta Q_R \ll 1$, a transducer on YZ LiNbO$_3$ will have about 14 dB lower loss than the same transducer on ST quartz, based on their relative k^2 (see Table 3.1). Furthermore, doubling β on any substrate leads to a 6 dB increase in the loss of each transducer.

Figure 3.18 is a plot of p_{13} from (3.26) that is useful for estimating transducer loss as a function of design variables β, Q_L, and Q_R.

When applying these principles to the design of surface wave filters, cases may arise for which $Q_L = 1/\beta$ does not guarantee that $\beta Q_R \ll 1$ (i.e., when $\beta < 2k/\pi$). In these cases, either Q_L can be reduced until $Q_L/Q_R \ll 1$, or Q_R may be

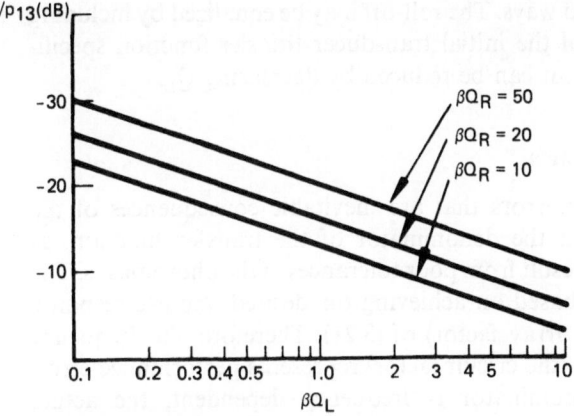

Fig. 3.19. Echo suppression for one transducer, p_{11}/p_{13}, versus circuit coupling parameter βQ_L (with βQ_R as a parameter)

increased to bring this about by using a substrate crystal with a *smaller* value of k^2. Thus, strong piezoelectric coupling, manifested by large k^2, is not necessarily advantageous when narrow band filters are desired.

In addition to controlling frequency-dependent circuit effects, the condition $Q_L/Q_R \ll 1$ also helps to suppress the reflection, or regeneration[6], of the wave incident upon a transducer. Such reflections cause multiple echoes (and passband ripple) which can be the major source of spurious signals. The efficiency of reflection at an acoustic transducer port is given by [3.1]

$$p_{11}(f)$$

$$= \frac{(Q_L/Q_R)^2}{\{1+(Q_L/Q_R)[G_a(f)/G_a(f_0)]\}^2 + \{Q_L(f/f_0 - f_0/f) + (Q_L/Q_R)[B_a(f)/G_a(f_0)]\}^2}.$$

or from (3.23),

$$p_{11}(f) = (Q_L/2Q_R)[G_a(f_0)/G_a(f)]p_{13}(f). \tag{3.30}$$

Hence, the efficiency of "regeneration" is proportional to the efficiency of coupling, but reduced by the factor $Q_L/2Q_R$. For the purposes of estimating echo level, the ratio $p_{11}(f)/p_{13}(f)$ from (3.10) is plotted in Fig. 3.19. A comparison of Figs. 3.18 and 3.19 shows that transducer loss can be traded off against echo suppression by decreasing Q_L such that $Q_L < 1/\beta$.

There is a second reason which can justify utilizing $Q_L < 1/\beta$, thereby accepting a concomitant increase in transducer loss. A comparison between Figs. 3.17 and 3.18 shows that with $Q_L = 1/\beta$, the circuit factor causes the loss at the passband edge to be 3 dB greater than at midband. Thus, the surface wave filter employing two transducers, tuned for $Q_L = 1/\beta$, exhibits a 6 dB roll-off at the edges of the passband. This roll-off is intolerable for many filter applications,

---

[6] These reflections are not the same as those produced by piezoelectric or mechanical discontinuities.

and it may be overcome in two ways. The roll-off may be equalized by including a roll-up as a *predistortion* of the initial transducer transfer function specification, or as discussed below, it can be reduced by decreasing Q_L.

3.4.2 Filter Error Characteristics

This section deals with filter errors that are inevitable consequences of the frequency-dependent terms in the denominator of the transfer function, as opposed to errors that might result from poor tolerances in the photomasks. The surface wave filter design is based on achieving the desired impulse response given by the numerator (the array factor) of (3.21). Therefore, the frequency variation of the denominator (the circuit factor) represents performance error. To the extent that the denominator is frequency dependent, the actual transducer transfer function departs in both amplitude and phase from the desired ideal response.

Under the condition that $Q_L/Q_R \ll 1$ [imposed to suppress the frequency dependence of $G_a(f)$ and $B_a(f)$ in the circuit factor $D(f)$ of (3.21)], the denominator reduces to

$$D(f) \cong 1 + iQ_L(f/f_0 - f_0/f), \quad Q_L/Q_R \ll 1$$

which, for $f \cong f_0$, becomes

$$D(f) \cong 1 + i2Q_L(f - f_0)/f_0. \tag{3.31}$$

As we have defined it, the transducer phase error results entirely from the angle of the denominator, namely

$$\phi_e(f) = -\arctan[2Q_L(f - f_0)/f_0]. \tag{3.32}$$

It is helpful to expand $\phi_e(f)$ about $f = f_0$ since the first term, which is linear, merely corresponds to a nondispersive delay error.

$$\phi_e(f) \cong -2Q_L(f - f_0)/f_0 + (8/3)Q_L^3(f - f_0)^3/f_0^3. \tag{3.33}$$

Defining the delay error δt by

$$\delta t \equiv -(1/2\pi)\frac{\partial \phi_e(f)}{\partial f}$$

gives a fixed-delay error δt_0 of

$$\delta t_0 = Q_L/\pi f_0. \tag{3.34}$$

Thus, even the fixed-delay error decreases with Q_L.

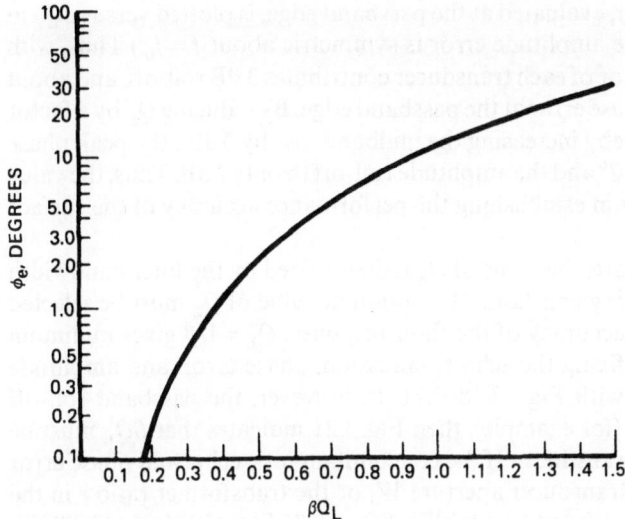

Fig. 3.20. Dispersive phase error for one transducer at upper passband edge versus circuit coupling parameter βQ_L

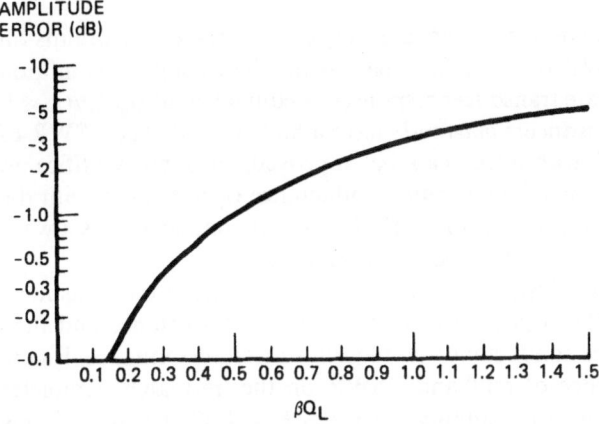

Fig. 3.21. Amplitude error for one transducer at passband edge versus circuit coupling parameter βQ_L

The higher order frequency terms in (3.32) represent dispersive error which, evaluated at the upper passband edge, is plotted against βQ_L in Fig. 3.20. (Note that the phase error is odd-symmetric about $f = f_0$.)

Similarly, the amplitude error is found from the frequency variation in magnitude of the denominator, namely

$$\delta |\hat{T}_{13}(f)| = -10\log_{10}[1 + 4Q_L^2(f - f_0)^2/f_0^2]\,(\text{dB}). \tag{3.35}$$

The amplitude error, evaluated at the passband edge, is plotted versus βQ_L in Fig. 3.21. (Note that the amplitude error is symmetric about $f = f_0$.) Thus, with $Q_L = 1/\beta$ the circuit factor of each transducer contributes 3 dB roll-off, and about $12°$ peak (nonlinear) phase error at the passband edge. By reducing Q_L by a factor of 2, however (and thereby increasing the midband loss by 3 dB), the peak phase error is reduced to only $2°$ and the amplitude roll-off to only 1 dB. Thus, the value of Q_L is very important in establishing the performance accuracy of the surface wave filter.

In summary, therefore, the value of Q_R is determined by the filter bandwidth and piezoelectric coupling constant. The optimum value of Q_L must be selected based on the required accuracy of the filter response. $Q_L = 1/\beta$ gives minimum insertion loss, thereby fixing the echo suppression, phase error, and amplitude roll-off in accordance with Figs. 3.18–3.21. If, however, the passband roll-off must be less than 1 dB (for example), then Fig. 3.21 indicates that βQ_L must be less than 0.35 for each transducer. If the concomitant loss, echo, and phase error are tolerable, then the transducer aperture W_0 or the transformer ratio r in the tuning circuit must be adjusted to establish $\beta Q_L = 0.35$ [see (3.18) and (3.28)].

3.4.3 Other Trade-Offs

In many cases, surface wave transducers are subjected to design constraints on input impedance or VSWR over the filter bandwidth. The coupling condition that $Q_L/Q_R \ll 1$, for accurate transducer responses, is equivalent to $G_a(f_0)/G_L \ll 1$. This implies that the transducers *inherently* have a high VSWR. The VSWR of the simple shunt tuned transducer can be improved, at some sacrifice to transducer efficiency, by including a shunt conductance G_c in parallel with the tuning coil. In any real transducer, some effective shunt conductance is always present corresponding to electrode resistive losses [3.13].

Using simple circuit analysis, it is straightforward to extend the transducer transfer function in (3.21) to include a shunt padding conductance and then calculate the effect of this conductance on the transducer performance. Figure 3.22 shows the dependence of midband VSWR on the mismatch parameter $X \equiv Q_L/Q_R$ for several values of a padding parameter $K \equiv G_c/G_a(f_0)$. In the region where the curves have negative slope, VSWR is improved by increasing the shunt conductance. Of course, VSWR could also be reduced by increasing Q_L, although, as we have seen, this would increase phase and amplitude errors. Figure 3.22 also shows that it is possible to increase shunt conductance to the point that the midband input VSWR is unity (i.e., the total conductance equals G_L). Further increases in G_c then begin to increase the VSWR, as indicated by the positive slope segments of the VSWR versus X curves.

Figure 3.23 plots the VSWR at the transducer passband edge as a function of βQ_L, with K and X included as parameters. From Fig. 3.22, we see that when $X \ll 1$, we require $K \cdot X \cong 1$ in order to achieve low VSWR. Figure 3.23 shows that even when these conditions are met, the low VSWR might not be maintained

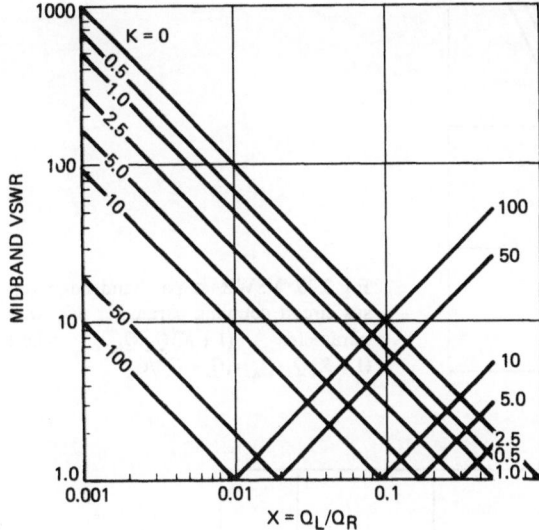

Fig. 3.22. Midband VSWR versus parameter $X = Q_L/Q_R$, with K as a parameter, where $K \equiv G_c/G_a(f_0)$

over a wide fractional bandwidth. Finally, Fig. 3.24 shows the transducer midband insertion loss as a function of X with K as a parameter. When $K = 0$ (i.e., no shunt conductance) the midband loss follows (3.27). For a given value of X, the curves show how transducer loss increases with increasing K. Figure 3.24 includes a line indicating a maximum practical value of X. For a given value of K, increasing X beyond this line produces only detrimental results, i.e., it increases transducer errors *and* increases loss. The procedure for using these design curves is described in the next section.

3.5. Summary and Examples

Let us now review the process of designing a surface wave filter.

1) Based on the filter specification, we must first select a substrate and transducer configuration. If high-temperature stability of the delay (i.e., in the phase response) is required, then a substrate with a low temperature coefficient of delay (TCD) such as ST quartz [3.34] is in order. If TCD is not critical, a strong piezoelectric substrate may be used.

2) Next, the transducer configuration must be determined. If the filter specification favors the use of a weak piezoelectric, then the apodized/unapodized configuration in Fig. 3.7 is required. If on the other hand, a strong piezoelectric such as YZ LiNbO$_3$ is used, then the more versatile MSC configuration in Fig. 3.8 becomes viable.

3) Having established the transducer configuration, the filter specification must be converted to a pair of impulse responses which will then be used to define the transducer geometries. In bandpass filter applications, a desired filter *transfer function is usually specified*. This can be converted to two factors,

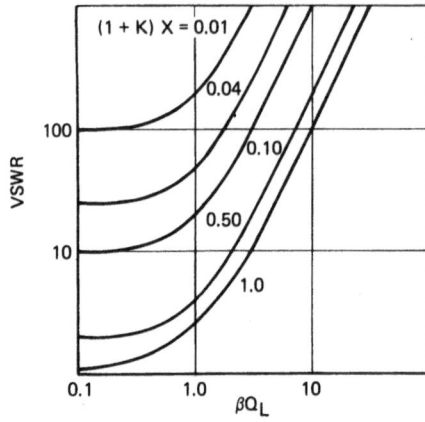

Fig. 3.23. VSWR at passband edge versus circuit coupling parameter βQ_L with parameter $(1+K)Q_L/Q_R$, where $(1+K)Q_L/Q_R = (G_a + G_c)/G_L$

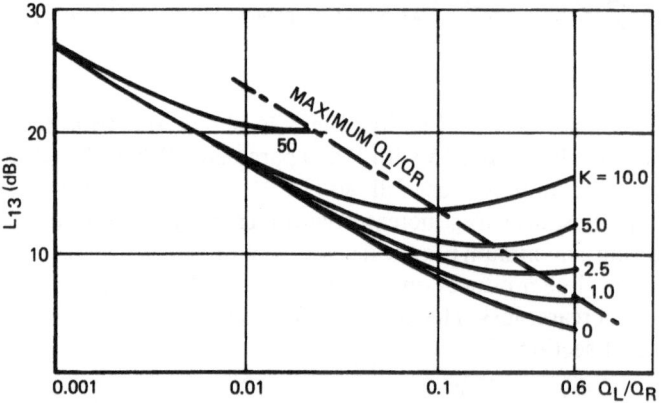

Fig. 3.24. Midband transducer loss versus parameter $X = Q_L/Q_R$, with $K = G_c/G_a(f_0)$ as a parameter

namely, a transfer function which approximates the specification and is produced by a synthesized unapodized transducer, and one that is given by the ratio of the specified to the unapodized transducer transfer function. When an MSC configuration is employed, the square root of the specified transfer function can be used as the transducer design transfer function. In performing the division of transfer functions or the square root operation, the resulting design transfer function may exhibit singularities near the edges of the passband. These singularities result from truncation of the impulse responses of the necessarily finite-duration transducers. Generally, a slight change in specification making the roll-off at the passband edge more gradual, i.e., using a "window function", removes the singularities [3.35].

Surface wave filters are not always specified in terms of desired transfer functions. In the case of tapped delay lines and dispersive filters (e.g., linear FM chirp filters), a desired *impulse response* is usually specified. This presents no problem since transfer functions and impulse responses are related through

Fourier transformation. The filter designer must simply remember that it is easiest to form the pair of transducer-design transfer functions by factoring a specified *transfer function*; similarly, it is easiest to design these transducers based on the individual *impulse responses*.

4) Once acceptable transducer-design transfer functions are derived, the design impulse responses are computed using the Fourier transform, given in (B.3).

5) Given the design impulse response in the form of (3.3), the electrode locations t_n are determined from the solution to the phase-weighting equation, (3.4). This also determines the required tap weight amplitudes $a(t_n)$.

6) The tap locations lead to instantaneous frequencies, from (3.5) and then to the required relative apodization w_n/W_0, given by (3.20).

7) Next, considerations of insertion loss, amplitude roll-off, phase error, and echo suppression, combined with an estimate of Q_R from (3.28), lead to an optimum choice for Q_L (see Figs. 3.18–21).

8) The total transducer capacitance, calculated for the appropriate apodization using (3.18), is then combined with (3.23) to determine practical values of maximum aperture W_0, load conductance G_L, and transformer turns ratio r to achieve the desired value of Q_L.

9) Finally, Fig. 3.22 shows the midband VSWR corresponding to the selected value of Q_L. For the particular value of $X = Q_L/Q_R$, this graph yields the value of $K = G_c/G_a(f_0)$ required to reduce the midband VSWR to a desired level. Figures 3.23 and 3.24 are then used to establish the value of VSWR at the edge of the band (of fractional bandwidth β) and the concomitant increase in midband insertion loss, respectively. If Fig. 3.23 indicates that the required VSWR is not maintained over the full design bandwidth, then a larger value of K, or a smaller of Q_L, must be tried. Both of these alternatives increase the midband insertion loss, shown in Fig. 3.24.

3.5.1 Examples

In closing, two examples of surface wave filters are presented, one implemented using the unapodized/apodized configuration in Fig. 3.7, and one using the MSC configuration of Fig. 3.8. The measured filter transfer function for the first case is shown in Fig. 3.25a and b. Midband insertion loss is 20.5 dB (at 132 MHz) and there is less than 0.2 dB peak to peak passband ripple. Wiggle in the horizontal dots indicates departure from linear phase on a scale of 10°/div. Thus, there is less than 1° phase ripple over the central 4.0 MHz. Figure 3.25b exhibits out of band rejection of 60 dB, with a 50 dB/3 dB shape factor of 1.86. This surface wave filter is representative of the outstanding performance that can be achieved up to frequencies around 1 GHz.

Figure 3.26 shows the amplitude response of a surface wave bandpass filter configured with two apodized transducers and an MSC. This device, centered at 140 MHz, has 20 MHz bandwidth and exhibits 30 dB midband insertion loss.

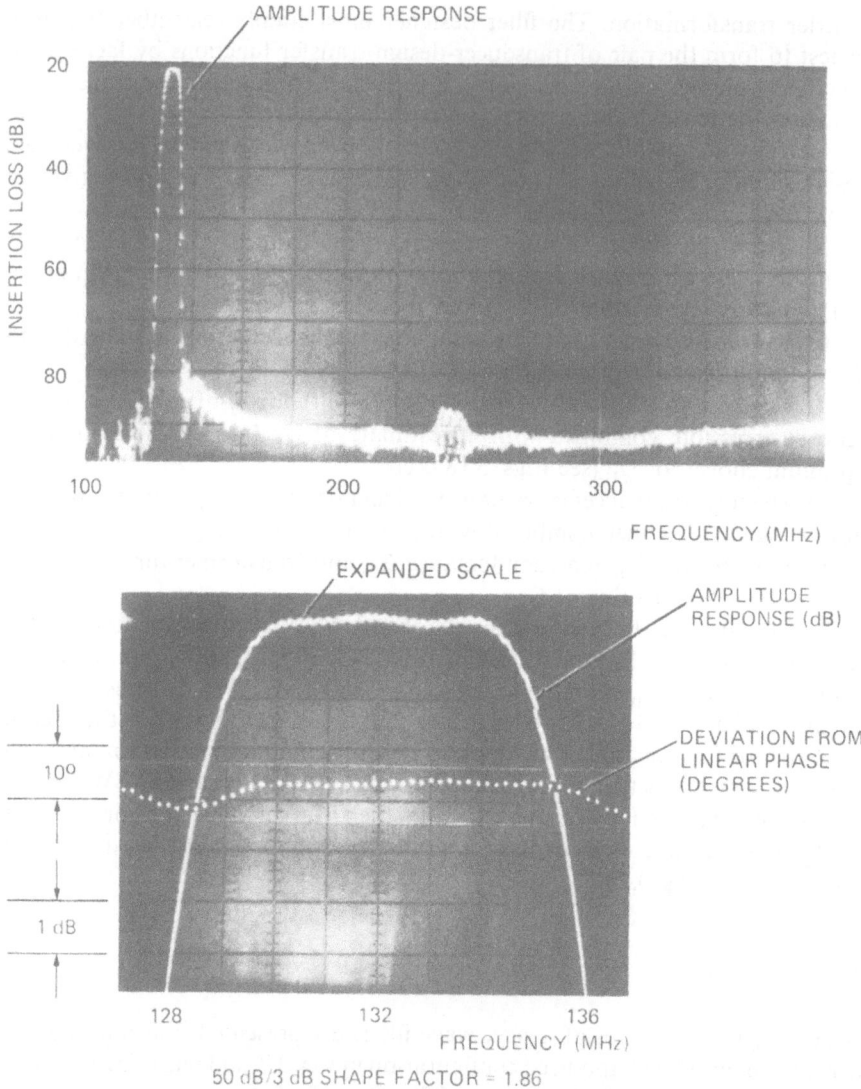

Fig. 3.25. Measured transfer function of surface wave bandpass filter with unapodized/apodized configuration of Fig. 3.7 (courtesy of *W. R. Smith*, Hughes Aircraft Company, Fullerton, Calif.)

The filter has a good shape factor, less than 10° peak to peak phase ripple over the band, and 50 dB out of band rejection.

Although the examples have focused on bandpass filters, the surface wave filter design principles presented here have been widely used in the development of dispersive filters for radar signal processing. Like the bandpass filters, above, the results for dispersive filters have been generally excellent. In most cases, chirp

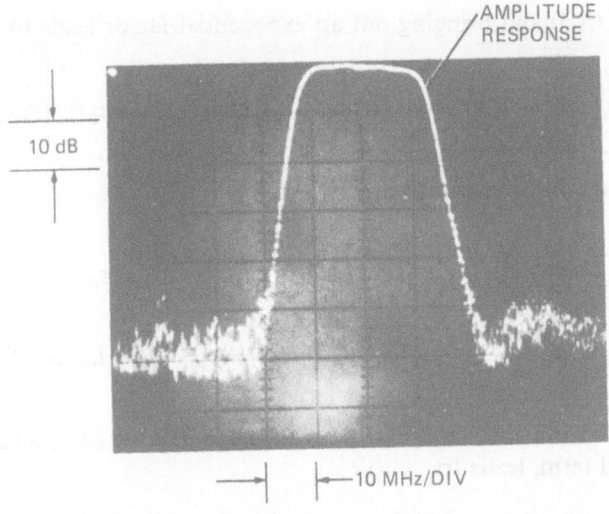

Fig. 3.26. Measured amplitude response of surface wave bandpass filter with multistrip coupled apodized transducers, as in Fig. 3.8. Center frequency is 140 MHz, and midband loss is 30 dB. (Courtesy of *W.R.Smith*, Hughes Aircraft Company, Fullerton, Calif.)

filters have been implemented with surface waves which could never have become practical with any other existing technology [3.2, 19, 36].

Based on the history of surface wave solutions to sophisticated filtering problems, one can project with confidence that the level of interest in surface wave filters will continue to increase as system requirements evolve to higher frequencies, to broader bandwidths, to more sophisticated desired transfer functions, to improved reliability, and to smaller size and lower power dissipation.

Appendix A: Symmetrical Transducers

The objective of this appendix is to show that if the impulse response $h(t)$ has a center of symmetry then the transducer has linear phase.

Assume the transducer impulse response is given by

$$h(t) = a(t - t_0) \cos \phi(t - t_0) \tag{A.1}$$

where symmetry in the transducer about $t = t_0$ leads to

$$a(t - t_0) = a(t_0 - t), \quad \text{and} \quad \phi(t - t_0) = \phi(t_0 - t). \tag{A.2}$$

The transfer function $H(\omega)$ of the symmetrical transducer is found by taking the Fourier transform of (A.1),

$$H(\omega) \equiv \int_{-\infty}^{\infty} dt\, h(t) \exp(-i\omega t). \tag{A.3}$$

Substituting (A.1) into (A.3) and bringing out an exponential factor leads to

$$H(\omega)=\exp(-i\omega t_0)\int_{-\infty}^{\infty} d(t-t_0)a(t-t_0)[\cos\phi(t-t_0)]\exp[-i\omega(t-t_0)]. \quad (A.4)$$

Now, separating the interval of integration gives

$$H(\omega)=\exp(-i\omega t_0)\int_{-\infty}^{t_0} d(t-t_0)a(t-t_0)[\cos\phi(t-t_0)]\exp[-i\omega(t-t_0)]$$

$$+\exp(-i\omega t_0)\int_{t_0}^{\infty} d(t-t_0)a(t-t_0)[\cos\phi(t-t_0)]\exp[-i\omega(t-t_0)]. \quad (A.5)$$

Substituting $u=(t_0-t)$ in the first term on the right-hand side of (A.5) and $u=(t-t_0)$ in the second term, leads to

$$H(\omega)=\exp(-i\omega t_0)\int_0^{\infty} du\, a(-u)\cos\phi(-u)\exp(i\omega u)$$

$$+\exp(-i\omega t_0)\int_0^{\infty} du\, a(u)\cos\phi(u)\exp(-i\omega u). \quad (A.6)$$

Therefore, making use of (A.2) in (A.6) leads to

$$H(\omega)=2\exp(-i\omega t_0)\int_0^{\infty} du\, a(u)\cos\phi(u)\cos\omega u. \quad (A.7)$$

Since the integral in (A.7) is real, the phase of $H(\omega)$ is $\phi(\omega)=-\omega t_0$. Consequently, $\tau(\omega)$, the delay associated with $H(\omega)$, is

$$\tau(\omega)\equiv(-1/2\pi)\frac{\partial\phi(\omega)}{\partial f}=t_0 \quad (A.8)$$

thereby illustrating that the transfer function associated with a real, symmetric impulse response corresponds to a simple delay (and a spectral shape factor).

Appendix B: Spectrum of Time-Sampled Signal

The objective of this appendix is to derive the frequency domain representation of a particular time-sampled replica of the general desired transducer impulse response $h(t)$, given by

$$h(t)=a(t)\exp[i\phi(t)]. \quad (B.1)$$

Guided by the electrode positioning law and the assumption that the sample points within a transducer lie in the center of the gaps, we consider the time-sampled replica of $h(t)$ given by

$$h_s(t) = \sum_{n=1}^{N} h(t)\{\delta[\phi(t) - (n+1/2)\pi] + \delta[\phi(t) - (n-1/2)\pi]\}. \tag{B.2}$$

Each term in (B.2) corresponds to two samples whose times are given by the solutions of

$$(t_{n\pm1/2}) = (n\pm1/2)\pi.$$

Now, substituting (B.1) into (B.2) and taking the Fourier transform to arrive at the frequency domain representation, $H(f)$ gives

$$H(f) = \text{F.T.}\{h_s(t)\} = \int_{-\infty}^{\infty} dt\, h_s(t)\exp(-i2\pi ft), \tag{B.3}$$

or

$$\begin{aligned} H(f) = \sum_{n=1}^{N} &[1/\dot{\phi}(t_{n+1/2})]a(t_{n+1/2})\exp\{i[(n+1/2)\pi - 2\pi f t_{n+1/2}]\} \\ &+ [1/\dot{\phi}(t_{n-1/2})]a(t_{n-1/2})\exp\{i[(n-1/2)\pi - 2\pi f t_{n-1/2}]\} \end{aligned} \tag{B.4}$$

where the dot notation indicates a time derivative.

Now, assuming that $a(t)$ and $\phi(t)$ are reasonably smooth, we make the approximation that

$$t_{n\pm1/2} \cong t_n \pm 1/4f_n, \tag{B.5}$$

where f_n, the instantaneous frequency, is defined by

$$f_n \equiv (1/2\pi)\frac{\partial\phi(t)}{\partial t}\bigg|_{t=t_n} \tag{B.6}$$

which is roughly equal to

$$f_n \cong (1/2\pi)\frac{\partial\phi(t)}{\partial t}\bigg|_{t=t_{n\pm1/2}}.$$

Finally, substituting t_n for $t_{n\pm1/2}$ and $2\pi f_n$ for $\dot{\phi}(t_{n\pm1/2})$ in (B.4) gives

$$H(f) = (1/\pi)\sum_{n=1}^{N} (-1)^n [a(t_n)/f_n]\sin(\pi f/2f_n)\exp(-i2\pi t_n). \tag{B.7}$$

References

3.1   W.R.Smith, H.M.Gerard, J.H.Collins, T.M.Reeder, H.J.Shaw: IEEE Trans. MTT-**17**, 856–864 (1969)
3.2   W.R.Smith, H.M.Gerard, W.R.Jones: IEEE Trans. MTT-**20**, 458–471 (1972)
3.3   G.L.Matthaei, D.Y.Wong, B.P.O'Shaughnessy: IEEE Trans. SU-**22**, 105–114 (1975)
3.4   D.A.Leedom, R.Krimholtz, G.L.Matthaei: IEEE Trans. SU-**18**, 128–141 (1971)
3.5   R.Krimholtz: IEEE Trans. SU-**19**, 427–436 (1972)
3.6   H.E.Kallman: Proc. IRE **28**, 302–310 (1940)
3.7   W.D.Squire, H.J.Whitehouse, J.M.Alsup: IEEE Trans. MTT-**17**, 1020–1040 (1969)
3.8   W.R.Smith: Proc. Ultrasonics Symp., IEEE Cat. CHO 1120–5SU, 547–552 (1976)
3.9   R.H.Tancrell, M.G.Holland: Proc. IEEE 393–409 (1971)
3.10 H.B.Dwight: *Tables of Integrals and Other Mathematical Data* (McMillan Company, New York 1947)
3.11 T.L.Szabo, A.J.Slobodnik: IEEE Trans. SU-**20**, 240–251 (1973)
3.12 J.D.Maines, G.L.Moule, N.R.Ogg: Electron. Lett. **8**, 431–433 (1972)
3.13 K.M.Lakin: IEEE Trans. MTT-**22**, 418–424 (1974)
3.14 J.J.Campbell, W.R.Jones: IEEE Trans. SU-**15**, 269–217 (1968)
3.15 W.S.Jones, C.S.Hartmann, T.D.Sturdivant: IEEE Trans. SU-**19**, 368–377 (1972)
3.16 W.R.Smith, W.F.Pedler: IEEE Trans. MTT-**23**, 853–864 (1975)
3.17 T.W.Bristol, W.R.Jones, P.B.Snow, W.R.Smith: Proc. Ultrasonics Symp., IEEE Cat. CHO 708-8SU, 343–345 (1972)
3.18 A.J.DeVries, R.L.Miller, T.J.Wojcik: Proc. Ultrasonics Symp., IEEE Cat. CHO 708-8SU, 353–358 (1972)
3.19 H.M.Gerard, W.R.Smith, W.R.Jones, J.B.Harrington: IEEE Trans. MTT-**21**, 176–186 (1973)
3.20 H.M.Gerard, G.W.Judd, M.E.Pedinoff: IEEE Trans. MTTT-**20**, 188–192 (1972)
3.21 F.G.Marshall, C.O.Newton, E.G.S.Paige: IEEE Trans. MTT-**21**, 206–215 (1973)
3.22 R.H.Tancrell: Electron. Lett. **9**, 316–317 (1973)
3.23 T.W.Bristol: Proc. Ultrasonics Symp., IEEE Cat. CHO 708-8SU, 377–380 (1972)
3.24 C.S.Hartmann: Proc. Ultrasonics Symp., IEEE Cat. 73 CHO 807-SSU, 423–426 (1973)
3.25 C.F.Vasile: IEEE Trans. SU-**21**, 7–11 (1974)
3.26 R.F.Mitchell, N.H.C.Reilly: Electron. Lett. **8**, 329–331 (1972)
3.27 W.P.Mason: *Electromechanical Transducers and Wave Filters*, 2nd ed. (Van Nostrand, Princeton, N.J. 1948) pp.201–209, 399–409
3.28 H.Engan: IEEE Trans. ED-**16**, 1014–1017 (1969)
3.29 K.A.Ingebrigsten: J. Appl. Phys. **40**, 2681–2686 (1969)
3.30 H.Skeie: J. Acoust. Soc. Amer. **48**II, 1098–1109 (1970)
3.31 H.M.Bode: *Network Analysis and Feedback Amplifier Design* (Van Nostrand, New York 1945) pp.360–371
3.32 R.M.Fano: J. Franklin Inst. **249**, 57–84, 139–154 (1950)
3.33 T.M.Reeder, W.R.Sperry: IEEE Trans. MTT-**20**, 453–458 (1972)
3.34 M.B.Schultz, M.G.Holland: Proc. IEEE Lett. **58**, 1361–1362 (1970)
3.35 R.H.Tancrell: IEEE Trans. SU-**21**, 12–22 (1974)
3.36 G.W.Judd: Proc. Ultrasonics Symp. IEEE Cat. 73 CHO 807-8SU, 478–481 (1973)
3.37 R.W.Wagers: Proc. Ultrasonics Symp., IEEE Cat. CHO 1120-5SU, 536–539 (1976)
3.38 W.R.Smith: Wave Electronics **2**, 25–63 (1976)

4. Fundamentals of Signal Processing Devices

E. A. Ash

With 54 Figures

4.1 Introduction

Our primary aim in this book is to portray fundamental surface acoustic wave (SAW) device physics and further to provide a basis for the design of the elements which in combination lead to integrated signal processing subsystems. We have already seen in Chapter 3 how the basic theory of transduction and the associated frequency domain analysis lead directly to techniques for synthesizing a large class of filter structures. Frequency filters indeed represent one of the most important areas of application for SAW technology. In this chapter we now want to embark on a description of other signal processing functions which are capable of implementation using SAW technology.

The basic concept from which most of the device applications flow is extremely simple: it is the fact that the slow speed of acoustic propagation enables one to display a time-varying signal in space; that during its passage from transmitter to transducer, one can sample and even manipulate small sections of that signal. Consider a very simple example: given a signal $f(t)$, we would like a record of $f(t) + f(t + \tau)$. The time interval corresponds to a spatial displacement of $v\tau$. The simple arrangement shown in Fig. 4.1a will therefore accomplish the task. It is worth noting that the conventional digital electronic alternative, Fig. 4.1b, involves a rather complex system. Moreover, if $f(t)$ represents a sufficiently high frequency record, let us say a burst of a 500 MHz waveform, and if we want to perform the addition with an accuracy of around 1 %, current silicon-based digital technology is unable to perform the task in real time. The extraordinarily simple device of Fig. 4.1a therefore has the ability to perform what would otherwise be at best a complex and at worst an unattainable task. It is this basic capability which has led to such a rapid development of SAW signal processing technology and such a rapid infiltration into a wide variety of systems.

To give some account of these developments we shall have to touch on a large number of devices and concepts. Within the confines of a single chapter the treatment is therefore necessarily broad and descriptive. It is nevertheless hoped to convey the dominant features of the terrain and to reveal some of the key considerations which dictate the design and the range of applicability of various devices.

The simplest function with which we are concerned is that of delay (Sect. 4.2), with or without intermediate sampling points. One of the problems we shall

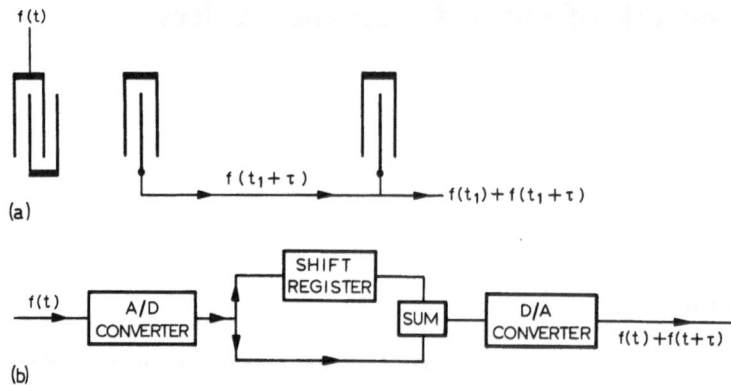

Fig. 4.1a and b. Simple example of signal processing capability arising from capability of displaying a time varying signal in space. (a) SAW processor. (b) Equivalent conventional electronic processor

encounter is how to obtain long delays—"long" used in the sense of exceeding L/v, where L is the maximum convenient sample length. One, though by no means the only, reason for requiring long delay is to realize an element having a large $d\phi/df$, where ϕ is the transfer phase. An alternative means of obtaining such large phase frequency gradients is by the use of surface wave resonators (Sect. 4.3), which have recently become of great importance. Any system—be it a delay line or resonator—which provides a large $d\phi/df$, can be used as the frequency determining element in an oscillator. Such surface wave oscillators (Sect. 4.4), are rapidly becoming among the most widely used SAW devices.

Without any doubt, the first SAW devices to have reached operating systems are coded time-domain filters for radar application. It turns out that the basic concept of a coded transducer structure (Sect. 4.5) is an extremely powerful technique for compressing and expanding signals. The impact of this concept extends well beyond radar systems—indeed to any communication system where the generation and recognition of complex coded signals can improve system performance.

We have seen in the illustration of Fig. 4.1 that the source of the power of SAW signal processing devices can be identified as the facility with which two analog signals can be added. There are, however, signal processing functions which require multiplication instead of or as well as summing operations. The nonlinear properties of elastic solids provide the means for effecting such functions (Sect. 4.6). The nonlinear phenomena lead to interactions which are by normal standards "inefficient". But it is just this low efficiency which ensures that the interactions lead to extremely pure products.

Most acoustic surface wave structures that have been studied are two-port devices, in the sense that there is a single non-branching acoustic transmission path, even though there may be numerous taps to sample the signal along this path. There are, however, applications where one is concerned with multiport acoustic paths (Sect. 4.7). These additional paths may be "internal" in the sense that they do not end in additional electric output ports, but are included purely to improve the performance. The multistrip coupler is an important component which is frequently used in this manner. In addition, however, there are systems in which the additional acoustic paths do lead to additional outputs which are germane to the operation of the device. An example in this class is a channel dropping filter.

Even in broad survey we shall have to omit some significant areas of endeavor, a notable example being that concerned with amplifications of acoustic surface waves. The performance of such amplifiers has continued to improve—but so has the performance of transducers, so that amplifiers become less necessary. Furthermore, the competition offered by conventional amplifiers is unlikely to be overcome.

4.2 Delay Lines

The simplest and perhaps the widest used signal processing function is that of pure delay. The applications range from those requiring delays of a few tenths of a microsecond, to others which would require in excess of 10 ms—from applications with minimal constraints on phase errors, temperature stability, or spurious response to others in which these "secondary" specifications are central to the purpose. We shall begin by briefly considering the simplest from of delay line, shown in Fig. 4.2. The bandwidth is determined by that of the transmitting and receiving transducers T and R, respectively. The insertion loss is determined first by the bidirectional loss of 3 dB per transducer—i.e., 6 dB—to which one must add losses arising from the material, from diffraction effects, from beam steering effects attributable to small errors in the orientation of the transducers, and from any electrical mismatch conditions.

4.2.1 Insertion Loss and Spurious Signals

It is natural to think first about the loss. However, for delays of the order of 10 μs or less, at center frequencies below 500 MHz, one is rarely concerned with the insertion loss as such, which with careful design can be less than 10 dB. Loss is readily compensated electronically and becomes an embarrassment only when it detracts excessively from the total available dynamic range. The only exception to this rule arises in the use of a filter at the front end of a receiver; however, delay functions would normally be postponed to a later stage in the receiver system. The dynamic range of a delay line may be set by the breakdown voltage on the transmitting transducer, or by the onset of spurious signals arising from nonlinear interactions; it will depend on choice of center frequency, on the bandwidth, and on the material, but it is not uncommon to approach 100 dB. Systems requirements on available dynamic range vary over very wide ranges but, as a general rule, an insertion loss of the order of 30 dB is not too damaging.

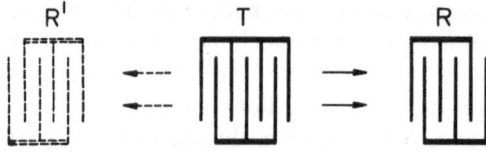

Fig. 4.2. Basic delay line configuration. The symmetrical disposition arising from an additional receiving transducer R' reduces insertion loss and increases triple transit suppression

The main problem is to ensure that the available dynamic range is not degraded excessively by the presence of spurious signals. The first to consider is the "triple transit echo", arising from the reflection of the signal at the receiving transducer R followed by a second reflection at the transmitting transducer T. It then arrives again at R with a delay of 3τ, where τ is the single path delay. If T and R are perfectly matched bidirectional transducers, the acoustic loss at each reflection is only 6 dB. In this case then the triple transit echo is at a level only 12 dB below that of the wanted signal. It is possible to make a dramatic improvement in this situation [4.1] by using two receiving transducers R and R' symmetrically disposed with respect to T, as in Fig. 4.2. The outputs from R and R' are combined, so that T is no longer subject to the 3 dB bidirectional loss. More importantly T is now—in principle—a perfect absorber of the acoustic signals reflected from R and R', so that the triple transit signal disappears. The perfect cancellation assumes that T is perfectly matched—which it can be only at the center frequency. It also assumes that T is perfectly symmetrical and symmetrically located and that R and R' are identical—fabrication targets which can be approached but not perfectly realized. Nevertheless, the additional suppression which can be thus obtained can be of the order of 20 dB.

This system of triple transit suppression is a circumvention of the bidirectionality of the transducer T. Alternatively we can incorporate unidirectional transducers based on the use of a pair with appropriate phasing [4.2], on a three-phase design [4.3], or on the use of a multistrip coupler [4.4] (see Sect. 4.7). These approaches are more elaborate, but also inherently more effective in that both T and R can be made nonreflecting. They also give more scope for broad banding the nonreflection property, by stagger tuning the two transducers.

The simplest suppression method of all is deliberately to mismatch the transducers. Here one is involved in trading triple transit suppression against increased insertion loss. Fortunately, the terms of trade are rather favorable; thus, increasing the basic bidirectionality insertion loss from 6 to 12 dB raises the triple transit suppression from 12 to 33 dB [4.5].

In addition to the triple transit echo, there are other sources of spurious signals. These include the bulk waves which are inevitably launched by a transducer; one seeks to suppress them by making reflecting surfaces, such as the base of the sample, rough, by the use of various absorbing compounds, and by a judicious choice of the crystal cut [4.6, 7]. One can also redirect them, out of harm's way, using multistrip coupler techniques [4.4] (Sect. 4.7). An example of a 5 µs delay line using such multistrip coupler techniques at a center frequency of 60 MHz had an insertion loss less than 10 dB and a maximum spurious level of −40 dB, both over a 25 % bandwidth. We see that, with sophisticated design

INSERTION LOSS (DB)

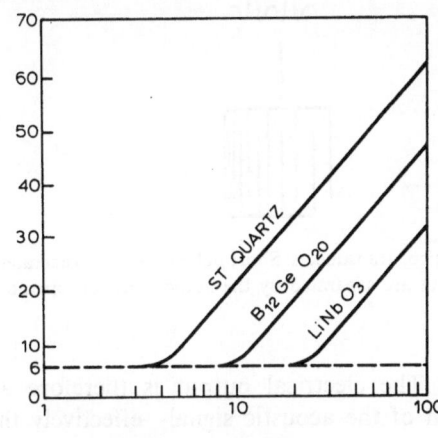

Fig. 4.3. Minimum insertion loss as a function of bandwidth (after [4.2])

FRACTIONAL   BANDWIDTH   (%)

techniques, delay lines with low insertion loss *and* having at the same time a low spurious level can be realized. It is relatively easy to obtain even better performance in either of these parameters at the expense of the other.

4.2.2 Delay Line Bandwidth

The bandwidth is controlled by the design of T and R (see Chap. 3). The maximum attainable bandwidth depends on the coupling constant of the material used and on the acceptable insertion loss. This situation is summarized, for a number of materials [4.5] in Fig. 4.3. In the case of LiNbO$_3$, one can obtain a bandwidth of 25 % without significantly exceeding the 6 dB bidirectional loss (itself, as we have seen, removable if desired). It is, however, important to appreciate that one can still achieve this same bandwidth on ST quartz if one can tolerate an additional insertion loss of 34 dB. There are many situations in which the temperature stability of ST quartz may be an attraction which outweighs such additional loss.

In the discussion of bandwidth we have not so far addressed ourselves to the question of the nature of the signal which it is intended to delay. If it is an analog signal with a fractional bandwidth well below unity, the transducer bandwidth requirements for a given permissible frequency distortion are readily specified. However, in some situations one will use T not merely for transduction to the acoustic wave but simultaneously as a filter; one will use R not just to transduce back to an electronic signal but as a matched filter. An extreme example of such usage is for an input signal in the form of a video pulse. If it is applied to a variable pitch transducer, as in Fig. 4.4, the filtering action is best portrayed in time domain; the acoustic wave launched will approximate the impulse response of the transducer, which may be many tens or even hundreds of times longer than the *duration* of the *video* pulse. On arrival at R, the signal encounters a

Fig. 4.4. Delay line using variable period of "chirp" transducers. Short pulse excitation generates a chirp train. Several such trains can overlap but are separated by the autocorrelation operation carried out by the receiving transducer

transducer which is identical to T. The electrical output is therefore an approximation to the autocorrelation of the acoustic signal—effectively the signal is recompressed into a band-limited version of the input. It is important to appreciate that the rate at which one can enter digital impulses is of the order $(2\Delta f)^{-1}$, where Δf is the transducer bandwidth, and is not determined by the length of impulse response of T. There is no need to avoid the superposition of many slightly displaced trains of signal (see Fig. 4.4). In principle, one can achieve the same result by using a short simple interdigital transducer having the same effective bandwidth. However, for large bandwidths, and hence transducers with very few finger pairs, it will not then be possible to match the transducer to a $50\,\Omega$ system; the variable period transducer provides bandwidth as well as presenting a more suitable impedance to the source. The variable period disposition is one example of a coded transducer to be discussed in more detail in Section 4.4.

4.2.3 Temperature Stability

A critical performance characteristic of a delay line is the temperature stability $(1/\tau)/(d\tau/dT)$. This is directly determined by the material and by the chosen direction of propagation. The only way in which the delay line design as such can bear directly on the temperature performance arises if there is an opportunity for having more than one direction of propagation in a composite path. This possibility has been successfully exploited in several types of design [4.8, 9]. More generally, one is concerned with the choice between a high coupling-constant material such as $B_{12}GeO_{20}$ or $LiNbO_3$ which have unhappily high temperature coefficients of the order of $10^{-4}\,°C^{-1}$ or ST quartz whose temperature coefficient is zero (at 28 °C) but which has a much smaller coupling constant. If the low coupling-constant material cannot meet the other delay line specifications, one can resort to the use of transducers on the low coupling-constant ST quartz, with an overlay of a thin film of a high coupling-constant material. Success has also been achieved with modifying the temperature coefficient of $LiNbO_3$, by deposition of a nonpiezoelectric film over the whole of the delay line. By depositing a layer of quartz onto $LiNbO_3$ the temperature coefficient has been decreased by an order of magnitude [4.10]. The use of such films will of course have other effects, on loss, and, probably of greater importance, also on dispersion. Moreover, such a technique represents a step away from the essential simplicity of surface wave devices, with an inevitable bearing on yield and costs.

In many systems, the critical issue is not so much the constancy of the total delay as the need to retain knowledge of its precise value. In such cases one can define a number of delay lines on a single substrate and use one of them to measure the actual delay. The delay of the others will be identical to that of the test line to a very high degree of precision. An example is seen in a surface wave delay line memory application [4.11]. This consisted of an array of seven parallel lines and used the variable period transducers shown in Fig. 4.4, providing a band from 40 to 110 MHz. To minimize the temperature problem, the transducers were formed on short lengths of $LiNbO_3$ but jointed to ST quartz, the latter providing most of the propagation delay of 67 µs. Each line stored 1500 bits, with a bit rate of 75 Mbits. Six of the lines were used for storage; the seventh was used to control the clock frequency, so that the tolerance to temperature-induced delay variations was thereby greatly increased.

4.2.4 Long Delay Lines

Substrate crystals are not very often used in lengths greater than 10 cm. Beyond this the cost rises much faster than proportionally with the length, and it is in any event difficult to obtain lengths of quartz, $LiNbO_3$, or $B_{12}GeO_{20}$ in lengths which exceed 25 cm. Quite apart from procurement, it becomes quite difficult to handle such long samples. There is therefore an economic and practical limit to the maximum delay which can be obtained on a single propagation path, which one might set at 50 µs for $LiNbO_3$ and at 100 µs for $B_{12}GeO_{20}$. Yet there are requirements for much longer delays, in both radar and communication systems. There is also considerable interest in the use of surface wave delay lines for storing a complete frame of a television display, an application which leads to a requirement for delay times in excess of a millisecond. The need therefore exists for delay lines with propagation path lengths which greatly exceed the dimensions of available crystals, i.e., for "long" delay lines.

The problem can be approached by stitching together a number of relatively short lines, as in Fig. 4.5a, with amplifiers for isolation and gain. Though unsophisticated, it is an approach which may in some circumstances prove to be a most effective engineering solution. The periodic use of amplifiers makes relatively large insertion losses for the individual lines permissible—insertion loss which can be "used" to improve the suppression of the triple transit, and other spurious signals. The arrangement also allows scope for stagger tuning to control the bandwidth and phase characteristics.

A number of purely acoustic solutions have been developed. As long ago as 1958 a delay line was described [4.12] in which surface waves propagated along a helical path on the outside of a metallic cylinder; see Fig. 4.5b. A delay time of

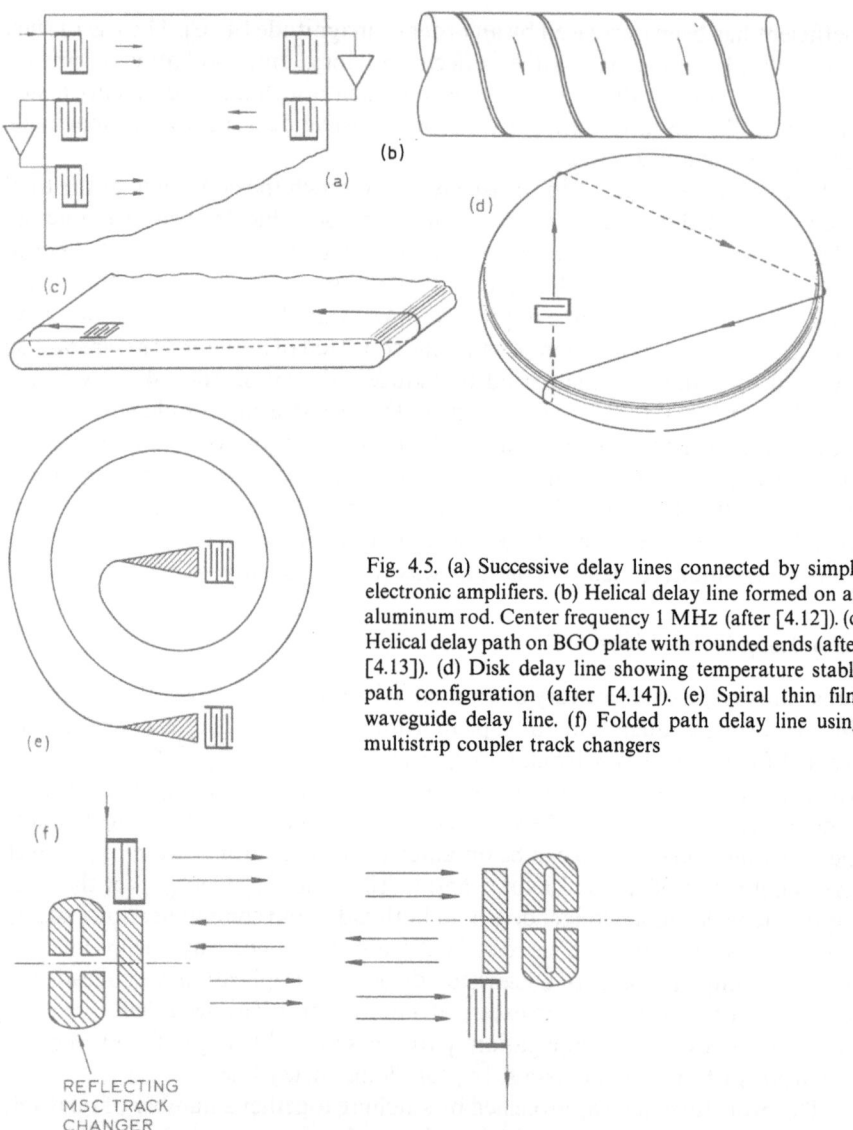

Fig. 4.5. (a) Successive delay lines connected by simple electronic amplifiers. (b) Helical delay line formed on an aluminum rod. Center frequency 1 MHz (after [4.12]). (c) Helical delay path on BGO plate with rounded ends (after [4.13]). (d) Disk delay line showing temperature stable path configuration (after [4.14]). (e) Spiral thin film waveguide delay line. (f) Folded path delay line using multistrip coupler track changers

REFLECTING
MSC TRACK
CHANGER

2 ms was achieved at a center frequency of 1 MHz. It is this basic scheme which in recent times has been redeveloped in the context of single crystal substrates [4.15, 16]. To obtain the longest possible path length for a given volume of material, the configuration shown in Fig. 4.5c was been adopted by a number of workers. In a particular example of such a delay line [4.13] constructed on $B_{12}GeO_{20}$ with a center frequency of 83 MHz, the overall delay time was 907 µs. The results obtained are shown in Fig. 4.6. The insertion loss of 60 dB is remarkably constant over the 35% band, a result achieved by designing

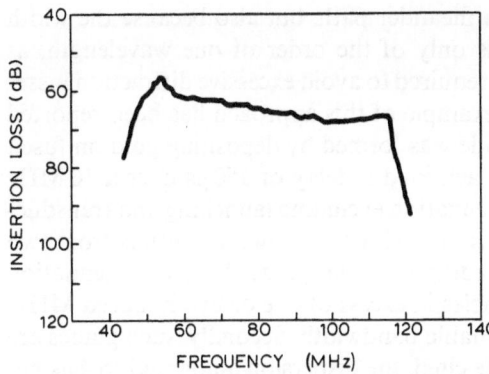

Fig. 4.6. Performance of a 0.9 ms delay line of the form shown in Fig. 4.5c (after [4.15])

transducers having an insertion loss characteristic which compensates that of the other, frequency-dependent losses encountered. Delay lines of this form rely on the propagation of surface waves around carefully polished ends of the plate. It has been found [4.17] that even at frequencies as high as 300 MHz it is possible to avoid seriously enhanced losses in the propagation from one side of the sample to the other.

A radically different conceptual design, in the form of a circular flat disk structure, has recently been described [4.14, 18, 19]; this also relies on a system of propagation paths on alternate sides of the sample, as seen in Fig. 4.5d. However, there is now an additional feature which comes into play as a consequence of the circular symmetry. It is easy to see that, for an isotropic material, a disturbance launched at the center of the disk on the top side will propagate circular waves which will come to a focus at the center on the bottom side of the disk—i.e., the curved edge of the disk acts as a geodesic lens. If the source of the disturbance is no longer at the center of the disk, it turns out that there is still a focusing action, albeit with aberrations, which under some specific conditions will persist even when the disk is anisotropic. The fact that the propagation paths are now no longer parallel provides an opportunity to choose them in such a manner that the temperature-induced phase error in one path is cancelled by one of opposite sign in another. This specific condition has been found for Y cut quartz, and for the beam pattern shown on Fig. 4.5d. Using a modified form of this propagation pattern, a delay time of 220 μs was obtained at 40 MHz. A notable feature of this form of delay line is the low spurious levels achieved—below -40 dB in the above example. This is attributable in part to the fact that the input and output transducers are not parallel.

Although it turns out that the polishing of the edges required is not necessarily an expensive operation, there is nonetheless some reluctance to abandon the simplicity of reliance on a single polished plane surface. To make effective use of such a single surface, some means have to be found for bending the propagation path so that it utilizes as large a portion of the available area as possible. This immediately suggests the use of acoustic surface wave waveguides (described in detail in Chap. 5), not only because they provide a means for steering the wave into a spiral or a meander path, but also because the width required for a propagation path is only of the order of one wavelength, as compared with tens of wavelengths required to avoid excessive diffraction losses in freely propagating beams. One example of this approach has been reported [4.20] in which a thin film waveguide was formed by depositing gold on fused quartz, as shown in Fig. 4.5e. This achieved a delay of 250 μs over a 10 MHz bandwidth. The total delay path attenuation excluding launching and transduction losses was only 7.5 dB. This form of delay line, however, suffers from two disadvantages: first, metal mass-loading thin film guides have an attenuation which becomes excessive at frequencies in excess of one or two hundred MHz; this represents a constraint on attainable bandwidth. Secondly, such guides are inevitably dispersive. In the example cited, the dispersion amounted to 4 μs per MHz—bad enough, if not compensated in some manner, to reduce the attainable bit rate by almost one order of magnitude below that implied by the bandwidth. Both these problems can be alleviated using more sophisticated waveguide technology. Waveguides formed by ion implantation [4.21] or by metal indiffusion [4.22] can have losses which are only marginally greater than that of the substrate material. By controlling the guide profile [4.23], one can achieve guiding with very little dispersion. These waveguides are discussed in greater detail in Sections 5.4.1 and 5.3. It seems likely that these developing techniques will lead to practical long delay lines having a delay per unit area of substrate which is greater than that attainable using competing techniques.

All of the forms of long delay lines so far discussed depart in some major respect from the basic simplicity of SAW technology—single mask photolithography on a single polished surface. One of the devices which has emerged from the multistrip coupler concept is a "track changer", to be discussed in Section 4.7. For the moment it is sufficient to appreciate the action as indicated in Fig. 4.5f. It allows a surface wave beam to be reflected into a meander path. The construction of such a delay line remains totally within the basic concept of SAW technology, and thereby lies its great attraction. The main disadvantage is that it requires the use of high coupling-constant substrate material, thus incurring the penalty of a high temperature coefficient of the delay. Further, although one is only concerned with single-stage photolithography, the complexity of the structure is such that it tends to lead to yield problems. A delay line built on this principle has recently been described [4.9], in which a total delay of the order of 100 μs was obtained by the use of two track changing couplers. The basic construction involves the use of three transducers at one end and four at the other. By suitable electronic switches it was possible to bring any pair into play; in this way, the delay could be varied from 94 to 102 μs in 1 μs steps. A frequency band of 60 MHz centered on 130 MHz was obtained by a modified single finger pair transducer. The midband insertion loss was 60 dB, but still allowed a dynamic range of 50 dB. Wedge-shaped metal films were used to correct small misorientations of the wavefront. A complete delay line, including the switching circuitry, and amplifiers to make the overall insertion loss zero, were packaged in a volume of approximately 80 cm^3.

A second example of a delay line built on this principle [4.24] used four track changers to realize a total delay of 130 μs over a 15 MHz band centered on 74 MHz with an insertion loss of 23 dB. The spurious level was -25 dB, attributed primarily to small photolithographic defects. A notable feature of these delay lines is the very small departure from phase linearity, which in this instance did not exceed $\pm 25°$ over the band. It seems probable that the basic track changer technique can be extended to delays of several hundred microseconds and with increased bandwidth.

Before leaving the topic of long delay, brief reference should be made to a radically different approach—using glass capillary tubes and propagating a modified Rayleigh wave down the inside surface [4.25], or, in another version using bulk waves in what is in effect a clad acoustic glass fiber [4.26]. The performance characteristics of these waveguides are presented in Section 5.4.3. In both cases there will be an upper limit to the attainable frequency on account of the losses associated with the glass. However, the losses below 100 MHz are small, it is very easy to obtain long path lengths, and the dispersion can also be small. The transduction is effected by means of thin film couplers, which have a very wide bandwidth. The prospects for achieving very large time bandwidths at relatively low frequencies therefore appear favorable.

4.3 Acoustic Surface Wave Resonators

The classical use of acoustic waves in electronics is in the form of a bulk-wave resonator—the ubiquitous "quartz crystal". Such devices are used both as frequency control elements and also as circuit elements for narrow band filters. Their operation is based on multiple reflection of bulk waves from two opposed surfaces of a crystal. The bulk-wave reflection obtainable from a free surface departs from unity only on account of air loading, and this too can be removed by encapsulation in a vacuum. Such crystal resonators require very accurate parallelity of the two reflecting surfaces, but are otherwise inherently simple. The Qs attainable are limited primarily by bulk losses, and reach values of the order of 10^5. The thickness is normally one-half wavelength, so that the devices become increasingly difficult to make, and fragile when made, at frequencies above 50 MHz. At higher frequencies, one normally resorts to harmonic operation where the thickness is equal to an odd number of half wavelengths. However, the electrical performance suffers in most respects, with increase in the harmonic number.

This leads naturally to the thought that it would be useful if one could make a SAW resonator by reflecting surface waves. One can, however, see at once that a discrete reflector is not realizable: if we consider two surface waves travelling in opposed directions, we find that there are, as one would expect, nodes of the vertical displacement. There are also nodes of the horizontal displacement. *Unfortunately, their loci are displaced by a quarter of a wavelength. Thus, even if*

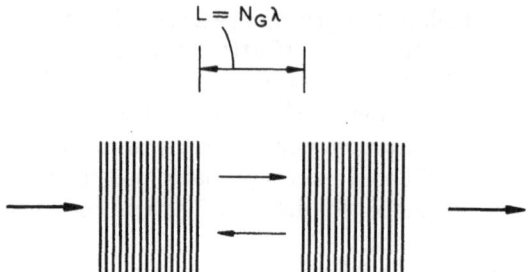

Fig. 4.7. Basic transmission resonator configuration

we could envisage an infinitely rigid reflector, the boundary condition cannot be satisfied by two contra-propagating waves. While a surface wave incident on a perfectly sharp edge of a sample will be partially reflected, a significant portion of the energy is transformed into a bulk wave. If a surface wave is incident on the straight boundary between a free surface and a surface plated with a "fast" material at a sufficiently large angle of incidence, the surface wave will be totally reflected, with none of the power being propagated away in bulk-wave form [4.27]. This effect has been proposed as the basis for several SAW devices, but the angle is too close to grazing to offer a practical solution for a resonator structure.

These considerations lead to the question as to whether one might be able to reflect a surface wave if one does so gradually—at a series of relatively small perturbations rather than at a single large one. The basis of this approach does not depend on the expectation that a small discontinuity will transform a smaller portion of the reflected wave into bulk waves. Rather, it relies on the fact that the bulk-wave sources represented by each individual discontinuity form an array which will not be phased matched to a bulk wave over the frequency range in which it is to be used. It is of course fundamental to resonator operation that the relevant frequency operation is small; the almost complete suppression of bulk waves can therefore be assured. Experiments based on this concept were carried out some time ago, using suitable loaded interdigital transducers as a reflecting element [4.28] and using mass loading gratings [4.8]. The experiments led to the attainment of Q values of the order of 100. More recently, a number of workers have developed this theme [4.29–32], and with improved technology, mass loading, or by using piezoelectric shorting strips on a high coupling-constant material, have achieved Qs in the range of several tens of thousands [4.31]. With such Q values, the performance which can be attained both as frequency control elements and as narrow band filters becomes most attractive. The basic structure, seen in Fig. 4.7, is a transmission cavity which will have periodic narrow pass bands. In this form the device is, in almost every way, comparable to a laser cavity. The mirrors take the form of a grating with alternating values of impedance—again reminiscent of the optical element with multiple dielectric coatings. There are, however, some significant differences. The total thickness of the dielectric coating on a laser mirror amounts to only a few wavelengths, while N_G, the number of wavelengths in the gap, may well be of the order of 10^6. In contrast, the SAW mirror may be quite thick, typically several tens or even hundreds of wavelengths, while N_G is normally of the same order, or even smaller. One consequence which flows from this is that while diffraction losses represent a key consideration in the case of laser resonators, one can in the case of SAW resonators work with beam apertures which are sufficiently wide to make diffraction losses negligible. Thus, the fact that in Fig. 4.7 we consider planar resonators which are known to lead to mode configurations on the verge of instability [4.33] is quite harmless in this situation.

4.3.1 Surface Wave Reflectors and Transmission Cavities

If then we can discount mode conversion—particularly to bulk-waves—the action of a grating can be completely represented in terms of a stepped impedance filter, in which we associate a characteristic impedance Z_0 with the unperturbed surface, and an impedance Z_1 with a grating line. If the width of each section is a quarter of a wavelength, one can readily show [4.34] that the amplitude transmission coefficient T and the reflector coefficient R are given by

$$T = \frac{2\varrho}{1+\varrho^2} \quad \text{and} \quad R = (1 - T^2)^{1/4} \tag{4.1}$$

where

$$\varrho = \left(\frac{Z_1}{Z_0}\right)^{N_R}$$

and N_R is the number of complete grating periods.

Even for small impedance ratios, the reflectivity approaches unity rapidly as N_R is increased, as shown in Fig. 4.8. It should, however, also be appreciated that the realization of high reflectivity using such small impedance steps imposes tight fabrication tolerances. The permissible random error in the positioning of the individual grating lines decreases rapidly as the impedance ratio is reduced, and hence the total number of grating periods required to effect the reflection is increased.

The periodic impedance change required can be implemented in a number of ways. The simplest is undoubtedly the use of thin metal strips as piezoelectric shorting elements. The impedance discontinuity so obtained, in the case of YZ lithium niobate, is 1.2%. This suffices to produce good reflectors with N_R of the order of 200. Experience at relatively low frequencies suggests that the positioning tolerance does not impose a major problem. It might well do so at higher frequencies. Furthermore, there will be applications where the temperature instability of $LiNbO_3$ would prevent its use. There is therefore interest also in the use of mass loading techniques, where larger impedance ratios are attainable, and hence larger errors can be tolerated. Resonators have been successfully realized using sputtered ZnO on ST quartz [4.30].

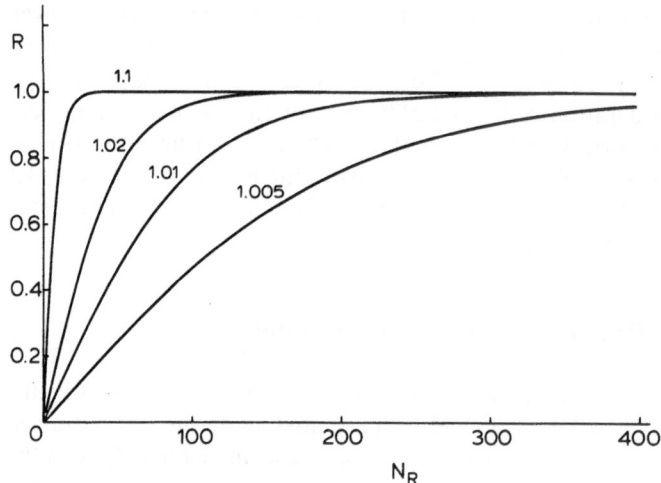

Fig. 4.8. Reflectivity of a single grating as a function of the thickness in wavelengths, N_R. The parameter is the impedance ratio ϱ

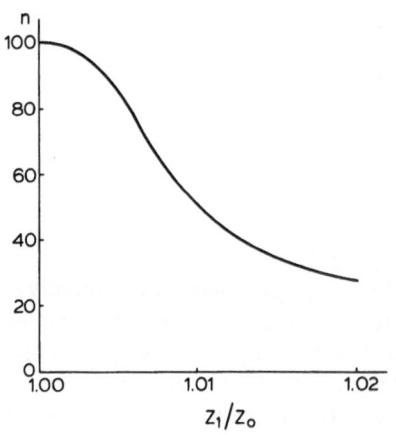

Fig. 4.9. Effective depth of penetration for a 200 period reflector. The reflected wave appears to come from the nth period (after [4.30])

The technology for ion etching grooves into the surface of the substrate, developed in the first instance for pulse compression filters (Sect. 4.5) has also been adapted to the fabrication of reflectors for resonators [4.31], and has led to encouraging results at relatively low frequencies. Success has also been achieved with ion implantation techniques [4.35] as with metal diffusion [4.36]. The diffusion technique seems very attractive, though so far it has been reported for $LiNbO_3$ only, and so does not bypass the temperature stability problem.

Wherever temperature stability is not the prime problem, the piezoelectric shorting techniques are likely to be employed—again because they alone fully retain and exploit the basic simplicity of SAW technology. However, there are, as we have seen, a number of other fabrication techniques which will allow the realization of reflectors on low-temperature coefficient materials.

The theory of resonators made by two opposed reflectors as in Fig. 4.1a is given by the classical relationships for a Fabry-Perot cavity [4.37], with one additional complication: the mirrors are "soft" in the sense that the wave penetrates a substantial distance into the grating structure. If, as is normally the case, the gap width $N_G\lambda$ is fairly small, the total effective mirror spacing of the resonator, $N_T\lambda$, can receive a considerable or even dominant contribution from penetration into the mirrors. This penetration depth $N_P\lambda$, defined as the distance into the reflector where the wave amplitude has dropped to $1/e$ of its value in the cavity, is shown for a specific case in Fig. 4.9. Since the reflection proceeds gradually, there is no unique way of specifying the effective length N_T. Following *Staples* et al. [4.30], we can define L in terms of the transmission Q of the cavity

$$Q \equiv 2\pi N_T/T^2. \qquad (4.2)$$

One can readily relate N_T to N_G and N_P by noting that the Q is the ratio of energy stored to energy lost (in this case the loss being attributed solely to leakage through the reflectors). The effective penetration is therefore obtained by taking the average of the distribution of the square of the amplitude. If we are dealing with a large number of grating elements we can write the amplitude of the wave at the Nth grating period as

$$A_N = A_0 e^{-N/N_P}$$

$$\frac{1}{A_0^2} \int_0^\infty A_N^2 = \frac{N_P}{2}$$

so that the effective length is simply given by

$$N_T = N_G + N_P. \qquad (4.3)$$

In the case of the more familiar electromagnetic Fabry-Perot, the mirrors are normally very broadband as compared with the frequency selectivity imposed by the spacing between them. As a result, one observes periodic transmission peaks, which, if losses are negligible and if the mirrors are identical, will approach 100 %.

These transmission peaks are separated by stop bands, as shown in Fig. 4.10a. In the case of the acoustic resonator, N_G and N_P are typically of the same order of magnitude. The selectivity of the overall device is therefore determined both by the grating mirrors and by the spacing between them. If the spacing is such $L = n\lambda/2$ at the frequency at which the grating reflectivity reaches its peak, the transmission characteristic will be again as in Fig. 4.10a. If, however, at this frequency $L = (n/4)(2\lambda + 1)$, there will be a central notch in the transmission characteristic, as appears in Fig. 4.10b.

It is clear that the behavior shown can form the basis of filter design or of a control element for a frequency stabilized source. Used in conjunction with multistrip couplers (Sect. 4.6), it could be the basis of a channel dropping filter. This approach of utilizing the properties of the resonator from the "outside"— i.e., by transmission through the structure—is however beset by one significant

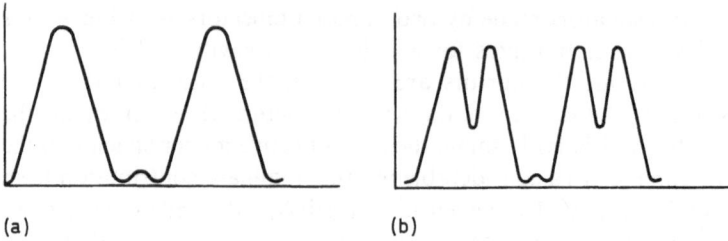

Fig. 4.10a and b. Transmission response of basic resonator. Effective spacing at grating resonance is (a) an even number of quarter wavelengths. (b) An odd number of quarter wavelengths

technological difficulty. As we have seen from Fig. 4.8, the reflectivity of each grating reflector is a very sensitive function of the impedance ratio Z_1/Z_0. If the reflectivity of each mirror is high, the transmission peak is significantly reduced if they are not identical. In the case of the piezoelectric shorting technique, Z_1/Z_0 is largely determined by the material constants. But even here there will be a small, and potentially variable, degree of mass loading. There will also be variations in the widths of the stripes and the gaps between them. The tolerances imposed by these considerations are therefore likely to be troublesome.

It is possible, to a very large extent, to escape from these difficulties by working from the "inside" of the resonator—by loading the resonator with one or more interdigital transducers. In this situation, which we shall now consider, one can make the thickness of the reflectors $N_R \lambda$ sufficiently great to ensure that the field has effectively decayed to zero at the outside edge. Small variations in Z_1/Z_0 will then still lead to small changes in N_P, but will not produce such violent changes in the first-order characteristics of the device.

4.3.2 The Loaded Surface Wave Resonator

We consider the surface wave resonator into which we have now inserted an interdigital transducer, as in Fig. 4.11a. In this basic system, the selective properties of the acoustic resonator are directly translated into electrical characteristics. Systems of multiple resonators can still be assembled by making electrical connections. Although one could at the same time also envisage acoustic connections by making the mirrors deliberately semitransparent, for the reasons discussed in the last section this may not be the preferred approach.

One can analyze the behavior of the interdigital transducer using the established model [4.38], but modified by the addition of an acoustic transmission line terminated by a reflector, attached to each of the two acoustic ports. The analysis can be substantially simplified by the fact that resonators are very narrowband devices, so that to a good approximation one can assume that the interdigital transducer is operated throughout near its synchronous frequency f_0. This means, for example, that one can neglect the susceptance arising from the acoustic radiation at frequencies away from synchronism. The

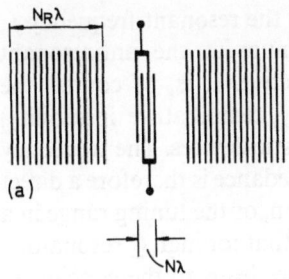

(a)

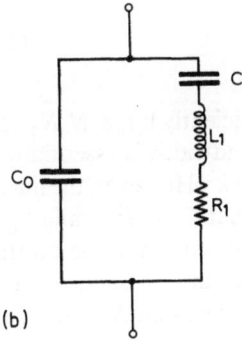

(b)

Fig. 4.11. (a) The basic configuration of a resonator loaded with an interdigital transducer. (b) Equivalent circuit

system of Fig. 4.5a can be represented by the equivalent circuit of Fig. 4.11b. This circuit is identical to that used to represent bulk-wave acoustic resonators. The capacity C_0 is the static capacity of the system, while C_1 and L_1 represent the dynamic reactances, and R_1 the radiation resistance as modified by the cavity environment. With the indicated approximations the elements of the equivalent circuit for an interdigital transducer with N finger pairs can be represented as follows [4.30]:

$$C_0 = NC_s \tag{4.4}$$

where C_s is the static capacity per finger pair.

$$R_0^{-1} = 8N^2 f_0 k^2 C_s .$$

$$R_1 = R_0 \frac{1 - |R|}{1 + |R|} \tag{4.5}$$

where

$$L_1 = \frac{N_T}{4} \frac{R_0}{f_0} \tag{4.6}$$

$$C_1 = 1/4\pi f_0^2 L_1 . \tag{4.7}$$

The impedance of the device will be minimum at the resonant frequency f_r, given by $f_r^2 = (L_1 C_1)^{-1}$, and then reach a maximum at the antiresonant frequency f_a. For $f_r < f < f_a$ the impedance is inductive. It is, of course, the deficiencies of electrical inductances (low Qs, bulky, temperature instability) which one wishes to circumvent by resorting to acoustic devices. The frequency range over which the transducer has an inductive impedance is therefore a direct measure of the usable bandwidth in a filter application, or the tuning range in a frequency control application. One can readily show that for high Q resonators, the circuit of Fig. 4.11b with the above element values leads to the conclusion that

$$f_a - f_r = \frac{f_0}{2} \frac{C_1}{C_0} = f_0 \frac{4k^2}{\pi^2} \frac{N}{N_T}. \tag{4.8}$$

Clearly N cannot exceed N_T though, if we make N_G sufficiently large, N/N_T can approach unity. Beyond this, we see that the usable bandwidth is essentially a materials property—not too surprisingly proportional to k^2. However, the linear relationship between C_1/C_0 and N arises from the fact that $C_1 \propto N^2$, while C_0 is directly proportional to N. This suggests [4.32] that one can improve on this basic situation by dividing the interdigital transducer into p sections and connecting them in series, such that acoustically is still acts as an N finger pair transducer, but that electrically the effective static capacitance is reduced by p^2, and the usable bandwidth increased by the same factor. This procedure is one which is not available in the design of bulk-wave resonators, where the available bandwidth is strictly determined by the material constants.

The quantity $f_0/(f_a - f_r)$ can be regarded, at least for certain applications, as a figure of merit. It is not, however, useful to think of it as a Q value—a parameter which should always retain the physical significance of a ratio of the energy stored to that dissipated per cycle. Provided the resonant and antiresonant frequencies are well resolved one can, however, define the Q in a form applicable to any single resonant circuit.

$$Q_{a,r} = \frac{f_{a,r}}{2} \left| \frac{d\phi}{df} \right|_{f_{a,r}} \tag{4.9}$$

where ϕ is the phase of the input impedance.

This definition will, for $Q \gg 1$, be approximately equivalent to the normal circuit definitions (e.g., $Q = \omega L_1/R_1$ for the series resonance). Since R_1 is independent of the gap width N_G, while L_1 increases linearly with N_G, it is clear that the Q can be raised simply by making the gap between the mirrors larger—until ultimately limited by diffraction losses. In this context it should, however, be appreciated that the theory of the resonator which we have outlined is based on the assumption that the fields have a pseudo-planar distribution. The width of the resonators is typically tens of wavelengths. It is clear therefore that modes with higher order transverse distributions can be excited. There is indeed experimental evidence [4.39] which shows that in some situations relating to piezoelectric shorting gratings, such modes are observed. The relationship between the Q and the gap spacing N_G may therefore be considerably more complex than would be predicted by elementary diffraction calculations based on pseudo-planar transverse distributions. Nevertheless, the increase in Q with N_G can be exploited—but at the cost of an ever increasing number of resonances. There will be applications, notably in the field of frequency control, in which the presence of adjacent, even fairly closely spaced resonances is not deleterious, and can even be advantageous. For use in filters it is desirable to work with resonators which have a single, or at most very few resonances.

A number of workers have reported results for the Qs obtainable in the multiresonance case for piezo-shorted $LiNbO_3$. Using interdigital transducers with three or four finger pairs, and N_G of several tens, the measured parallel resonance Q values are of the order of 2000.

To design resonators with single or at least low order resonance response, it is necessary drastically to restrict the bandwidth of the interdigital transducer. One can again analyze the complete system consisting of the transducer, with acoustic ports connected to grating reflectors [4.40]. In this case it is, however, no longer justifiable to make the approximations which led to equations (4.4) to (4.7), based on operation very close to the transducer synchronous frequency. The frequency response is characteristic of the overall coupled system; one can nevertheless obtain a useful insight into the problem, by comparing the frequency response of the empty resonator with that of the unobstructed interdigital transducer. This comparison immediately shows that by placing the desired resonance at the synchronous frequency of the transducer, the adjacent resonator responses will be at the first zeros of the transducer response if $N = N_T$. However, since N_T is necessarily larger than N by at least $2N_P$—the case when the transducer just fills the gap—this condition cannot be achieved [4.32]. However, if N_G is not too small, one can still ensure that the adjacent resonator responses come fairly close to the transducer zeros. With a design on this basis established, one can then exploit the inductive behavior of the device in the frequency range given by (4.8).

One can, for example, realize the well-known quartz crystal filter structures [4.41] so widely used in critical communication systems. An example which has recently been demonstrated [4.32] is the bandpass behavior of a lattice section, Fig. 4.12a implemented as shown in Fig. 4.12b. Here the A elements are surface wave resonators, and the B elements simple capacitors. The interdigital transducer had 60 periods, but was divided into two portions electrically connected in series to reduce $C_0(p = 2)$. The finger overlap width was 20λ, and N_T was 108. The capacitors were realized on the same substrate in the form of an interdigital electrode system but facing the X rather than the Z crystal direction. The X direction has a substantially smaller coupling constant, and in any event the period was designed so that these structures could not radiate bulk or surface waves over the frequency range of interest. The experimental filter response is

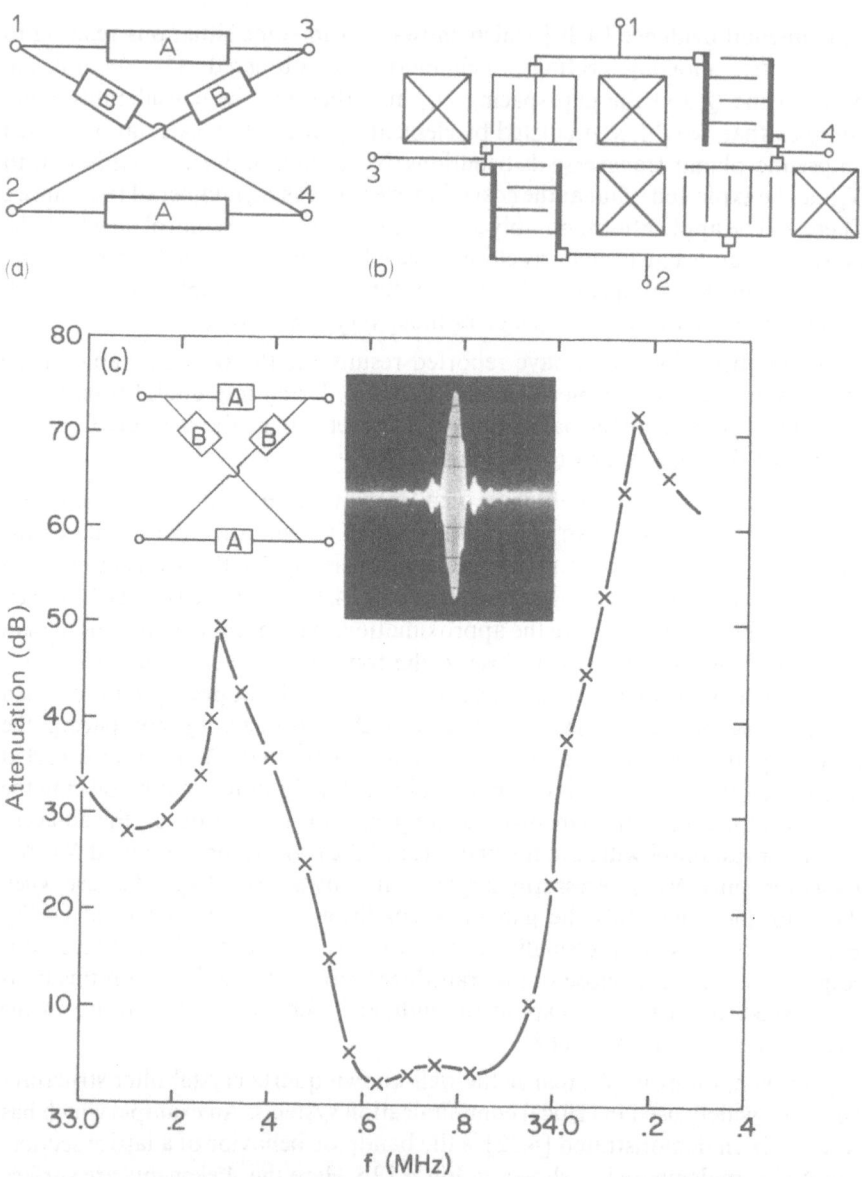

Fig. 4.12a–c. Lattice filter structure. (a, b) Equivalent circuit, and its realization using SAW resonators. (c) Filter response (after [4.32])

shown in Fig. 4.12c. The extension to higher order filter circuits is straightforward, and would still be realized with a single-mask photolithographic process. Resonators are also finding increasing application in very narrow band filters [4.42], where they offer major advantages both in size reduction and in performance. The exploitation of these techniques on a large scale can be anticipated, though there remains the problem of temperature stability which is inevitably a central consideration in the use of narrow band communication filters.

4.4 SAW Delay Line Oscillators

The SAW resonators discussed in the previous section can be used as the frequency control element to form a stable signal source. These resonators derive from our ability to reflect a surface wave using grating reflectors. It is, however, quite possible to construct a surface wave resonator without the use of surface wave mirrors, by resorting to a recirculating transmission path. This approach is familiar in the case of electromagnetic waves where, for example, one uses waveguide ring resonators at microwave frequencies. Recirculating paths are also used at optical frequencies as cyclic interferometers, and in the form of laser gyroscopes. Such recirculating paths for acoustic surface waves can also be devised using suitable grating structures. However, it is simpler to retain a linear propagation path for the acoustic waves, and effect the feedback electrically as shown in Fig. 4.13a. This circuit will have resonant frequencies whenever the total phase shift around the loop amounts to an integral number of 2π.

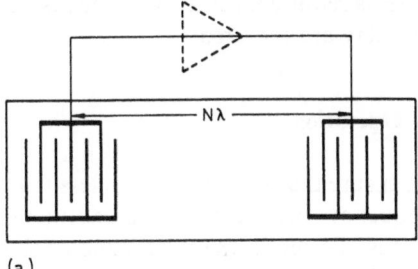

(a)

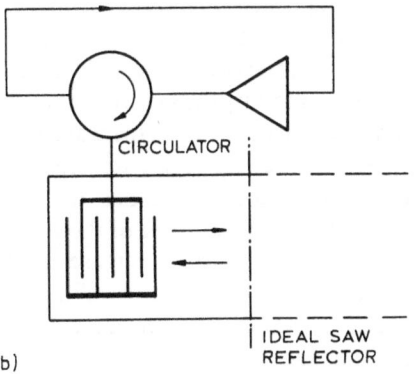

(b)

Fig. 4.13. (a) Basic SAW feedback system. Insertion of the amplifier produces oscillations. (b) Oscillator configuration, equivalent to that of (a)

The ring resonator of Fig. 4.13a becomes an oscillator if we include in the feedback path an amplifier with sufficient gain to make up the overall loop loss [4.43]. The frequencies of oscillation will satisfy the equation

$$\frac{\omega L}{v} + \Phi_E = 2n\pi \tag{4.10}$$

where L is the acoustic path length and Φ_E is the total phase change associated with the transducers and with the amplifier. To obtain a very precise frequency control, an oscillator must contain an element having a phase characteristic which varies very steeply with frequency. In the SAW oscillator, this element is of course the delay line which has a transfer phase slope $d\Phi/df$ proportional to length. This point will be further discussed in Section 4.4.2. For the moment we note simply that the total delay path will be long—typically hundreds or even thousands of wavelengths. Now, the spacing of the resonant frequencies given by (4.10) is f_0/N, where f_0 is the central frequency of the transducers, and N is the number of wavelengths contained in L. In general, the mode spacing will be much narrower than the width of the gain characteristic of the amplifier, so that oscillations can be obtained at a number of different frequencies, separately, or even simultaneously. In this respect, the situation is very similar to that encountered in the case of gas lasers. In practice, one will normally wish to have a single frequency output, though there are also applications in which the simultaneous presence of a set of frequencies is required. Either way, means for effecting control of the mode or modes in play are required.

4.4.1 Mode Control for Single-Frequency Operation

To obtain single mode operation, we must ensure that the loop gain is less than unity for all but the desired oscillation frequency—we must include additional selectivity to attenuate all but one frequency. To retain the advantage of the compactness and hopefully temperature stability of a SAW device, it is at once clear that this additional selectivity must be built into the device itself rather than included in the external loop. This is easily accomplished (Fig. 4.14a) by making one of the transducers sufficiently long and hence sufficiently narrow band. Specifically, if we make the length $N_T\lambda$ and then space the center of this transducer from the center of the other short transducer by the same distance, we can ensure that all of the solutions of (4.10) will fall on the zeros of the transducer response. This method of mode control is effective; however, we see from Fig. 4.14a that one-half of the average acoustic path lies under the long transducer. For the usual mark to space ratio of 1:1, one-quarter of the path is covered by a metallic thin film. This leads to a significant increase in the attenuation, and hence a reduction in the effective Q. Moreover, the well-known secondary effects [4.44] arising from acoustic reflections and from regeneration can become very serious in such a long transducer, and lead to unwanted

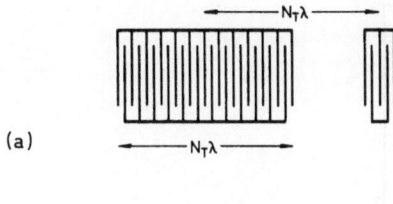

(a)

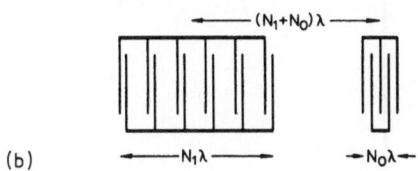

(b)

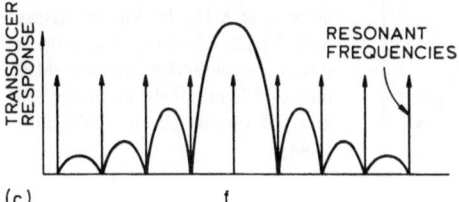

Fig. 4.14a–c. Strategy for choosing mode propagation. (a) Use of long transducer. (b) Use of long thinned transducer. (c) Response of transducer shown in (b)

responses. Since we are only concerned with the location of the zeros of the long transducer, it is possible, while retaining its overall length, to remove most of the finger pairs [4.45]—to work, in the language of electromagnetic radiators, with a "thinned array". The disposition shown in Fig. 4.14b still leads to zeros which coincide with all the unwanted resonant frequencies (Fig. 14c) but very greatly reduces the fraction of the surface covered by a metallic film. It is the system which has been most widely used in oscillator design; recently though, the replacement of the delay line by an acoustic resonator has been shown to offer some significant advantages [4.46].

4.4.2 Oscillator Stability

The frequency stability of the oscillator is of primary importance; it is determined by a number of physical phenomena, including thermal and electronic noise, random temperature fluctuations, and, largely determining the long-term stability of the oscillator, small changes in the transmission medium itself. The short-term stability, measured in intervals of a second or less, will be almost entirely determined by noise.

If for the moment we discount the Johnson noise associated with the acoustic and *transduction losses*, we can identify random phase changes in the amplifier

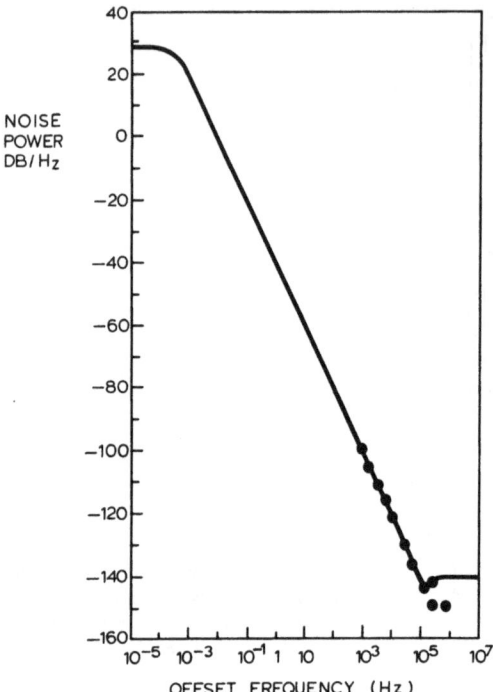

Fig. 4.15. Single sideband FM noise, relative to carrier, for a SAW oscillator at 60 MHz, having an acoustic path length of 100 λ. The total insertion loss of the delay line was 24 dB, the noise figure of the amplifier 3 dB, and the output power 1 mW (after [4.48])

as the main source of FM noise in the oscillator output. From (4.10) we can see that the frequency deviation δf associated with an amplifier phase change $\delta \phi_E$ is given by

$$\frac{\delta f}{f} = -\frac{\delta \Phi_E}{2\pi N} \tag{4.11}$$

where N is the number of wavelengths in the acoustic transmission path. We can identify the quantity $2\pi N$ as playing the role of a Q factor,

$$Q' \equiv 2\pi N. \tag{4.12}$$

Thus we see that Q' increases indefinitely as we increase the path length and hence N. While Q' bears no direct relationship to the normal definition in terms of energy storage and dissipation, it is nevertheless the parameter which governs the oscillator frequency drift arising from amplifier phase changes, such as those due to temperature changes.

It is physically apparent that the short-term stability, or, using frequency domain description, the spectral width of the oscillator output, cannot be indefinitely improved by increasing N. Beyond a certain length, the effect of the acoustic losses will more than outweigh the benefits of the steeper phase characteristic. Perhaps the easiest way to appreciate this is to show the oscillator of Fig. 4.13a in a folded version, as in Fig. 4.13b. We postulate a perfect reflector at the midpoint of the acoustic delay line. The circulator separates the incident and reflected waves, as shown. In this configuration the delay path appears as a one-port. If we were to make the delay path progressively longer, we would see that the input impedance of the line becomes purely resistive. The classical theory of oscillators using transmission line resonators is directly applicable.

Assuming that the dominant noise source is Johnson noise from the input resistor of the amplifier, one can calculate the power spectrum of the oscillator [4.47, 48]. A typical result is shown in Fig. 4.15. It is seen that, as expected from classical theory, the noise power density decreases at a rate of 20 dB per octave over a frequency range of over 8 decades. A useful single parameter which is a measure of the width of the power spectrum is the coherence time τ_c—roughly the period over which phase deviations remain within π. This coherence time can be expressed in the form

$$\tau_c = \frac{Q'^2}{(\mathrm{NF}) \cdot G^2} \left(\frac{1}{4kT\omega_0^2} \right), \tag{4.13}$$

where NF and G are the noise figure and the voltage gain of the amplifier. In this formulation, one can readily grasp the effect of losses in the feedback path. To maintain the oscillations in the face of such losses, it is clear that the voltage gain G will have to be increased by the overall amplitude loss factor D. It follows that (4.13) for the coherence time will remain valid provided that we replace Q' by the effective Q, Q_{eff}, defined by

$$Q_{\mathrm{eff}} \equiv \frac{Q'}{D}. \tag{4.14}$$

Here D can embrace all the losses in the feedback path—including those arising from mismatches and losses in the transducers. However, as we make the delay path progressively longer, the dominant factor will be due to delay line losses. If α is the real part of the acoustic propagation constant,

$$D = e^{\alpha L}. \tag{4.15}$$

In this limiting case, we find that the definition of Q_{eff} of (4.14) coincides very nearly with the conventional definition of the Q in terms of the ratio of energy stored to the energy dissipated per cycle. Since D is an exponential function of L, while Q' depends linearly on L, it is clear that Q_{eff} will reach a maximum value at a specific length and fall rapidly with further increase in length. This variation of Q_{eff} with L has been calculated [4.48] for ST quartz, for a number of different frequencies; these results are shown in Fig. 4.16.

The fact that the propagation loss increases as the square of the frequency leads to a rapid variation in the optimum length of the delay path, of the order of

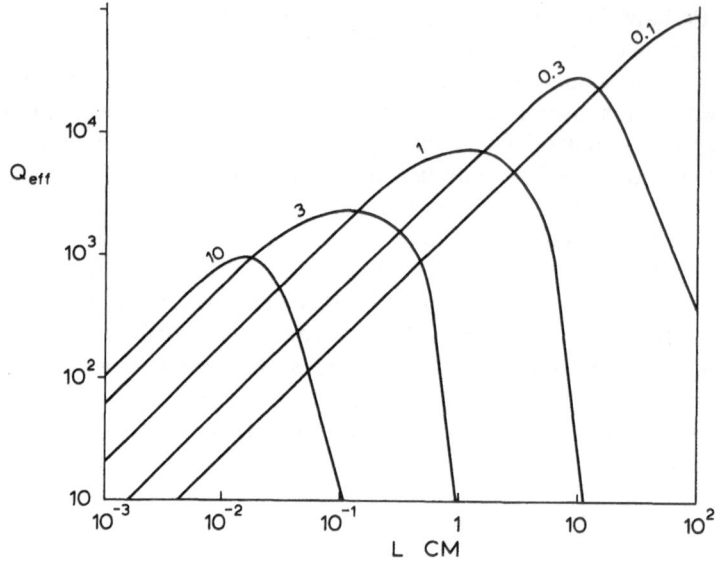

Fig. 4.16. Effective Q, Q_{eff} as a function of the propagation length L on ST quartz. The parameter is the frequency in GHz (after [4.48])

1 mm at 3 GHz to 1 m at 100 MHz. To achieve the latter, one could resort to some form of folded delay path. An alternative is to use an acoustic surface wave resonator [4.39], where the effective path length is increased by a factor of the order of $1/(1 - R)$ where R is the grating reflection coefficient.

In many applications, one is concerned with the frequency stability over extended periods of minutes or hours, as well as with the spectral purity or "short-term stability" discussed above. Such frequency drifts arise predominately from temperature effects. By using quartz, and a cut close to ST, the incremental variation of delay with temperature can be made zero at a temperature which one can choose over a fairly wide range. There then remains a quadratic dependence on velocity, which leads to a frequency deviation of Δf ppm, when the temperature is displaced by $\Delta T\,°C$ from the turning point, such that

$$\Delta f = 0.03(\Delta T)^2. \tag{4.16}$$

Even crude temperature stabilization will ensure a frequency deviation—from this cause—of less than 1 ppm. Considerably more troublesome is the frequency drift arising from amplifier phase changes. In principle, one can incorporate the linear phase temperature curve of the amplifier in the whole oscillator design, by choosing a slightly different cut—and so obtaining an overall zero linear temperature coefficient. However amplifier characteristics are hardly as reproducible as those of a sample of quartz. Since one can readily measure the amplifier transfer phase shift, one can in principle compensate for any

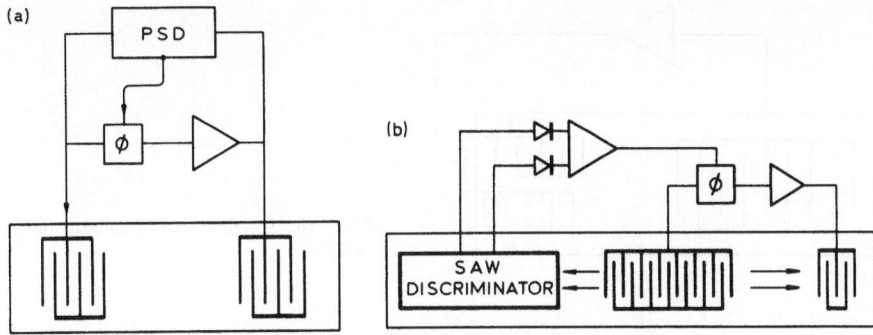

Fig. 4.17a and b. Improvement of medium-term stability of SAW oscillator. (a) Direct control of delay path transfer phase. (b) Use of separate discriminator with long time constant control circuit

temperature-induced phase deviations. A simple technique for effecting this has recently been described [4.49] and is shown in Fig. 4.17a. A phase-sensitive detector is used to provide an error signal proportional to the change in the amplifier transfer phase. The error signal controls a varactor phase shifter to complete the control loop. Experimental results show a reduction by two orders of magnitude in the sensitivity of the frequency to amplifier temperature changes.

An even more direct attack on the problems arising from temperature effects in the amplifier has been proposed and successfully tested. It uses an acoustic frequency discrimination [4.50, 51] to monitor the frequency of the SAW oscillator. One form of such a discriminator consists essentially of a pair of staggered filters with an approximately triangular frequency response, so that the rectified output is proportional to the frequency deviation. This output can then be used, as in Fig. 4.17b, to control the phase and hence frequency of the oscillator. The discriminator is built on the same substrate as the delay line, and is therefore also subject only to quadratic temperature effects. At first sight one might think that one has merely transferred the problems to another device—the discriminator—itself subject to the deleterious effects of noise in its detection circuitry. However, it is important to appreciate that one is relying on the discriminator control only to cope with the relatively long-term temperature drifts. One can therefore use time constants of the order of a second or more. This means that the effective bandwidth of the control loop can be less than one hertz, with a correspondingly low noise. Effectively, this allows one to locate an operating point on the discriminator characteristic with very great precision—it is not necessary for the discriminator itself to operate over very narrow bands. The system of Fig. 4.17b has in fact led to dramatic improvements, amounting to perhaps two orders of magnitude not only in the temperature stability but also in a reduction to the effects of supply voltage variations. Since the discriminator requires only a somewhat more complex mask, it seems likely that it will be widely adopted in SAW oscillators.

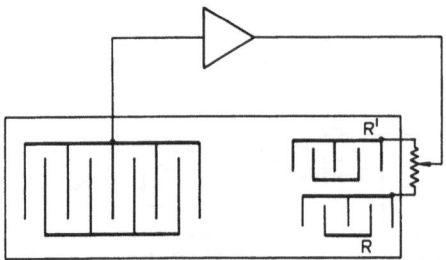

Fig. 4.18. Tunable SAW oscillator (after [4.47])

A tacit assumption in all these methods which endeavor to stabilize the frequency is that the stability of the substrate material is assured. However, as is well known from decades of experience with bulk quartz crystals, there are long-term "ageing" effects, presumably due to the movement of dislocations through the material. In the case of the best bulk crystal oscillators, such ageing effects can be as low as 1 ppm yr^{-1}. Moreover, the ageing characteristic is linear and accurately predictable, so that the prediction of the oscillator frequency can be substantially better than implied by this figure. Such ageing effects have also been observed in the case of SAW oscillators, and initially they appeared to be several orders of magnitude worse than their bulk-wave counterparts. This observation is not perhaps entirely surprising, since the surface has inevitably been subjected to work damage and strain. It has been known for some considerable time that the velocity of surface waves shows a slow, but measurable, decrease with frequency as a result of the different properties of the surface layer [4.52]. It is therefore clear that the long-term ageing characteristics can be expected to depend sensitively on the treatment to which the surface has been subjected before and during polishing, and in subsequent stages of fabrication. Recent developments suggest that it is now possible to achieve an ageing rate of the order of 10 ppm yr^{-1}, and further improvements are to be anticipated.

Finally, it should be appreciated that the application dictates the stability characteristics required—that it is by no means necessary for a particular device to excel in all respects. For example, a radar signal must have a spectral width which is less than that imposed by Doppler effects of the slowest moving objects whose speed may be of interest. This imposes a stringent requirement on spectral width, while the long-term ageing effects might here be almost totally irrelevant. This is particularly so if the oscillator frequency can be adjusted over a small range.

4.4.3 Tuning and FM Capability

If we include a phase shifter in the external loop, it is possible to tune a SAW oscillator. To ensure that tuning does not lead to a jump to a different mode, the total additional phase change should not exceed $\pm\pi/2$. From (4.11), we see that for an oscillator with a delay path of $N\lambda$, we can obtain a tuning range $\Delta f/f$ of the order $1/4N$. Since the short-term stability is largely determined by N, a large tuning range is therefore at the expense of stability: the balance between these conflicting considerations will be determined by the nature of the application. It is worth noting that even for $N = 250$, the tuning range is enough to compensate for small variations in the photolithographic fabrication and allow the frequency to be set to a precise, prescribed value.

A varactor phase shifter can also be used to frequency modulate the oscillator output. The maximum deviation which one can obtain is comparable to the tuning range. The rate at which the frequency can be changed is necessarily determined by the transit time of the acoustic wave. Experiments have suggested [4.47] that it may take up to 10 circulation times to establish a new frequency, so that the maximum modulation frequency is of the order of $f_0/10N$. Here again we face a compromise between short-term stability and modulation capability. In any event, there are many applications where a modulation band of 0.1 % is entirely adequate.

Instead of effecting the tuning by using a voltage-controlled phase shifter, one can build in a fixed phase shift in the acoustic circuit, as shown in Fig. 4.18. The transducers R and R' are staggered by $\lambda/4$ and so give the maximum phase shift which can be used. By varying the two resistors, one can then change the phase over the complete $\pi/2$ range. If the variable resistors take the form of PIN diodes, one can use this system for frequency modulation.

4.4.4 Multifrequency Operation

In Section 4.4.1 we saw how it is possible to design transducers such that only a single mode of oscillation can be observed. If, however, the design is directed towards achieving approximately equal insertion losses for a set of modes, it becomes possible to observe oscillations at more than one frequency. The situation is quite analogous to that of a laser where the atomic transition has a much bigger bandwidth than the mode separation. The question then arises as to whether the system can simultaneously support several oscillation modes. It is an issue which can only be discussed in terms of the nonlinear behavior of the amplifier. One can, however, say that amplifiers normally behave more like homogeneously broadened rather than inhomogeneously broadened amplifying media. That is, once a particular mode has reached the threshold of oscillation there is a tendency for the nonlinear behavior to be such as to reduce the gain at other frequencies. If, therefore, the gain of the amplifier is adjusted to be above, but not too far above, the value required to achieve threshold for one of the set of modes, there is a tendency for single-mode operation to occur.

When an oscillator so designed is switched on, the mode having the lowest insertion loss is the one which will grow fastest, and which, when approaching saturation, will cause the available gain for competing modes to be suppressed below the threshold. The reason is simply that the oscillations build up from

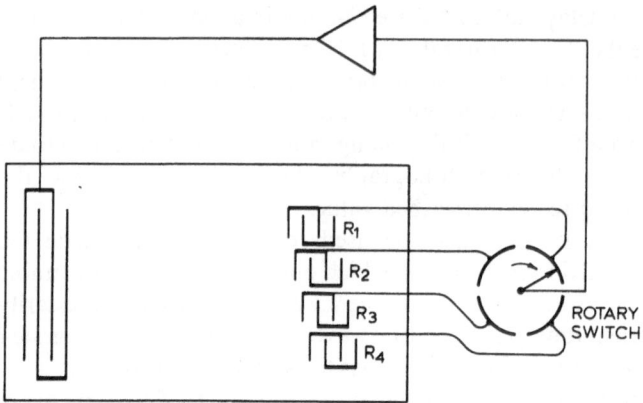

Fig. 4.19. Basic arrangement of a SAW frequency synthesizer (after [4.53])

broad-band noise, which does not favor any particular mode. Suppose now that before switching on the amplifier we inject a low level signal at a frequency f_n, corresponding to the nth mode. This mode will therefore grow not from noise but from the much larger base of the injected signal. It will now be the first to reach saturation, and will suppress competing modes. The nth mode, under these conditions, can win through, even if its insertion loss is substantially larger than that of some other mode. The permissible variation in insertion loss is of course determined by the detailed nonlinear characteristics of the amplifier.

This is the physical phenomenon which forms the basis of operation of a new class of device—the SAW frequency synthesizer [4.53]. In its simplest form it represents an extension of the tunable oscillator of Fig. 4.18. However, this time, in Fig. 4.19, we have four receiving transducers, staggered by intervals of $\lambda/4$. A rotary switch, which can of course be realized in electronic form, is at any instant connected to one of the four receiving transducers. If initially it is connected to R_1 and then moves to R_2, the frequency will change as before by $\delta f = 1/4N$. The decaying wave train associated with operation at R_1 will have sufficient power in the adjacent frequency range to ensure that it is this mode which is favored. In particular, it will discriminate adequately against the nearest other modes which can operate using R_2, which differ from this frequency by at least $3\delta f$. Proceeding to R_3, R_4, and then again to R_1, one can obtain progressive frequency changes, exploring the whole range of the comb spectrum for which the insertion loss of the delay system is sufficiently small. A device of the kind shown in Fig. 4.19, with a center frequency of 60 MHz has been used to demonstrate this principle [4.53]. In this design, δf was 25 kHz, and it was possible to cover the complete range from 59 to 61 MHz, in 80 distinct steps. In its simplest form, the device can, in some situations, give rise to signals which are frequency modulated at a rate corresponding to the intermode spacing frequency of f_0/N. The spectrum of such a wave will show components at frequencies corresponding to adjacent modes (i.e., spaced by $4\delta f$). One can see this phenomenon as essentially due to the fact that the nonlinear saturation suppression of the gain is not quite adequate to shut out adjacent modes. These adjacent modes are not however free running—they have a well-defined phase relationship to the central oscillation mode. It turns out that this difficulty can however be overcome by some elaboration of the feedback circuitry [4.53]. It seems probable that this device will prove of considerable importance in the realization of frequency synthesizers, and related systems.

The device which we have just discussed allows operation at one of a large number of frequencies. If we raise the gain of the amplifier sufficiently, we can achieve a situation in which many of the modes oscillate simultaneously, providing a comb spectrum of frequencies. In this form the device is, however, very hard to control. The distribution of the power among the modes is determined by "second-order" nonlinear effects. Quite small changes in the conditions, including temperature changes, can lead to marked variations in the modal powers—even the extinction of some and the appearance of others. If, however, we introduce additional nonlinear controls in the feedback circuit we can mode-lock the various modes so that they have a fixed phase relationship to each other. In this we are essentially seeking to exploit the very coupling phenomenon referred to above, which is a source of some embarrassment in the frequency synthesizer.

Such mode-locked systems are familiar from laser technology. The phenomenon has recently been demonstrated for SAW oscillators [4.54] using several different techniques, one of which is shown in Fig. 4.20. To understand its operation one must appreciate that, when mode-locked, the temporal output of the oscillator takes the form of narrow pulses having a width of the order of the total inverse bandwidth separating the extreme participating modes, and a repetition period corresponding to the delay time τ. If then we imagine the system in operation, the detector and video amplifier will operate the modulator at the frequency $1/\tau$. One can then arrange matters such that the modulator provides a low loss just when the rf pulse appears, and a high loss at other times.

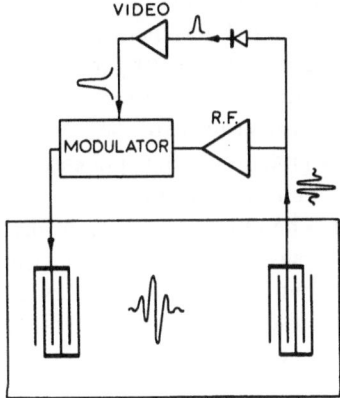

Fig. 4.20. Mode-locked SAW oscillator (after [4.54])

The mode-locked pulse is therefore favored in competition with other, free running, multimode operation. A mode-locked system of this kind has been used in conjunction with a 100 MHz center-frequency SAW oscillator designed so that modes over a 20 MHz bandwidth could participate. The participation of all the modes was demonstrated by showing that the temporal output of the device is indeed in the form of 50 ns pulses separated by the delay time of 1.8 μs.

4.5 Coded "Time-Domain" Structures

The action of simple transducers can be described in terms of a transfer function. In Chapter 3 we saw how, by varying the finger spacing, and by the use of various apodization techniques, it is possible to synthesize a wide class of frequency filters. The feature that characterizes most classical filter applications is a concentration on the amplitude characteristic; beyond that one must satisfy a more or less stringent requirement on the constancy of the group delay over the relevant bandwidth—i.e., one seeks a phase transfer function which is at least approximately linear with frequency. In this section, we shall now discuss a series of devices which are distinguished from such filters in a number of respects, but perhaps most significantly in the abandonment of the nondispersive requirement. Indeed for one important class of devices, the pulse expansion and compression filters used in chirp radar [4.55], the group delay dispersion is central to the functional purpose.

It is entirely possible to discuss the characteristics of dispersive and related structures in terms of their transfer functions, i.e., in terms of their behavior in the frequency domain. The methods of synthesis developed in Chapter 3 are immediately applicable. However, in use these devices are often excited by short digital pulses so that their impulse response comes into play in a direct, literal manner. Moreover, the systems applications which the devices serve impose requirements which are most readily expressed in terms of the impulse response, i.e., in the time domain. It is in the recognition of the requirements imposed by the application that we choose to regard this class of devices as "time-domain" filters. They are distinguished from frequency-domain filters neither by the underlying physics nor even by the design techniques but primarily by the nature of the most convenient language which characterizes the applications.

We can at best sample a number of the generic devices which are contributing to the rapidly growing field of SAW time-domain signal processing applications, beginning in Section 4.5.1 with what has been and is likely to continue to represent the most important single application—the pulse compression filter for chirp radar. In Section 4.5.2 we shall discuss a class of phase-coded devices which are of growing importance in new telecommunication techniques. In Section 4.5.3 we shall return to the theme of pulse compression filters, to explore some of the more general processing functions of which they are capable.

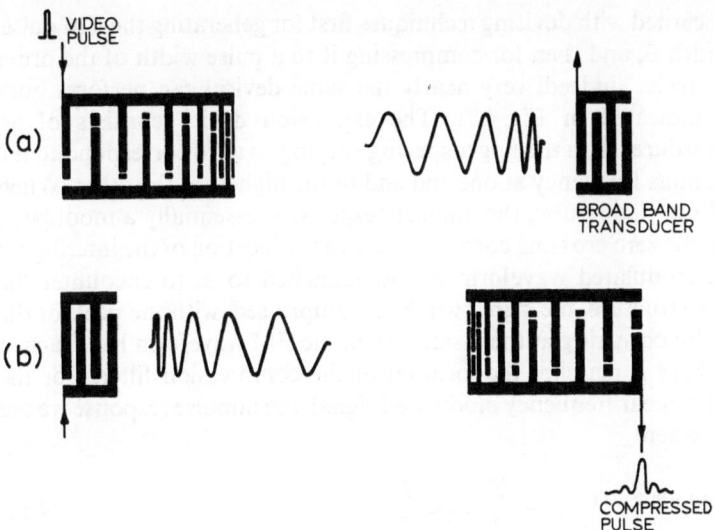

Fig. 4.21a and b. Chirp signal processing. (a) Expansion of a narrow video pulse into a frequency modulated wave train. The bandwidth has been greatly exaggerated for clarity but normally does not exceed 25%. (b) Compression of the chirp pulse by the complementary structure. Note that the low-frequency portion of the wave has to pass underneath the high frequency portion of the transducer

4.5.1 Pulse Compression Filters for Chirp Radar

The basic motivation for the use of chirped (i.e., FM) signals in radar systems derives from two basic facts [4.55]: that the detection sensitivity of a radar is proportional to the transmitted energy, and that the resolution capability of a simple modulated pulse radar is determined by the pulse width. These twin requirements influence the design towards transmitters with very high peak power capability, and hence towards high cost. By the use of a coded pulse, one can obtain a resolution corresponding to a subsection of the transmitted pulse, a section which very roughly we can identify with the smallest distinguishable feature within the pulse waveform. The most widely used code is that of a linearly frequency modulated pulse. One can show readily that the effective time resolution corresponds to B^{-1}, where B is the bandwidth. Thus, for example, if we have a pulse which is frequency modulated over a range of 20 MHz, the resolution which we obtain is that which, with a simple pulse radar, would have required a pulse width of 50 ns, even though the actual FM pulse might extend over 50 µs. An important measure of performance is the *time-bandwidth product* TB, where T is the duration of the FM pulse. In the above example $TB = 1000$; the effective pulse length, judged on the criterion of attainable resolution, is the actual pulse length divided by the time bandwidth product. The information about the target is embedded in the return echo and has to be decoded, before it can be presented in a useful form such as, for example, the range.

We are concerned with devising techniques first for generating the FM pulse, with a bandwidth B, and then for compressing it to a pulse width of the order B^{-1}. A SAW device (indeed, very nearly the same device) can perform both functions, as indicated in Fig. 4.21. The expansion device consists of an interdigital transducer with the finger spacing varying so as to correspond to the lowest synchronous frequency at one end and to the highest at the other. When excited by a short video pulse, the impulse response is essentially a modulated sine wave, with the zero crossing corresponding to the location of the interdigital fingers. If this modulated waveform is now launched so as to encounter the complementary structure, the signal will be recompressed, with the peak of the compressed pulse occurring at the instant when the FM waveform has reached the position where it matches the location of the compression filter. For the specific case of a linear frequency modulated signal, the impulse response we are seeking is $h(t)$, where

$$h(t) = e^{2\pi i f_0 t (1 + Bt/2f_0 T)} \qquad -\frac{T}{2} < t < \frac{T}{2}. \tag{4.17}$$

The instantaneous frequency f_i defined in terms of the rate of phase is therefore

$$f_i = f_0 + \frac{Bt}{T} \tag{4.18}$$

and varies over the range B,

$$\left(f_0 - \frac{B}{2}\right) < f_i < \left(f_0 + \frac{B}{2}\right).$$

On the simplest delta-function model of the interdigital transducer developed in Chapter 3, we can then immediately write down the location of the nth finger x_n in the structure

$$x'_n + \frac{B'^2}{2(BT)} x'^2_n = \frac{n}{2} \tag{4.19}$$

where $x'_n \equiv x_n/\lambda_0$ is the location of the nth finger normalized with respect to the center frequency wavelength λ_0, and where $B' = B/f_0$ is the fractional bandwidth.

In most cases the total number of fingers will be large, so that we can differentiate (4.19) to obtain the separation between fingers dx'_n/dn. For $x'_n = 0$ this yields $dx'_n/dn = 1/2$, corresponding to the appropriate finger spacing at the center frequency f_0. By direct inspection of (4.17), the total number of phase changes of π over the pulse width is $2f_0 T$, which gives N_0, the total number of fingers required in the transducer. Of these, a number n_1 form the transducer for frequencies above the center frequency, and n_2 the transducer for frequencies below f_0

$$n_{1,2} = f_0 T(1 \pm B'/4). \tag{4.20}$$

Equation (4.19) therefore has to be solved for all integers n in the range $-n_2 < n < n_1$.

When the signal is recompressed using the complementary structure, the output is approximately the autocorrelation function of $h(t)$, which takes the form $g(t)$,

$$g(t) = \frac{\sin(\pi Bt)}{\pi Bt} \qquad (4.21)$$

as indicated in Fig. 4.21. The central peak occurs at $t = 0$, with the first pair of zeroes at $t = \pm 1/B$, and the 3 dB pulse width is $0.89/B$.

The main defect of the combination of the basic system so far developed lies in the fact that $g(t)$ has subsidiary maxima where $2BT$ is an odd integer larger than unity. Specifically, the innermost "side lobes" occur at $t = \pm 3/(2B)$ such that

$$\frac{g(\pm 3/2B)}{g(0)} = 0.212.$$

Expressed in power terms, this implies that the innermost side lobes are only 13.46 dB below the peak of the compressed pulse, which is unacceptable for almost all applications. The problem derives essentially from the fact that we have designed a filter which, to a first approximation, has a sharp cut-off in its frequency response at $f_0(1 \pm B'/2)$. As in the analogous case of a bandpass filter, this leads to truncation ripples in the frequency response which show up as side lobes in the compressed temporal response. The remedy is to extend the frequency response a little beyond the band covered by the signal, and taper off the response gradually at either end [4.56]. In theory it is in this way possible to reduce the time side lobes as much as may be desired; the penalty is an increase in the width of the main lobe. It turns out that one can obtain a dramatic reduction in side lobe level for a modest increase in main lobe width. For example, for the widely used Hamming [4.57] frequency weighting, for $BT = 1000$, the time side lobe can be reduced to 42 dB at a cost of doubling the width of the main lobe. The spectral weighting is implemented by one of the apodization techniques already discussed in Chapter 3. In fact, quite apart from spectral weighting for reduction of side lobes, it is in any case necessary to include some apodization to take account of the fact that the power radiated by a pair of transducer fingers is frequency dependent. As already shown in Chapter 3, to remove this dependence one must vary the finger overlap according to a $(f/f_0)^{-3/2}$ law. A typical form of the apodization is indicated in Fig. 4.21.

When the structure is excited by a cw signal at a frequency f, the energy is radiated mainly from that region where the interdigital period is v_r/f. The effective number of finger pairs involved, N_e, depends on the dispersion characteristics, and can be readily estimated using the criterion that the total phase error over the active region must not exceed π.

$$N_e/N_0 = 2(BT)^{-1/2}. \qquad (4.22)$$

The frequency change Δf required to move from one set of activated fingers to the neighboring set is simply $B\cdot(N_e/N_0)$.

$$\Delta f = 2(B/T)^{1/2}. \tag{4.23}$$

This is *not*, however, a measure of the frequency resolution limit of the device, as we will find in further discussion in Section 4.5.3.

The extremely simple considerations which we have advanced can take us a considerable distance towards a viable design; for relatively undemanding applications they might suffice in themselves. However, for systems requiring large TB and low side lobe levels, we must consider a series of second-order effects and means for reducing their deleterious impact.

As in the case of frequency filters, one must account for diffraction losses which, of course, depend on the location of specific fingers, their spacing, and, strongly on the overlap length. The diffraction effects are well understood [4.58], and can be allowed for by making appropriate changes in the "zero order" apodization function.

We have so far tacitly assumed that a surface wave once launched ignores the remainder of the transducer under which it has to travel. In fact there are a number of effects which come into play. The metallization perturbs the velocity of the wave, both by field shorting and by purely topographic effects. This leads to some reflection which is particularly severe in those parts of the transducer which are synchronous for the particular wave component with which we are concerned. A second effect, closely related to the first, is due to interaction of the wave with the electrical circuit, which will give rise to regeneration; this can enhance the reflection, and also perturb the phase of the transmitted wave. Both of these effects [4.59] can be shown to be proportional to k^2, the square of the effective electromechanical coupling constant. They can be assessed quantitatively; they can also be allowed for in the design of the finger placement so as to largely eliminate their effect [4.57]. There is, however, another approach—at first sight less elegant—which is however much easier to adopt. It is to use materials, such as quartz, having a relatively low electromechanical coupling constant [4.60]. Beyond that, there is a further technique which can be used with great advantage—to incline the dispersive transducers as shown in Fig. 4.22. If the angle θ is chosen so that the number of finger pairs in any horizontal slice $W\cdot\cos\theta$ is of the order of N_e (4.22), the propagation path between sets of fingers which can communicate will not encounter any other part of the transducer. The price one has to pay is in the increased area of substrate, which makes it difficult to use this approach for the highest TB products. The efficacy of this technique is illustrated by a particular example where the pulse compression loop performance was computed for the normal as well as, the inclined geometry (Fig. 4.23). In this case there is a 10 dB improvement in the level of the far-out side lobes.

The use of inclined transducers is also largely effective in overcoming problems associated with the generation of bulk waves. Any finite length

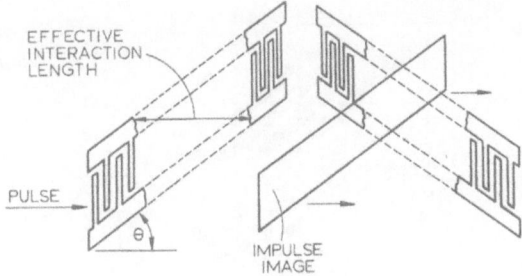

Fig. 4.22. Compression filter using inclined geometry. The interaction length is much less than the overall length of the dispersive transducers

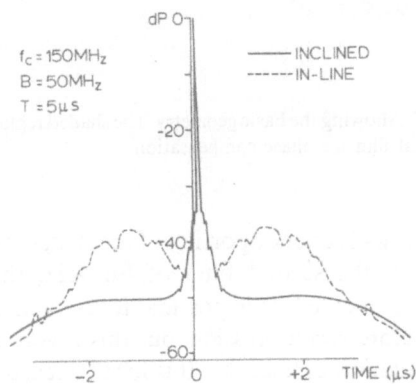

Fig. 4.23. Simulation of compressor performance comparing in-line with inclined geometry (after [4.60]). The improvements deriving from the reduced interactions in the latter are clearly revealed

transducer will generate bulk-wave components. In normal operations a narrow band transducer will not, however, be phase matched to a bulk wave with a propagation velocity v_b, since in all cases $v_b > v_R$. In a wide band transducer, and specifically in a wide band dispersive filter, sections of the large-period portion of the transducer may phase match to bulk waves at the upper frequencies of the band. If a surface wave at frequency f_1 encounters a portion of the structure whose period is synchronous for waves of frequency f_2, phase matching to the bulk waves will occur if

$$f_1 > f_2 \frac{2v_b}{v_b + v_R}. \tag{4.24}$$

The worst case arises when f_1 is the highest frequency in the band and f_2 is the lowest. If even in this situation we wish to avoid the possibility of phase matching to the bulk waves, we must restrict the relative bandwidth B' to a value which is approximately given by

$$B' < (v_b - v_R)/2v_b. \tag{4.25}$$

The necessity for this restriction is avoided if we wish to generate a down chirp, *as in the first of the transducers shown in Fig. 4.21;* here the high-frequency

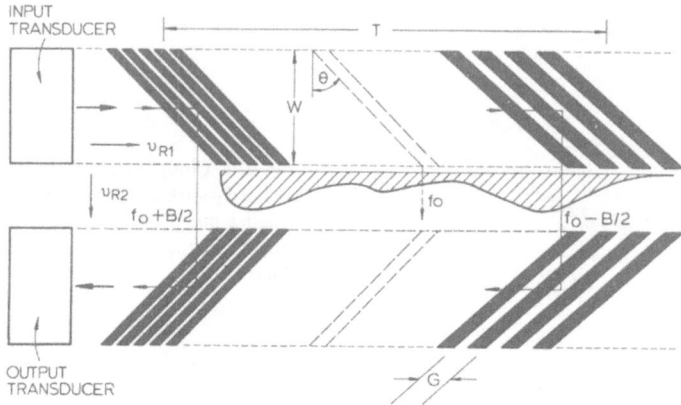

Fig. 4.24. The reflective array compressor or "RAC", showing the basic geometry. The shaded region between the two tracks indicates a deposited metal film for phase compensation

components do not propagate under the low-frequency portion of the structure. However, in recompression, as shown in the second filter of Fig. 4.21, the problem cannot, in any direct manner, be avoided. It is possible to use active methods spectrally to invert the chirp before recompression, but this involves additional complexity and possible degradation of linearity. It is for this reason that the inclined structure of Fig. 4.22 is so attractive. Various frequency components of the complete band encounter only sections of the structure, and these are approximately synchronous. The bandwidth restriction of (4.25) is applicable to the subsections of the complete transducer of length $W \cdot \cos \theta$.

The use of the inclined structure, particularly on quartz, leads to a performance which for TB products of up to a few hundred can be very close to the theoretical optimum. An analysis of the permissible errors indicates that the amplitude characteristic must be rendered with an accuracy which typically is less than 1 dB, and the phase characteristic, with an accuracy of a few degrees. Nevertheless, for bandwidths up to perhaps 100 MHz, refined but essentially standard photolithographic techniques are able to define the metallization with the requisite accuracy. In common with most SAW technology and indeed most microelectronics, we find a situation in which we can produce a highly sophisticated product, rather inexpensively—once the photolithographic mask has been designed and realized.

It is therefore with some reluctance that one would be prepared to abandon the basic manufacturing simplicity of a single metallization SAW device. However, for sufficiently large TB product devices, the difficulties of the metallized interdigital structures become formidable and it is here that the advent of structures which make use of arrays of reflecting slots has led to remarkable advances in attainable performance. The concept has its origin in bulk-wave devices. However in its surface wave garb it takes the form [4.61, 62] shown in Fig. 4.24. The path followed by any given frequency component of the input wave is U shaped. It propagates to the region where the slot spacing is synchronous, is there reflected to the other leg of the U, and once again reflected towards the receiving transducer. We can at once identify a number of advantages for this system.

1) The required substrate length for a given total dispersion is halved by comparison with a conventional transducer-based in-line structure.

2) The correct surface wave path is controlled by two reflections and leads to a complicated route. Bulk waves and other spurious waves generated by the transducers or by the slots have a minimal chance of reaching the receiving transducers with the wavefront parallel to the interdigital fingers. Spurious responses due to such waves are therefore very effectively suppressed.

3) The reflecting slots are not interconnected by an electrical circuit. The regeneration phenomenon, with the attendant distortion problems, vanish.

4) Since the frequency components follow a unique path between the two legs of the U, it is possible to apply phase compensation in this region, by the use of an additional thin film.

5) The amplitude taper required in a compression filter can be introduced by varying the depth of the grooves.

The emergence of this concept owes much to the development of a new technology—ion beam machining [4.63]—which has proved to be capable of highly controllable fabrication. It is in this sense that "(5)" can be regarded as an advantage. The ion beam process involves complex equipment but, with computer control, can permit a variation of the amplitude weighting function, from one sample to the next. It remains true that devices so produced are likely to remain inherently expensive. The cost is, however, justified not only for sophisticated radar systems but also as the basic component in the signal processing systems which we shall discuss in Section 4.5.3.

One can envisage a device operating on the same basic principle in which the slots are replaced by a series of metallic stripes. The reflection of such a stripe, unless very thick, will be predominantly due to piezoelectric shorting effects. It is therefore difficult to devise a method of apodization. As has recently been shown [4.64], it is, however, possible to replace each stripe by an array of distinct dots. The effective reflection of each line of dots can then be controlled by varying the spacing between the dots. In this way it is possible to realize a reflective dot array (RDA) which parallels the performance of the reflective array compressor (RAC) but without requiring the expensive ion beam technology. The RDA technique may well prove to win a much wider scope for this class of compressive filter.

The first step in the design is to establish the appropriate groove angle θ, and the groove period G. In this we have to remember that except in the case of isotropic materials, the Rayleigh wave velocity will differ in the longitudinal direction v_{R1} from that which obtains in the transverse direction v_{R2}. From the requirement that the reflected wave shall have a plane wavefront normal to the direction of propagation, we can immediately write down the conditions,

$$\frac{\beta_1 G}{\cos\theta} = \frac{\beta_2 G}{\sin\theta} = 2\pi \tag{4.26}$$

where

$$\beta_{1,2} = \frac{2\pi f}{v_{R1,2}}.$$

The first equality gives

$$\theta = \tan^{-1}(v_{R1}/v_{R2})$$

and the second that

$$G = \lambda_{R1}\cos\theta.$$

Since θ will be of the order of 45°, we see that $G \sim 0.7\lambda_{R1}$. Thus the line spacing is almost three times greater than in the case of an interdigital transducer, where it is 0.25λ, which represents a further advantage in the realization of large bandwidth devices.

In using (4.26), we have tacitly assumed that the reflection from each slot is weak, so that multiple reflections can be discounted. By making N_{eff} sufficiently large, we could nonetheless ensure that most of the energy centered on the frequency f was reflected so as to minimize the insertion loss. However, one must appreciate that the waves reflected by the slot array encounter a number of other slots before they reach the free surface between the two arms of the U. The energy reflected at the center of the slot will have to encounter approximately $W/2\lambda_R$ slots before finally emerging. Clearly if N_e slots can reflect most of the energy, we would have to ensure that $W/2\lambda_R \ll N_e$.

This in turn would force a design with W very small, so that diffraction effects would become very significant. In practice, it is preferable not to attempt to utilize most of the input energy, but rather to incur an insertion loss, arising from this cause, of the order of 20 dB. This will ensure that the weak reflection theory is applicable, and that multiple reflections can be taken into account as a perturbation of the basic design which neglects them [4.65].

In order to obtain the complete amplitude characteristic of the device, one needs to know the reflectivity of each groove, $r(f)$. On the weak coupling theory, we then attribute the reflection of the spectrum centered around f to N_e grooves. Since each wave is reflected twice, the total amplitude response $H(f)$ takes the form

$$|H(f)| = [(N_e \cdot r(f)]^2 \cdot g(f)$$

where g is a geometric factor. The need for such a factor arises from the fact that a particular groove, the nth on the right-hand side of the array, will communicate not only with the nth slot in the left-hand half, but also with all those slots in the range $n \pm N_e/2$ which *partially* overlap the nth slot. Clearly, if $W/N_e\lambda_R \gg 1$, the nth slot will in fact overlap almost totally with all the slots with which it is able to communicate. In this case $g(f) \sim 1$. The opposite limit, when $W/N_e\lambda_R \ll 1$, also permits a simple intuitive assessment for $g(f)$: in this case the nth slot will be able to communicate with a fraction $W/N_e\lambda_R$ of the synchronous portion of the array; this fraction is therefore an appropriate approximation for $g(f)$. An approximate form of the complete function $g(f)$ has been calculated [4.66].

It has been shown theoretically [4.67] and demonstrated experimentally [4.68] that the reflection arising from a slot of depth h is proportional to h/λ — provided $h/\lambda \ll 1$. Since N_e is proportional to the frequency, we see that, quite apart from the frequency dependency arising from $g(f)$, $H(f)$ will be proportional to the fourth power of the frequency. This, together with the desired amplitude taper, can be incorporated in the slot depth profile. There are also alternative methods of amplitude weighting by varying the length of the reflecting slots [4.62]; in this situation one must adopt some means for compensating for the distortion of the wavefronts arising from the small difference in propagation velocity associated with the slotted as compared with the unperturbed surface. Techniques entirely analogous to the use of dummy fingers in ordinary transducer structures have been shown to be effective [4.69].

The bandwidth is limited by the attainable resolution, precisely as is the case with conventional transducers, but with the benefit of a line spacing of 0.7λ as compared with 0.25λ. The maximum total delay is limited by the length of available crystals. For high center frequencies—required to realize large bandwidths—the variation of propagation loss with frequency may also set a limiting obstacle. Very shortly after the first demonstration of a successful reflective array compressor (RAC), a device with a compression ratio in excess of 5000 was demonstrated [4.70]. Figure 4.25a shows the amplitude response compared with the required Hamming function. Figure 4.25b shows the phase response. After measurement, it was possible to compensate the error by calculating the profile of a thin metallizing film between the two legs of the U. The film depresses the velocity by piezoelectric shorting, which in the case of LiNbO$_3$ amounts to approximately 2%. At 1 GHz, the differential phase shift per micron path length is around $2°$, and this is in fact the order of remanent error. Recently, a device with a bandwidth of 250 MHz and a delay of 40 µs has been successfully demonstrated [4.71]. It is likely that the compression ratio of 10000 will be exceeded in the near future.

4.5.2 Phase-Coded Transducers

The pulse compression filters discussed in the last section can be regarded as transversal filters, in which each finger pair samples a portion of the wave, the *magnitude of the sample being determined by the apodization*; the timing—and

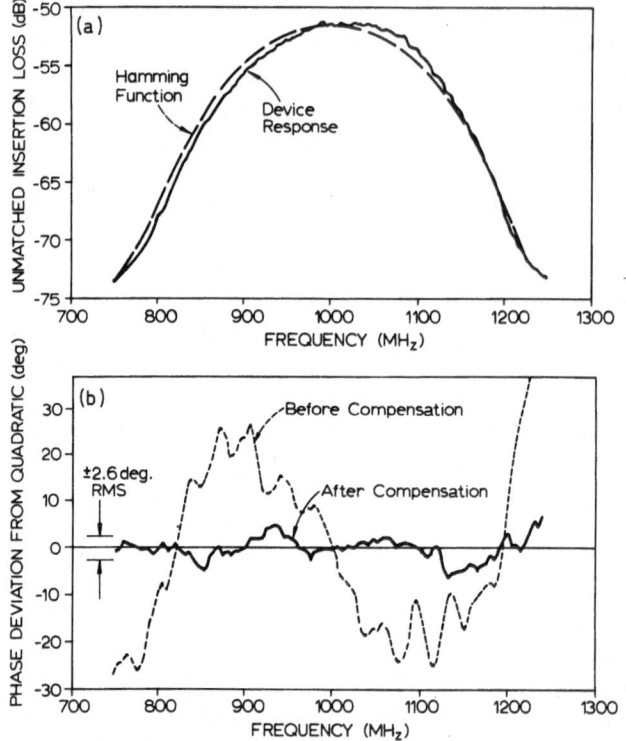

Fig. 4.25. (a) Amplitude response. (b) Phase response of a reflective array compressor with a bandwidth of 500 MHz and a dispersion of 10 μs.

The amplitude response is within 1 dB of the desired Hamming weighting function. The phase response before and after compensation by evaporation of suitably shaped thin metallic film is indicated

hence the phasing—by its location. The frequency filters discussed in Chapter 3 answer to the same description. The set of realizable transforms which can be implemented by a transversal filter is infinite; nevertheless, in view of its importance in communication systems signal processing applications, it is worthwhile to single out yet one more specific type—the biphase coded transducer. The biphase coded transducer is concerned with the generation and processing of waves having a specific form of modulation—binary phase shift keyed (binary PSK) modulation. Figure 4.26a shows an example of a wave modulated with a particular code; Fig. 4.26b shows how this çode can be generated using a SAW coded interdigital structure. Each section (chip) of the wave consisting of a sine-wave pulse represents a binary digit, and is generated by the corresponding section of the transducer. The phase change of π which differentiates a binary one from a zero is effected very simply by changing the sequence of the finger connections. The importance of SAW devices in the context of PSK waveforms derives just from the simplicity of implementation of

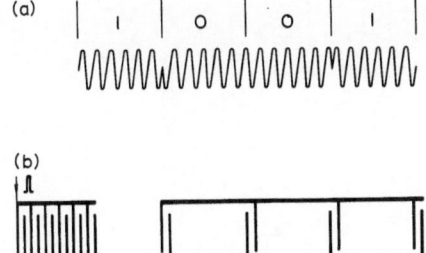

Fig. 4.26. (a) Form of binary PSK code-illustrated for a chip length of 6 cycles, showing a 1001 sequence. (b) Phase-coded filter that will generate the code of (a). Note that the input transducer has a length occupied by one complete chip

the phase changes. Furthermore, since the difference between a 0 and a 1 depends only on whether a set of alternate fingers is connected to the upper or the lower bus bar, it is at once clear that one can effect changes electrically, and thus realize a programmable device.

The basic design of the biphase filter of Fig. 4.26 is relatively simple: the input transducer bandwidth is such that it produces the desired chip length. Since the output from each of the taps is the convolution of the impulse response of the input transducer with that of each tap, it is clear that to avoid distortion the latter should be as close to a delta-function as possible. In practice, as long as its bandwidth is substantially larger than that of the input transducer, this condition is adequately fulfilled. In many cases, the tap transducer consists of two or even of a single finger pair. In the absence of spurious signals this basic design will launch the required PSK code; the complementary filter will compress the code, and give a peak to side lobe discrimination determined by the code itself.

For relatively modest compression gains—approximately equal to the total number of chips in the sequence—these design aims are translated into practical performance figures without undue difficulty. Thus, a 13 chip device using a Barker code has given close to theoretical performance [4.72]. However, in filters with many tens or hundreds of chips, it has been found by many workers that the effect of spurious signals becomes the performance limiting factor. As in all other SAW devices, we are concerned with acoustic reflections, re-radiation induced phase changes and the effect of bulk-wave generation. Much of the recent work on phase-coded devices has been concerned with the reduction of these spurious effects.

The acoustic reflections are caused primarily by the change in acoustic impedance engendered by the short circuiting of the electric field. The effect is very much weaker on a low-k^2 material such as quartz than it is for $LiNbO_3$. It is nonetheless sufficiently damaging to require remedial action for designs with several tens of chips.

One remedy which has been widely used for this application, as for pulse compression filters, is to split each electrode into two, so that there are now four electrodes per period instead of two [4.73]. The result is that the reflection peak, instead of occurring at the center frequency f_0, is translated to $2f_0$. At the same time, the larger number of fingers per tap, as well as their higher spatial

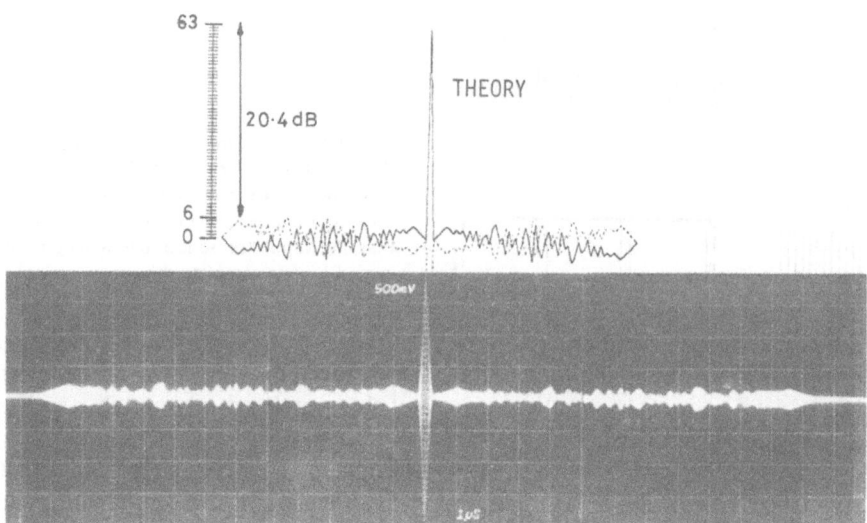

Fig. 4.27. Theoretical and measured response of a 63 chip coded split electrode filter (after [4.70])

frequency, substantially reduces the bulk-wave generation. In fact, bulk waves cannot phase match to a long transducer with the half period finger structure—though of course a component arising from end effects remains. The end effects—whether or not split fingers are employed—can be greatly reduced by filling in the gaps between the taps by additional dummy fingers, not connected to any electrical circuit. Other than small differences arising from the different electrical loads on the taps and on the dummy fingers, the whole structure is now homogeneous, and will not radiate to bulk waves until the bandwidth exceeds the expression given in (4.25).

The split fingers do not prevent the re-radiation from the transducer structure; an answer to this problem is supplied by the use of a complete set of dummy taps [4.74], displaced by a quarter wavelength from the "real" set. The dummy taps are connected to bus bars in a way which completely mimics the real set, even to the use of the same electrical load. The reradiated component is therefore completely suppressed, at least at the center frequency, as is the acoustic reflected component. The bulk-wave generation is not improved, and indeed, if the bandwidth exceeds the limit imposed by (4.25), it will be increased. The problem of bulk waves is not only that they may be reinjected into the proper signal path, but also that the implied progressive losses lead to a taper on the impulse response. In principle, one can overcome this taper by the use of a suitable apodization of the taps. The perfection of the response which can be achieved with a 63 chip device is illustrated in Fig. 4.27, which compares the computed and observed autocorrelation response (i.e., expansion by one device followed by recompression with its complement). Using standard techniques it is now possible to achieve performances not too far from the theoretical optimum

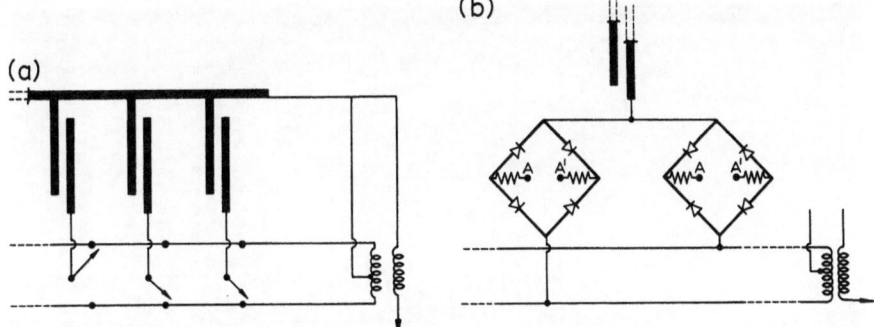

Fig. 4.28. (a) Programmable coded filter. The position of the switches determines the sign of the contribution from each individual tap. (b) Implementation of the configuration using ring diode circuits

with codes of up to 500 chips [4.75]. One result, using a quartz substrate in which the surface waves propagated from one side to the other via a curved end section, achieved good results with a 1010 chip code, with a chip length of 0.1 μs [4.76].

One of the most important uses for binary PSK coded devices is in spread spectrum communications, in which a single bit is encoded in a pseudo random sequence. The bandwidth required by the signal is thus increased by a factor which is approximately equal to the chip code length. There is, of course, a consequent increase in the system noise by the same factor. However, the signal to noise ratio is ultimately recovered by the processing gain of the compression phase-coded filter—again approximately equal to the chip code length. One of the possible reasons for using such spread spectrum techniques lies in the security which can thereby be achieved. Messages cannot be intercepted without knowledge of the pseudo random code used. The difficulty of "breaking the code" is of course a function of its length. For this reason, there are applications where one wishes to use a code length which is very much longer than the 500 or even 1000 chip devices which can be realized using current SAW technology. Further improvements in this technology might win another factor of two, but will not yield the increase of several orders of magnitude which are sought, e.g., for the implementation of a 24-h code. It is here that the use of programmable filters, in which the code used can be rapidly changed, is of such great importance.

In principle, all that we need to do to make the device programmable is to provide each tap with a two-pole double-throw switch, as shown in Fig. 4.28a. The switch can be implemented by a four-diode circuit, which in simplified form is shown in Fig. 4.28b. There are two main problems associated with this technique if one seeks to implement it using discrete switching components. The first emerges from the basic fact that one needs to make four connections to each tap, and find room for the discrete components in a space equal to the chip length. For a 100 MHz center frequency and a chip rate of 10 MHz, each chip will

Fig. 4.29. AlN/SOS monolithic integration tapped delay line and switches on common substrate (after [4.78])

be approximately 10 wavelengths, i.e., only 300 μ, in length. This forces the designer to adopt a fan-shaped layout which immediately limits the total number of chips which can be accommodated. A second problem arises from the finite reverse bias impedance of the diodes. An individual finger will be connected to the unwanted bus bar via the reverse bias capacitance of the switching diode. A typical value for the internal series capacitance of the finger might be 0.2 pF. If the reverse bias capacitance of the diode is of the same order, it is clear that the on-off ratio of the switch will be around 2 : 1, which is of course quite inadequate. It is therefore essential to use diodes with a capacitance which is substantially less than that associated with the individual finger. In practice, these constraints limit the implementation of such a hybrid technology to chip rates not exceeding 10 MHz and code lengths not exceeding 50.

An approach which is yielding good results, and which appears to be capable of substantial further expansion, is to base the acoustic system on a thin epitaxial film of AlN on sapphire, and the electronic switching system on a silicon on sapphire film. It turns out that the processing required is compatible, so that the whole system can be realized on one sapphire substrate [4.77, 78]. In this monolithic technology, the problem of forming the connections becomes a matter merely of producing the right masks for the photolithography. The diodes are formed by vertical diffusion through the $1\,\mu$ film. Thus the junction area is $w\,\mu^2$, where w is the width in microns. For $w = 20\,\mu$, the reverse capacitance is of the order of $0.02\,\mathrm{pF}$, a figure which it would be very difficult to achieve in conventional microelectronic technology. Using the combined AlN and Si on sapphire technology, a PSK device with 63 chips and operating at a chip rate of 20 MHz (center frequency 190 MHz) has been successfully demonstrated [4.77] and has given results which are not substantially worse than those obtained with fixed coded devices. A micrograph of the complete circuit is shown in Fig. 4.29.

The emergence of practical programmable binary PSK devices coupled with conventional storage and logic circuits opens up a very large class of new signal processing applications. It is already clear that these developments will have a major impact on spread spectrum systems and certain classes of radar systems.

4.5.3 Signal Processing and Linear Chirp Filters

In Section 4.5.1 we explored the use of linear FM filters for chirp radar applications. It has, however, become progressively clearer that the linear chirp filter is a generic device which can form the basis of a much wider class of signal processing applications. In this section, we shall give some examples of the applications which have emerged.

We can rewrite the impulse response of a linear chirp filter, (4.17) in the form

$$h(t) = e^{i2\pi f_0 t} \cdot e^{i\alpha t^2},$$

where

$$\alpha = \frac{\pi\beta}{T}. \tag{4.27}$$

The output signal $S(\tau)$, arising from an input signal $S_i(t)$, is therefore,

$$S_0(\tau) = \int_{-\infty}^{\infty} S_i(t)\, e^{i\alpha(t-\tau)^2}\, dt$$

where the factor $\exp(i2\pi f_0 t)$ has been suppressed. Expanding the squared term in the exponent,

$$S_0(\tau) = e^{i\alpha\tau^2} \int_{-\infty}^{\infty} [S_i(t)e^{i\alpha t^2}]e^{-2i\alpha t\tau}\, dt. \tag{4.28}$$

Equation (4.28) is close to the form of a Fourier transform for the quantity in brackets, the transform variable being

$$\omega \equiv 2\alpha\tau.$$

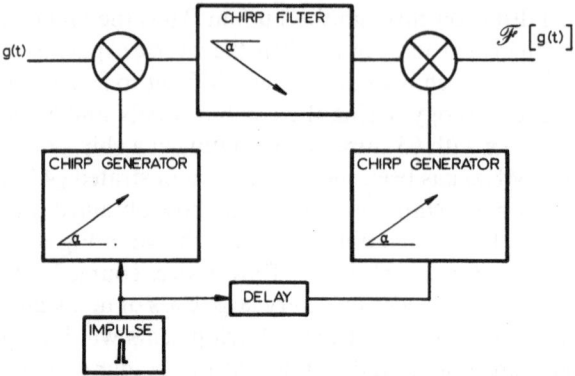

Fig. 4.30. Fourier transform processor. The sign and magnitude of the slope of the chirp are indicated by the arrowed lines

The transform is distorted only by the quadratic phase term outside the integral. In many situations, one is concerned not so much with the Fourier transform as with the Fourier spectral density, i.e., with $S_0(\tau)S_0^*(\tau)$, and in this the phase term, of course, vanishes.

The more serious problem which must be solved before we can hope to use the chirp filter as a Fourier transformer is to disentangle the actual input signal $S_i(t)$ from the factor $\exp(i\alpha t^2)$. This can be effected [4.79] very simply by premultiplying $S_i(t)$ by an exponential factor of opposite slope, i.e., by $\exp(-i\alpha t^2)$, using conventional mixer multiplier circuits. If desired, the phase factor outside the integral can also be removed by a further multiplication by $\exp(-i\alpha\tau^2)$. The complete arrangement to give effect to these transformations is shown in Fig. 4.30, where the sloping lines indicate the sign and magnitude of the slope of the chirp. The chirp signal required for the multiplication can be generated using a similar, but not the *same*, device as that used in the main chirp filter. The most immediate reason is that in simple mixers the product is recovered as an upper or lower sideband; the center frequency will be shifted by an interval equal to the center frequency of the chirp. One could of course overcome this problem by introducing a frequency translator between the chirp generator and the mixer.

There is, however, a more fundamental issue with which we have to contend. The dispersive filter must be able to handle the complete range of frequencies derived from multiplying the signal having a bandwidth B_s with the chirp having a bandwidth αT_c, where T_c is the chirp length. The available bandwidth of the filter is $B = \alpha T_f$, where T_f is the total filter delay. The maximum signal bandwidth which can then be used is therefore

$$B_s = \alpha(T_f - T_c). \tag{4.29}$$

It is clear that we must make the multiplying chirp length substantially less than the filter delay time. The length of the signal sample which we can handle, T_s, is clearly equal to the chirp time T_c. Thus, the available signal time-bandwidth product is

$$B_s T_s = \alpha(T_f - T_c) T_c . \tag{4.30}$$

We can maximize this by setting

$$T_c = T_f/2 \tag{4.31}$$

so that

$$(B_s T_s)_{max} = \frac{1}{4}(B_f T_f). \tag{4.32}$$

Since we will wish to make $T_c = T_s$, (4.31) immediately implies also that the useful signal bandwidth B_s is

$$B_s = B_f/2. \tag{4.33}$$

In a Fourier transformer, one is in addition concerned with the attainable frequency resolution Δf_r. Using (4.32) we find

$$\Delta f_r = \frac{B_s}{(B_s T_s)_{max}} = \frac{2}{T_f}. \tag{4.34}$$

Comparing this result with that obtained in (4.23) from simple considerations based on N_e, we find that f_r is substantially smaller—by a factor $(T_f B_f)^{1/2}$. The use of the premultiplier chirp allows us to resolve frequency very much more effectively than by the direct use of the filter as a simple discriminator.

These extremely simple results at once indicate that such Fourier transformers can have an impressive performance. It will, of course, inevitably be degraded by the appearance of side lobes in the transformed output. One can improve the peak to side lobe ratio by the use of amplitude weighting—again at the expense of resolution. An implementation of a Fourier transform has recently been described [4.80]; the center frequency of the chirp filter was 70 MHz, the bandwidth was 17.6 MHz, and the delay T_f was 88 μs. The amplitude Hamming weighting to improve the peak to side lobe ratio led to an effective signal time-bandwidth product of 260, which; very roughly, can be regarded as the number of distinctly resolvable frequencies. The corresponding frequency resolution was 37 kHz. This performance is attained using 44 μs samples of the signal and, of course, the results are available in "real time". Figure 4.31, after [4.80], shows an example of the use of this Fourier transformer to display the harmonics of a square wave.

With the development of compression filters with time-bandwidth products approaching, and perhaps exceeding, 10000, one can anticipate that the capability of such transformers will improve rapidly in the near future.

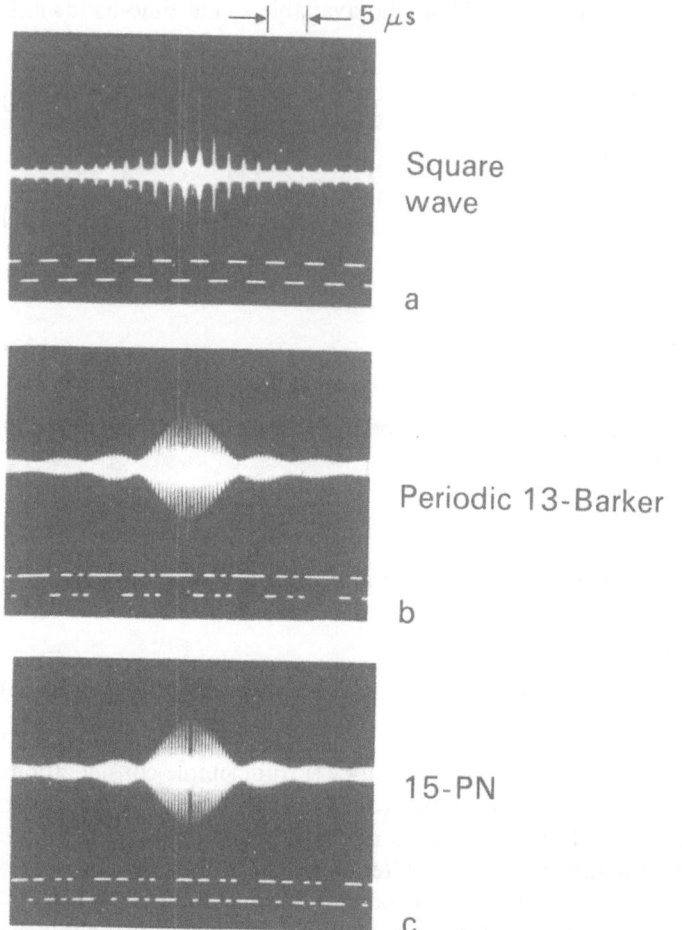

Fig. 4.31a–c. Fourier transforms obtained with a system having processing time-bandwidth product of just under 400. (a) Square wave. (b) Periodic 13 bit Barker sequence. (c) 15 chip PN sequence (after [4.80])

The most immediate application of a Fourier transformer is in spectral analysis. A particular form is known in radar usage as a "compressive receiver". It serves simply to display frequencies present in an input signal; it can do so very fast—in fact in a period of the order of T_f. It will succeed in detecting even single pulses which appear within the sampled signal. Moreover, since the signals of different frequency are, by the action of the Fourier transformer, separated in time, it is possible to make full use of the available dynamic range of the system, even when one is concerned with a relatively weak signal in the presence of others which are much larger. In this application one is concerned only with the spectral power, so that the final post-convolution chirp can be omitted.

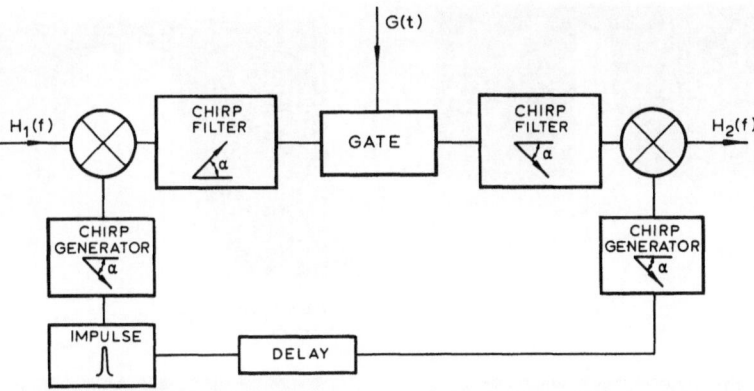

Fig. 4.32. Frequency filter implemented by means of two successive Fourier transformers. The timing and shape of the gating pulse $G(t)$ determines the filter response

An application which demands both amplitude as well as phase information is in the realization of a network analyzer. In a prototype instrument recently described [4.81, 82], the measurement is effected by Fourier transforming the output of the test network excited at the input by a periodic train of rf pulses, the modulation frequency being at the center of the test range, and the pulse width adjusted so as to give the required density of spectral test lines in the measurement interval.

Clearly, any transformation that can be expressed in terms of Fourier transforms can also be implemented by suitable combinations of the basic system of Fig. 4.30. As one example [4.80], it is possible to obtain the correlation between two signals $S_1(t)$ and $S_2(t)$ by making use of the well-known theorem,

$$\int S_1(t) \cdot S_2(t-\tau)\,dt = \mathscr{F}^{-1}[\mathscr{F}(S_1) \cdot \mathscr{F}(S_2^*)] \tag{4.35}$$

where the left-hand side is the required convolution and $\mathscr{F}$ indicates the Fourier transform operation. A direct implementation of the right-hand side would require six chirp generators, as well as the three chirp filters. In fact, it was shown that various operations could either be combined, or were redundant, so that only three chirp generators were needed.

A further exploitation of the basic principle which has recently been accomplished [4.83–85] is the realization of frequency filters. If a signal is successively Fourier transformed, and then transformed once more, one will recover the original signal—naturally with distortion arising from band limitation. In between the two transformations, the frequency components are displaced in time. This then offers the opportunity of affecting a *portion* of the spectrum while leaving the remainder untouched, by introducing a time-dependent attenuation, or, in the extreme case, a gate. The basic system is shown in Fig. 4.32. The first transform is effected using an up-chirp filter, the second *using a down-chirp.* In the absence of a time gate, the post-convolution chirp of

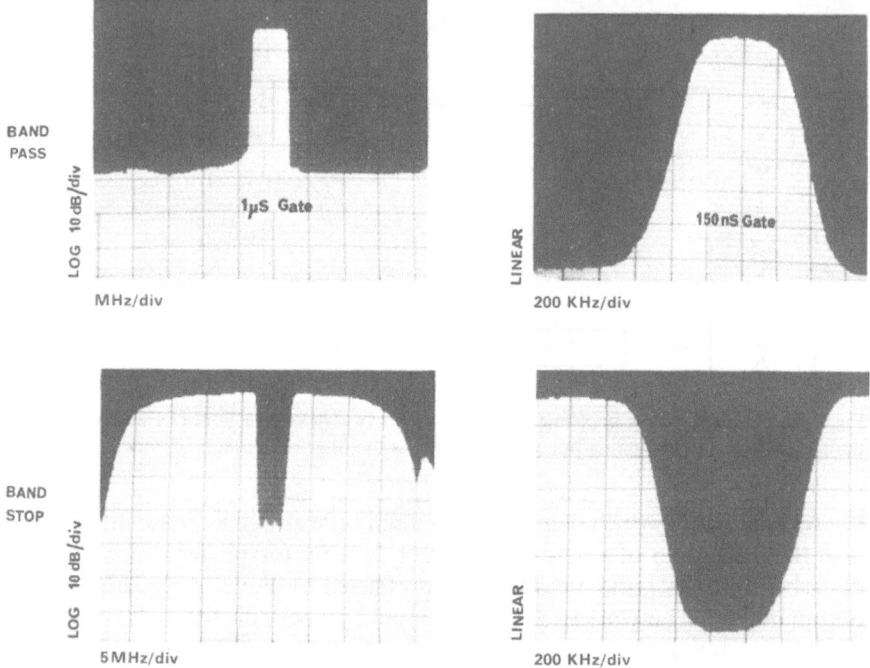

BAND PASS

LOG 10 dB/div

1μS Gate

MHz/div

LINEAR

150 nS Gate

200 KHz/div

BAND STOP

LOG 10 dB/div

5 MHz/div

LINEAR

200 KHz/div

Fig. 4.33. Frequency response of Fourier processor filters. The response is controlled by varying the length of the time the gate is closed during the transmission of the signal (after [4.83])

the first transform and the premultiplication chirp of the second cancel; nor will the presence of a time gate, which acts only on the amplitude of the signal, affect this basic phase cancellation. Both of these chirp generators can therefore be omitted. Since the signal sample time T_s is, according to (4.31), $T_f/2$, it is necessary to use two channels in parallel with the sampling periods interleaved, in order to obtain a continuous filtering action. Figure 4.33 shows the results which have been obtained using this system in the realization of both pass and stop band filters [4.83]. They were obtained with filters having a $T_f B_f$ product of 180, so that the available signal $T_s B_s$ product was only 45. Using large time-bandwidth RAC devices, high-performance filters with bandwidths of the order of 100 MHz would become realizable. It is important to appreciate that such a filter is fully programmable—that by using more elaborate gating signals it would be possible to synthesize a wide variety of filter responses.

It is also possible to use a compression delay line as a variable time delay element [4.86–88]. Given a narrow band signal, it is clearly in our power to change its frequency with the help of a mixer and a local oscillator of variable frequency. By changing the local oscillator frequency, we can then locate the signal in any part of the filter bandwidth B_f and hence vary the delay over the range of the dispersion, T_f. In this simple form the system would have two

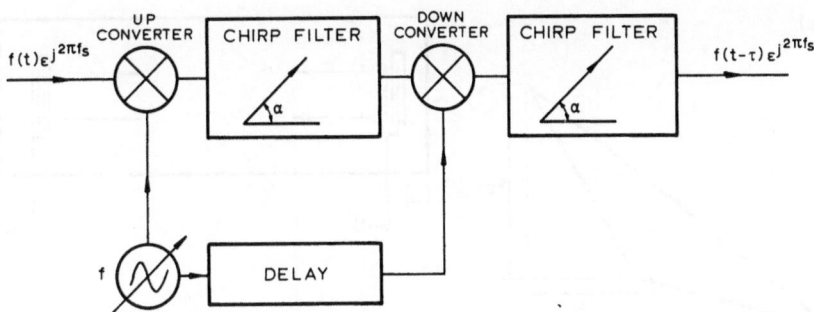

Fig. 4.34. Variable time delay processor. The output is at the same frequency as the input

defects. The first is trivial, though probably inconvenient: the output signal will have been shifted in frequency. The second defect is far more fundamental. We are dealing with a dispersive delay line, so that the signal will have been dispersed. Both these defects are overcome in the system shown in Fig. 4.34. The local oscillator frequency is variable about a central value f_0, which is such that (f_s+f_0) generated in the first upconverter mixer reaches the center of the available up-chirp filter band B. In the second mixer we chose the lower sideband and assigned a center frequency f_s to the dispersive filter. The frequency after the second mixer is then identical to the input signal frequency. The second filter will provide an additional variable delay equal to that derived from the first. However, owing to the spectral inversion of the second mixer, the sign of the dispersion will be reversed, so that the net dispersion of the overall signal is zero. The total variable delay τ is given by

$$\tau = 2(f-f_0)\frac{T}{B} \qquad (4.36)$$

with a maximum differential delay of $2T$.

In addition to variable delay, it is also possible to realize variable time compression, expansion, or inversion [4.60]. In all. such applications, the ultimate performance depends on the capability of the basic dispersive delay line. Such sophisticated signal processing applications make continually escalating demands on the device performance; at the same time, they represent applications for which the cost of the most sophisticated RAC devices can be justified.

4.6 Nonlinear Signal Processing

The basic capability of SAW signal processing devices derives from the facility with which one can implement summing operations; it is essentially a matter of devising the appropriate metallization pattern. However, as we have already

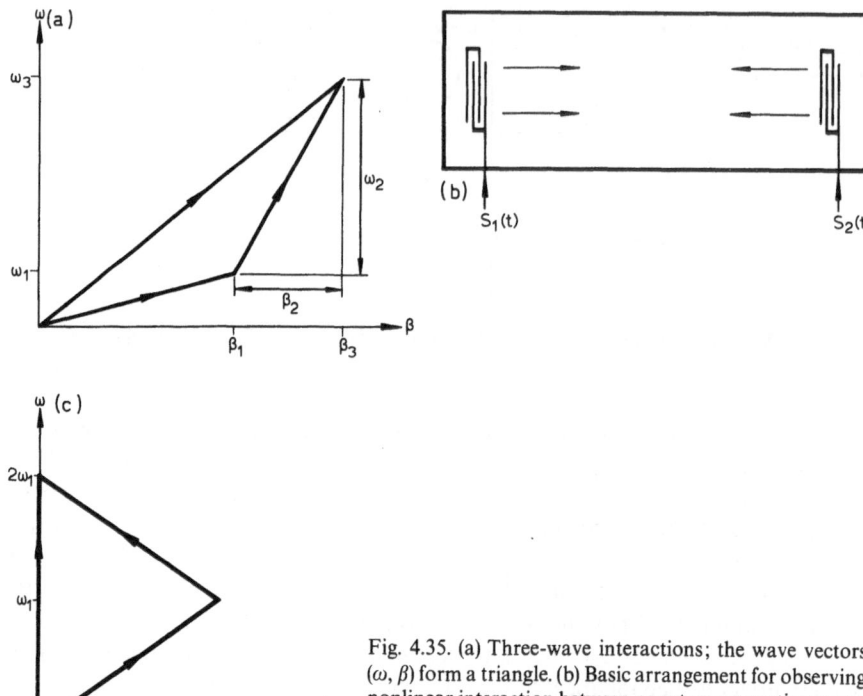

Fig. 4.35. (a) Three-wave interactions; the wave vectors (ω, β) form a triangle. (b) Basic arrangement for observing nonlinear interaction between counterpropagating waves (after [4.92]). (c) Wave vector diagram for degenerate case

seen in Section 4.5.3, the range of signal processing operations which can be effected is greatly increased if in addition we can perform multiplications. In the systems we have so far discussed, these have been based on the use of conventional mixer diodes; in practice, the multipliers have not usually proved to be the performance limiting element. There is, however, an alternative, which has been progressively explored over a number of years: the direct exploitation of the nonlinearity of the acoustic medium itself.

All material media will, of course, give rise to nonlinear effects at sufficiently high power levels. Specifically, if we have a wave of frequency ω propagating in a medium we shall expect generation of the second harmonic 2ω, and at sufficiently elevated power levels higher harmonics as well. One can derive the relevant results by developing the elastic equations but including second-order terms; this analysis will lead to the result—common to electromagnetic as well as acoustic wave propagation—that the power output at the second harmonic will be directly proportional to the square of the power density at the fundamental frequency. Without writing down any equations, this result is intuitively obvious if we look at the interaction in terms of phonons. The process is essentially a collision between two phonons of frequency ω generating one of frequency 2ω. The collision probability is clearly proportional to the square of the number of phonons present, and hence to the square of the power density. If the interaction takes place between two waves, one having a frequency ω_1 and power density $P(\omega_1)$, and another ω_2, $P(\omega_2)$, the nonlinearly generated wave at a frequency of ω_3 will have a power density $P(\omega_3)$ given by

$$P(\omega_3) = c_3 P(\omega_1) \cdot P(\omega_2). \tag{4.37}$$

The fact that it is the power density, rather than the power, that comes into play immediately gives surface waves a significant advantage over bulk waves in the implementation of nonlinear interactions: surface waves are guided by the surface; the depth of penetration is of the order of λ_R. If, in a given device we require a beam width of $N\lambda_R$ in order to avoid excessive diffraction, we find that, for a given power, the power density in a bulk device, which requires a beam having *both* lateral dimensions $N\lambda_R$, is N times lower than for the comparable surface wave device. Since a typical value for N is around 50, this represents a very substantial advantage.

The constant c_3 in (4.37) is a complex function of the third-order elements of the elastic and the piezoelectric tensors of the medium. In practice it is necessary to measure its value rather than attempting a computation from more basic data. However the *variation* of c_3 with the direction of the interacting waves and the cut of the crystal is amenable to analysis and can correctly predict some of the observations [4.89].

The nonlinear interaction which one can observe is weak, in the sense that the power transfer from one frequency to another *per wavelength of interaction* is small. To obtain useful transformation it is therefore necessary to have extended interaction lengths. The requirements for such extended nonlinear interactions between three waves are well known [4.90]

$$\omega_3 = \omega_1 + \omega_2 \tag{4.38}$$

$$\beta_3 = \beta_1 + \beta_2 \tag{4.39}$$

where β_i is the propagation constant of the ith wave.

Again, both equations follow at once from the phonon picture: (4.38) is an expression of the conservation of energy, (4.39), of the conservation of momentum. If we represent the waves as vectors (ω, β) in the $\omega - \beta$ plane, we see that (4.38) and (4.39) imply that the three vectors must form a triangle, as in Fig. 4.35a. For the particular case of second harmonic generation (i.e., $\omega_1 = \omega_2 = \omega_3/2$), we see from (4.39) that $\beta_3 = 2\beta_1$, i.e., the waves must be nondispersive. It is just because Rayleigh waves are indeed nondispersive that second harmonic operation is relatively easy to observe. The first results were reported by *Chaban* [4.91] and by *Svaasand* [4.92], who also described an experiment with two waves propagating in opposed directions (Fig. 4.35b). Figure 4.35c shows that in this case the nonlinearly generated wave at $2\omega_1$ has a zero propagation constant, and therefore represents a spatially uniform rf voltage. *If the amplitude of the two counterpropagating waves is amplitude*

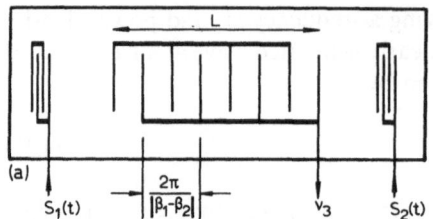

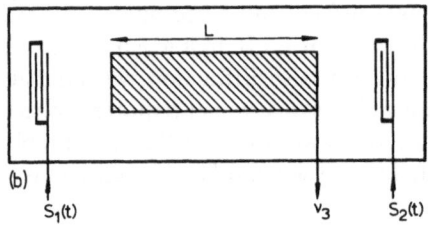

Fig. 4.36a and b. Electrode structures for detection of product wave. (a) Nondegenerate case. The period of the central transducer is determined by the difference between the two wave vectors. (b) The degenerate case. The convolution product is detected by a continuous metal plate

modulated, the generated wave will have a more complex spatial distribution, but still with very small phase change along the length of the structure. Specifically, if we write the input waves in the form

$$S_i(t) = S_i(t) \cdot e^{i(\omega_i t - \beta_i z)} \tag{4.40}$$

where the spectral width of S_i is very much less than ω_i, the distribution of the output voltage $v_3(z, t)$ can be written

$$v_3(z,t) = S_1\left(t - \frac{z}{v}\right) \cdot S_2\left(t - \frac{L-z}{v}\right) e^{i[(\omega_1 + \omega_2) - (\beta_1 - \beta_2)z]}$$

$$\equiv V_3(z,t) e^{i[(\omega_1 + \omega_2)t - (\beta_1 - \beta_2)z]}. \tag{4.41}$$

For the "degenerate" case, $\omega_1 = \omega_2$, $\beta_1 = \beta_2$ we find

$$v_3(z,t) = V_3(z,t) e^{2i\omega_1 t} \tag{4.42}$$

where $V_3(z,t)$ is defined by (4.41).

We can follow the now familiar SAW route to obtain the spatial integral of $v_3(z,t)$ by depositing a series of sampling taps and connecting them to a bus bar. In the general case of (4.41) we would have to locate the taps at the specific points where $(\beta_1 - \beta_2)z = n\pi$, as indicated in Fig. 4.36a.

However, for the degenerate case of (4.42), there is no fast phase variation of $v_3(z,t)$. Accordingly, the integration can be effected with extreme simplicity, by a continuous metallization as shown in Fig. 4.36b. The specific form of the output signal $v_3(t)$ becomes

$$v_3(t) = V_3(t) e^{2i\omega_1 t} = e^{2i\omega_1 t} \int_0^L S_1\left(t - \frac{z}{v}\right) \cdot S_2\left(t - \frac{L-z}{v}\right) dz. \tag{4.43}$$

We can put this into a slightly simpler form by writing

$$\tau \equiv \left(t - \frac{z}{v}\right)$$

$$V_3(t) = \int_0^L S_1(\tau) \cdot S_2\left(2t - \tau - \frac{L}{v}\right) d\tau. \tag{4.44}$$

If we assume that the length of the signal trains $S_1(t)$, $S_2(t)$ and their timing are such that the complete nonlinear interaction takes place under the continuous metal electrode, we can safely extend the limits on the integral of (4.44) to infinity; we can therefore omit the offset term L/v. Thus,

$$V_3(t) = \int_{-\infty}^{\infty} S_1(\tau) \cdot S_2(2t - \tau) d\tau. \tag{4.45}$$

In this form it is clear that $V_3(t)$ is the convolution of $S_1(t)$ and $S_2(t)$, time compressed by a factor of two (owing to the fact that it is $2t$ which appears in the integrand). The possibility of such signal processing using the nonlinear interaction of acoustic waves has been reported by a number of authors [4.93]. The first experimental demonstration of convolver action with bulk waves was made by *Wang* [4.94]. The development of the subject was given a particular impetus by the impressive experimental results with a bulk-wave convolver reported by *Thompson* and *Quate* [4.95]. Shortly thereafter, the basic phenomenon and some additional signal processing operations were demonstrated using surface waves [4.96], which, as we have already seen, present an inherent advantage in that the power level required to obtain strong nonlinear interaction is substantially lower.

4.6.1 The Piezoelectric Convolver

It is important to appreciate that the theory which we have outlined is based on "weak interaction" assumptions. Thus, for example, in (4.41) we have tacitly assumed that the amplitude and phase of the interacting waves S_1 and S_2 are totally unaffected by the process itself, and by the appearance of the distribution $v_3(z, t)$. In practice this assumption is beyond reproach: we can express the constant c_3 in (4.37), known as the "bilinear coefficient", in decibels. For a typical experiment using $LiNbO_3$ the value is of the order of -80 dB. Devices have been built [4.97] with an interaction time (L/v) of 10 μs and a bandwidth of 100 MHz at a center frequency of 400 MHz, so that the time-bandwidth product is 1000— a value which is quite comparable to that which can be realized with the transducer based time-domain filters discussed in Section 4.5. Even though the output power is 80 dB below the inputs into ports 1 and 2, a dynamic range, as limited by noise, of up to 60 dB is still attainable.

In a systems application, it is however not so much the dynamic range as defined by the noise level which comes into play. Rather it is the range down to the *largest interfering signal*, whether this be noise or some spurious response. In the case of most SAW devices, and the class of nonlinear devices is no exception, the *available* dynamic range is usually set by spurious signals rather than by noise. Specifically, if in the convolver of Fig. 4.36b, there is a spurious response which is at a level $-30\,\mathrm{dB}$ relative to the main signal output, it is this figure which determines the effective dynamic range rather than the much lower figure which one might deduce from noise considerations. The figure of $-30\,\mathrm{dB}$ cited is not entirely arbitrary; in practice it is far from easy to reduce the spurious signals to much lower levels.

It is not difficult to identify several of the mechanisms which can give rise to such spurious signals. The signals injected at port 1 can be partially reflected by the edge of the convolver plate, which represents a small impedance discontinuity. They can be further reflected by the transducer at port 2. In either case, one may see signals arising from the convolution of the reflected front end of a signal convolving with the tail end which is still moving in the original direction; one can also experience convolution between the signals injected at the two ports on their second transit. Quite apart from such spurious convolution interaction, each signal will generate its second harmonic. While the convolver plate is, of course, insensitive to this frequency, there are always edge effects which will lead to a finite second-harmonic signal emerging from port 3. There is also a troublesome spurious signal which can arise from plate resonances excited by the convolution signal $v_3(z, t)$.

A considerable amount of effort has been devoted to reducing the impact of these signals [4.98, 99] using more elaborate electrode structures, and with a considerable measure of success.

The ability to perform the operation of convolution immediately implies that one can also obtain the correlation between two signals if we can devise some means for time inverting one of them. If, for example, in (4.45), we have access to the time-inverted signal $S_1(-t)$, and apply this to port 1, we shall obtain $V_3(t)$ at port 3,

$$V_3(t) = \int_{-\infty}^{\infty} S_1(\tau) \cdot S_2(\tau - 2t)\, d\tau \tag{4.46}$$

which is the correlation of S_1 and S_2. If $S_1(t)$ happens to be symmetric in time, the inversion is, of course, redundant—convolution and correlation are then not distinguishable operations. The practical importance of these considerations lies in the identification of one of the two signals, say S_1, with a received radar or communications coded signal, and the other with the reference code. The device can then be used as a matched filter for the reception of the coded signal. If the code used is a frequency modulated chirp derived from a chirp filter, one can hope to have a reasonably accurate version[1] of its time inverse by placing a

---

[1]   The symmetry will to some extent be degraded by a number of second-order effects, such as the bulk-wave generation when the higher frequency portions of the chirp pass under the lower frequency portions of the transducer.

broad-band transducer on both sides of the chirped transducer. The easiest code to invert is undoubtedly a pseudo random binary sequence. There is no difficulty in inverting one widely used—the binary pseudo random PSK code. Here we can expect to have near-perfect time-inverted versions of the signal, and hence be in a position to perform the correlation of (4.46).

The importance of the nonlinear correlator derives from the fact that it is capable of correlating any signals that lie within the bandwidth and processing time of the device, provided only that it is possible to obtain a time-inverted reference. As we will find, the basic interaction scheme is capable of performing a number of other operations, including that of time inversion. In principle, therefore, this last restriction falls away. It leads to the possibility of using as a reference in a radar system, not a perfect sample of the transmitted pulse, but rather a return received from a nearby large target. It is this return signal which can then be correlated with that from a more distant and possibly much smaller target. By adopting this procedure, the performance of the radar is no longer degraded by distortions introduced in the transmitter. As long as both signal and reference suffer equivalent damage, the height of the correlation peak will be affected very little.

One example of the use of a correlator in a spread spectrum communication system has recently been described [4.100]. The signal consisted of a 70 MHz center frequency, 10 Mchip s^{-1}, binary PSK code. The length of the code was many times longer than the available interaction time, 30 µs, of the convolver. The code was therefore time inverted in sections of approximately this length, and the resulting correlation peaks summed in a recirculating delay line. It is important to appreciate that the appearance of a correlation peak does not depend on precise timing of the two interacting waves. As long as the encounter between substantial portions of the code arises under the convolution plate, there will be little reduction of the correlation peak amplitude. This feature greatly eases what is one of the main problems in a spread spectrum system: the need to acquire synchronization between the received signal and that from the local pseudo random sequence generator. It can be shown that, as compared with a correlator which has to be synchronous with an accuracy of a portion of one chip, the time required to obtain synchronization is reduced approximately by the time-bandwidth product of the correlator—in the present instance a factor of the order of 300.

If, however, we are concerned with a code such as the binary PSK, and if we are not concerned with transmitter distortion, then the nonlinear processing must be regarded as an alternative solution to a problem which, as we have seen in Section 4.5.2, can also be solved using phase-coded matched filters, if need be, programmable. The choice between these alternatives is therefore competitive. The transducer-based matched filter has the advantage of not requiring a time-inverted reference; the signal levels required at the input are substantially lower. Against that, the convolver of Fig. 4.36b is inherently a rather simple structure. Even when the precautions against various spurious signals are incorporated [4.98], one is still concerned with a single photolithographic operation. On the other hand, the monolithic programmable matched filter is a product of a highly sophisticated set of technologies. On other performance criteria, such as attainable time-bandwidth products, usable dynamic range, and maximum correlation length, there is not at present a very clear-cut advantage either way. It is likely that both systems will continue to be developed and find systems applications. However, it seems probable that it is in systems where the precise nature of the codes to be used is not established at the design stage that the balance will tend to turn towards the nonlinear convolver. The capability of introducing a memory function which we shall discuss in Section 4.6.2 will open up a completely new set of applications.

There is, however, no doubt that if it were possible to improve the efficiency of the device, i.e., to increase c_3 in (4.37)—and if one could do so without degrading, preferably improving, the dynamic range, that the nonlinear processor would greatly improve its competitive position. This has been one of the aims of resorting to acoustoelectric systems to be described in the following section. However, even within the confines of a purely piezoelectric system, one can make a substantial improvement. We have already seen that surface wave nonlinear interactions are stronger than their bulk-wave counterparts, simply by virtue of the fact that they are guided by the surface, that the width of the beam in a vertical direction is therefore only of the order of a wavelength. If one can also confine the surface wave in a lateral direction, by the use of a waveguiding structure, it is at once clear that one will gain a further factor—roughly comparable to the degree of compression achieved. The first experiments which were performed with this aim were based on the use of a wedge-shaped topographic guide, formed on a sample of PZT [4.101]. The bilinear coefficient for this configuration was of the order of $-50\,\text{dB}$, i.e., an improvement of around 30 dB as compared with the basic convolver on $LiNbO_3$. The experiments were carried out at low frequencies, so that the time-bandwidth product was small. However, it has since been shown that it is readily possible to fabricate a wedge structure on $LiNbO_3$ which operates successfully at 60 MHz [4.102]. There does not appear to be an insuperable obstacle to further extending the center frequency and in this way the attainable bandwidth.

Experiments were also carried out with thin film waveguides [4.103, 104] and at an input frequency of 100 MHz. Recently, very promising experiments were reported [4.105] using a piezoelectric shorting waveguide ("$\Delta v/v$ guiding") on $LiNbO_3$. The beam from the ports 1 and 2 transducers was compressed by a factor of 15 using multistrip couplers (Sect. 7.4) to excite the waveguide. The device had a bandwidth of 50 MHz centered on 156 MHz and an interaction time of 12 μs, so that the time-bandwidth product was 600. The bilinear coefficient referred to the terminals, i.e., including the effect of transduction losses, was 71 dB. The expected improvement arising from the compressed SAW beam was therefore achieved.

A feature of all of the waveguide convolver experiments which have been reported is the relative ease with which spurious signals can be reduced. The reason lies in the very fact that the interaction is in a waveguide—usually a

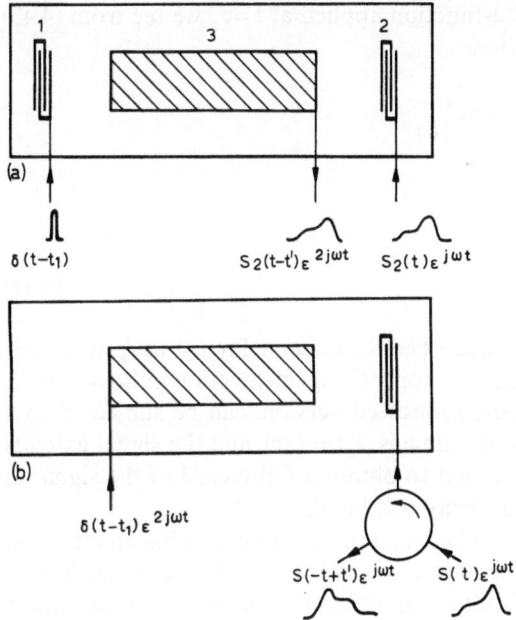

Fig. 4.37. (a) Variable time delay; the output on port 3 is determined by t_1 the instant of application of a pulse to port 1. (b) Time reversal; the delayed time reversed signal is separated from the input signal by a circulator

single-mode waveguide. The excitation of such a guide relies on very specific input field configurations which must be carefully designed. It is therefore not surprising that various spurious waves and in particular bulk waves will normally be only weakly coupled to the guide. In the piezoelectric waveguide structure referred to above, the spurious level was more than 40 dB down on the desired signal.

The use of a thin film waveguide does bring one additional problem—the fact that the system ceases to be totally free from dispersion. If one does not seek to work with the narrowest possible guides, the dispersion can be reduced—but at the expense of allowing more than one mode to propagate, a feature which can in itself give rise to distortion effects. The dispersion can be readily calculated, and its effect on the convolution fidelity is now also understood in detail [4.106]. It appears that dispersion effects would not prevent the attainment of time-bandwidth products in excess of 1000. There is every reason to anticipate that the waveguide interaction will prove to be the preferred configuration for the piezoelectric convolver.

We have so far confined the discussion to the case of the convolver of Fig. 4.36b and the correlator which differs from the convolver only in the nature of the fact that a time-inverted signal is applied to one of the ports 1 or 2. It was, however, appreciated [4.107, 108] that a number of other signal processing operations are possible using the same device. Figure 4.37 shows two examples which have potential practical importance. The first is the realization of a variable delay line; it is implemented by introducing the signal into port 2, and injecting a short pulse in port 1. If we can regard this pulse as sufficiently short so that we can represent it by a delta-function applied at $t = t_1$ we see from (4.43) that the output $V_3(t)$ can be written

$$V_3(t) = \int_0^L \delta\left(t - t_1 - \frac{z}{v}\right) \cdot S_2\left(t - \frac{L-z}{v}\right) dz$$

$$= S_2\left(2t - t_1 - \frac{L}{v}\right)$$

$$\equiv S_2(2t - t_1') \qquad\qquad\qquad (4.47)$$

where in the final version we have absorbed the offset delay L/v in t_1'. It is clear that by varying the moment t_1 when we apply the delta-function pulse to port 1, the output, which is again a time-compressed version, can be subjected to a varying delay. If the total interaction time is $T_0 (= L/v)$, and the signal extends over a time T_s, one can readily see that to obtain a full record of the signal at port 3, the maximum variation of delay will be $T_0 - 1.5 T_s$.

We have taken the impulse applied to port 1 to be a delta-function. In practice one wants this signal to be short so as not to degrade the resolution; at the same time, the demands of efficiency require a large spectral power density in the vicinity of the center of the signal frequency band. These requirements lead to the conclusion that the length of the pulse ought to be a little less than $1/B$ cycles of the signal center frequency; in practice it is possible to obtain good results with a video pulse having a duration of one-half cycle.

A final example of a signal processing function is indicated in Fig. 4.37b, where a signal introduced at port 2 is reflected by the nonlinear process arising from the application of a delta-function pulse to the convolver plate, port 3. We can immediately deduce that a time-reversed reflected signal will arise, by comparison with the case of correlation, and invoking the reciprocity principle. In the correlator we applied a signal to port 2, and its time-inverted complement to port 1. The two contrapropagating waves were then shown to give rise to the correlation of (4.46). The correlation will consist of a short pulse, limited in its sharpness only be the finite time-bandwidth of the device. If then we apply a short pulse to port 3, we will generate the time-inverted contrapropagating wave from the signal S_2. In this case the "delta-function" should, of course, have its maximum spectral power density centered on twice the signal center band frequency. The time-inverted signal is not time compressed or expanded. The system of Fig. 4.37b can therefore be used [4.109] to provide the time-inverted signal needed to effect an autocorrelation.

4.6.2 Acoustoelectric Nonlinear Signal Processing

We have seen that in spite of the inherent weakness of piezoelectric nonlinearity it is possible to achieve signal processing functions which are viable, in the sense of disposing of an adequate signal to effective noise ratio. An associated

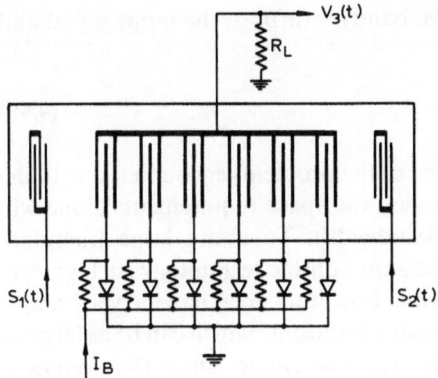

Fig. 4.38. The diode convolver (after [4.112])

inconvenience is the need to work with relatively high input powers, in many cases approaching the 1 W level. The inconvenience appears as a major barrier to usage in the case of certain equipment where weight and energy are at a premium. There is therefore a desire to improve the efficiency of nonlinear signal processing. Beyond this there is the further hope that one might eventually improve also the signal to effective noise ratio. Electronic devices are inherently nonlinear; indeed much of the first half century of electronics has been devoted towards finding means of achieving an acceptable degree of linearity in amplifiers and mixers in spite of this basic fact. It is therefore a natural endeavor to seek to exploit such strong nonlinearity to fulfill our aims. There is a vast literature which bears on this and the related topics of acoustoelectric amplifiers and image detectors, which has recently been portrayed in a detailed review by *Kino* [4.109]. Here we shall confine ourselves to a brief description of devices which show prospects of winning acceptance for systems usage.

The most direct means of introducing electronic nonlinearity [4.110, 111] is by abandoning the continuous convolver plate used in the degenerate interactions, and instead use a sampling transducer in which each finger is connected to a discrete diode, as in Fig. 4.38. The main impact on the theory of the device is that the integral in (4.43) becomes a discrete sum

$$V_3(t) = \sum_{n=1}^{N} S_1\left(t - \frac{z_n}{v}\right) \cdot S_2\left(t - \frac{L - z_n}{v}\right)$$

where   (4.48)

$$N = \frac{L}{\Delta L}.$$

If the sampling is to be sufficiently dense to avoid loss of information, we require two taps per wavelength of the modulation envelope of the output distribution—approximately equal to twice the time-bandwidth product. We can express this result in terms of λ_0, the wavelength corresponding to the center of the input signal band and the relative bandwidth B' of the input signal and find

$$\Delta L \sim \frac{\lambda_0}{2B'}. \tag{4.49}$$

This result immediately highlights one of the problems encountered with the diode convolver. If one uses discrete diodes, the space requirements alone will force the use of a relatively narrow bandwidth. To attain large fractional bandwidth, one will certainly have to use a monolithic technology such as AlN and Si on sapphire. The diode correlator, however, does have a very major attraction in its high efficiency, with bilinear coefficients which can be as large as $-30\,dB$, and are achieved with a large dynamic range. Since the nonlinear interaction takes place outside the acoustic system, the two counterpropagating acoustic beams need not be coincident. They can be arranged to propagate along parallel tracks, with the sampling electrodes spanning both beams. In this configuration it is easy to avoid the second transit effects, by providing an acoustic termination at the end of each acoustic beam. Using such a device [4.112], with 24 Si on sapphire diodes, it was possible to obtain a convolution peak which was free from spurious responses down to the $-60\,dB$ level, i.e., a true 60 dB dynamic range. A further basic merit of the diode correlator lies in the fact that the bias current to each diode can be separately adjusted so as to ensure a high degree of uniformity of nonlinear response along the length of the interaction region.

The separate accessibility of all of the nonlinear elements—the diodes—also leads to a further possibility. One can regard each of the diode terminals as a possible distinct signal port. In this way, one can realize devices which have some serial inputs—the interdigital transducers at each end of the device, and a set of ports which permit parallel access. Two applications which exploit this feature have been described. In one [4.113], a correlation operation is realized by the use of integration in *time*, using a conventional electronic integrator attached to each diode.

In this system, the length of the sequences which can be correlated depends only on the capability of the electronic integrator, not on the physical length of the acoustic device. The aim of a further example [4.114] of a device which makes use of the multiple diodes as ports is the realization of a Fourier transformer. The signal to be transformed is presented to the device in parallel form, with each sample controlling the bias of one of the diodes. Using this configuration with 32 diodes, a real-time transform with 10 MHz bandwidth was obtained. It may well prove that the parallel input feature of the diode correlator when assessed in the light of its high inherent efficiency will lead to increasing usage.

The diodes in the convolver can be operated in forward or reverse bias. In the latter case, the nonlinear behavior is reactive, comparable to that which one associates with varactor diodes. In the simplest terms, one can look on this nonlinearity as arising from a two-step process: if we apply a voltage V_1 to the

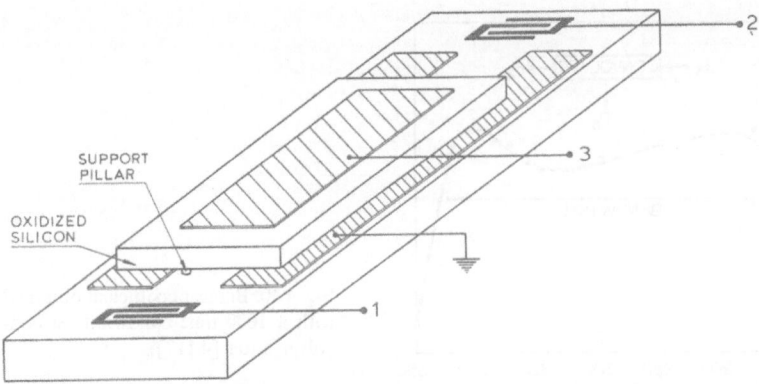

Fig. 4.39. The silicon convolver. The silicon slice is supported on a random array of pillars so that the LiNbO$_3$-Si spacing is a few tenths of a micron

diode circuit, a portion of this voltage, kV_1, will appear across the diode itself. The value of k will depend on the diode capacitance C, being the larger the smaller the value of C. The presence of the voltage kV_1 will lead to a proportionate change in the capacitance ΔC, thus resulting in a voltage component across the diode proportional to $\Delta C \cdot V_1$ and hence to V_1^2. This then is the source of the nonlinearity which is exploited in the diode convolver.

We have seen that the number of diodes which can be incorporated, and hence the time-bandwidth product, is severely limited by the difficulty of making the necessary connections. An escape from these constraints is provided by the use of a continuous silicon slice which is brought into close proximity to the piezoelectric surface. The piezo field can then act *directly* on the semiconductor—the need for connections is avoided. The basic configuration [4.109] is shown in Fig. 4.39. The normal component of the electric field is the effective interaction component, the thickness of the oxidized silicon slice being such that the longitudinal component is very nearly shorted out. If the local value of the electric field is $E_y(x)$, and if for the moment we assume that the surface is in the flat band state, a simple application of Poisson's equation leads to the well-known result that the voltage drop across the depletion region is $\Phi(x)$,

$$\Phi(x) = \frac{\varepsilon E_y^2(x)}{2qN_D} \tag{4.50}$$

where q is the electronic charge and ε and N_D the silicon dielectric constant and doping density. The effect of the uniform metal electrodes will be to produce an output voltage which is the integral of $\Phi(x)$, i.e., it will act very similarly to the uniform convolver plate in a degenerate piezoelectric convolver.

While it is intuitively clear that the depleted surface case is the one which will lead to the highest output voltage, it is in practice by no means easy to control the surface state densities so as to achieve this situation, and, even more importantly,

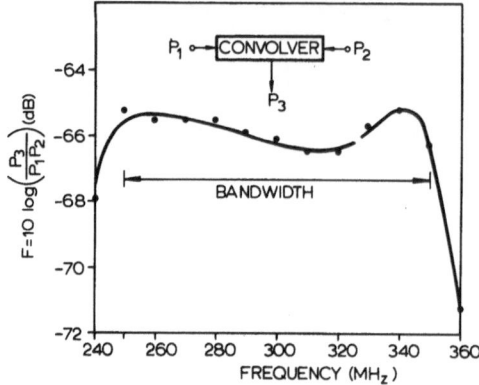

Fig. 4.40. Bilinear coefficient obtained with a 1000 time-bandwidth Si convolver (after [4.117])

to ensure that the conditions do not vary along the length of the convolver. Even if one should have achieved the condition by carefully processing of the silicon sample and control of the oxide growth, the situation can be perturbed by local static electric fields which can be generated during assembly of the structure. Again, it is the variation of such fields along the length which is particularly damaging to the performance. This problem can be avoided by the use of a very thin oxide layer under conditions which lead to moderately large densities of relatively deep, and hence slow, surface states. The effect of these states is that the system is now very insensitive to small local DC electric fields. The response to the high-frequency fields with which we are concerned in the operation of the convolver is, however, very little impaired [4.109]. Using this approach, it has become possible to build such silicon convolver systems with a remarkable degree of uniformity.

Detailed analysis shows that the silicon sample must be spaced by a distance from the surface of the piezoelectric material which is typically of the order of one twentieth of the Rayleigh wavelength. A very successful approach to solving this clearly formidable technological problem has been developed: the piezoelectric sample is ion beam etched so as to leave a series of pillars [4.115], whose position and size have been photolithographically defined. The height of the pillar is determined by the required spacing. The diameter is made less than a quarter wavelength, and the pillars are disposed in a random pattern with spacings which are several tens of wavelength. The inevitable scattering of the surface waves can be reduced to insignificance, while the spacing is maintained with great accuracy.

As an example of the performance which can be achieved with such a silicon convolver, a device with a time-bandwith product of 1000 has recently been demonstrated [4.117]. The center frequency was 300 MHz, and the interaction time 10 μs. The bilinear coefficient is shown as a function of frequency in Fig. 4.40 and is seen to vary by less than 1 dB from − 66 dB over the complete 100 MHz bandwidth. A noise-limited dynamic range of 50 dB was achieved, and the spurious signals were sufficiently low so as not to degrade this figure. These results are impressive; they are somewhat superior to the best which has been obtained using direct piezoelectric convolvers. However, the improvement in performance is not dramatic, particularly when compared with the waveguide-based piezoelectric convolvers [4.105]. Against this there is clearly a very large increase in the technological complexity involved. The role of the silicon convolver in competition with the simpler piezoelectric devices must therefore remain an open question, as long as the functional purposes are comparable. It is only where the acoustoelectric system offers unique operational features that one can confidently expect its continuing development. In fact, there is a function that the silicon-based device can perform which is in this class—that of combining memory with the other signal processing functions. It is, as we will see, a feature which can greatly widen the scope of the nonlinear processing operations which can be performed.

4.6.3 Acoustoelectric Processing with Inherent Memory

An intense acoustic surface wave propagating on $LiNbO_3$ generates piezoelectric potentials which can reach several hundred millivolts. Conversely, if one could find a means of depositing a charge distribution on the surface of the sample which gave rise to potentials of this order, the rapid removal of this charge would generate an acoustic wave whose initial distribution would correspond to the charge-induced stress pattern. These basic considerations inspired a remarkable experiment [4.116] involving the use of an electron beam in conjunction with a surface wave device. An electron beam of a few hundred volts energy floods the surface of the acoustic sample for a period which is short compared to the rf cycle. The secondary electrons which are released have energies which are comparable to the piezoelectric potentials produced by the acoustic surface wave, and are deflected towards the positively charged regions. This charge distribution is then a record of the acoustic field distribution at a particular moment. It will gradually leak away, but the time constants can easily be minutes. If the sample is then exposed to an electron beam pulse, the charge distribution will tend towards uniformity. The piezoelectrically induced stresses are relieved, and thereby launch a pair of waves, one of which proceeds in the original propagation direction, and provides a delayed output; the other will propagate in the reverse direction and hence provide a time-inverted version of the original signal. The device provides long-term memory as well as at least one important signal processing function.

In this form, where the acoustic device comes inextricably attached to the paraphernalia of vacuum electron beam apparatus, the concept is hardly likely to find acceptance in systems. However, it was realized by the original investigators [4.118], and by a number of other investigators [4.119, 120], that one could envisage solid-state systems embodying very similar concepts. The earlier experiments used the basic type of silicon convolver structure of Fig. 4.39, the storage mechanism being based on charge storage in surface states. While these experiments demonstrated the principle, it proved difficult to control the uniformity of the storage along the length of a sample. More recently, it has been found that one can obtain immensely better control over uniformity and other characteristics of the storage process by using arrays of discrete diodes [4.121, 122].

The basis of the storage mechanism in this case is inherently simple. Consider a Schottky barrier diode in the vicinity of the silicon sample. Under normal conditions there will be a depletion region which will have a potential drop whose magnitude depends on the contact alloy used, and can amount to almost a volt. Under these conditions the potentials which attend the propagation of the surface wave on the piezoelectric substrate will not lead to any charge transfer across the depletion region, though they will, of course, modulate the thickness of the depletion region. If, however, we apply a short pulse to the silicon sample so as to bias it positively, charge will flow to the diode contact, and the amount transferred will be modulated by the total instantaneous potential, which will include a component due to the acoustic wave. If the positive biasing pulse is short as compared with the inverse frequency of the acoustic signal, the amount of charge transferred will be proportional to the amplitude of that signal. When the biasing pulse has been removed, the diode reverts to its nonconducting state, but with a remanent bias which is a record of the acoustic wave sample. Eventually, the charge thus deposited on the diode contact will leak away. However, it is possible to have decay times of the order of 100 ms. In contrast, the writing time can be exceedingly short—certainly as low as 0.1 ns.

Since the nonlinear conversion of a diode will depend on its state of bias, it is clear that the record of the acoustic sample which has been imprinted can be subsequently recovered in the form of the nonlinear conversion efficiency of the diode. If we now incorporate a two-dimensional array of diodes on the silicon sample, the whole of the structure becomes capable of providing a spatial memory. In one set of experiments, *Ingebritsen* [4.123] has used an array of diodes with 4 μ contact diameter, spaced at 5.8 μ centers. The contact material was platinum-silicate on 25 Ω-cm material. Using this silicon system, a usable time-bandwidth product of 140 was achieved with a 30 MHz bandwidth centered on 145 MHz.

Suppose that we inject a signal, $F(t)\exp(i\omega t)$, into port 1 of Fig. 4.41. If we now apply a short forward-biasing pulse at a time t_1 to the silicon diode array, the charge stored will be a sampled version of $Q(z)$ where

$$Q(z) = F_0\left(t_1 - \frac{z}{v}\right) \cdot \cos \Phi \tag{4.51}$$

and $\Phi = (\omega t_1 - \beta z)$.

If the relative bandwidth of the signal is B', the writing pulse can be periodically repeated up to $1/B'$ times without loss of significant information, i.e., instead of a single short pulse, one can use a modulated rf pulse at frequency ω, with up to $1/B'$ cycles.

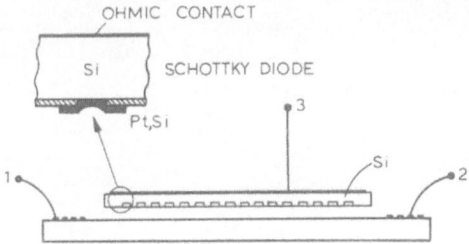

OHMIC CONTACT

Si   SCHOTTKY DIODE

Pt,Si

Si

1   3   2

Fig. 4.41. The Schottky diode array convolver system. The diodes are typically centered on 0.4 λ spacings

The $\cos \Phi$ term in (4.51) can be interpreted as a standing wave of zero frequency; in terms of vectors in the (ω, β) plane it can be regarded as being sum of two vectors $(0, \beta)$ and $(0, -\beta)$.

If, having stored the signal $F(t)$, we subsequently apply a signal $G(t) \exp i\omega t$ to the same port 1, following the formulation of (4.43), we deduce that the output signal from port 3 will take the form $S_0(t)$

$$S_0(t) = e^{i\omega t} \int F\left(-\frac{z}{v}\right) \cdot G\left(t - \frac{z}{v}\right) dz \tag{4.52}$$

where it is assumed that the silicon sample is long enough to encompass the complete nonlinear interaction between the two distributions, a condition which implies that the sum of their spatial extents must be less than the length L of the silicon sample. It is seen that the output is the convolution of the recorded signal $F(t)$ with the reading signal $G(t)$. The output in this case is *not* transformed in frequency. Moreover there is no time compression. Both conclusions derive from the fact that one of the two interacting distributions is static. The interaction which takes place is shown in the (ω, β) plane in Fig. 4.42a. Only the $(0, -\beta)$ component of the stored distribution is utilized; the other component will give rise to a distribution having a periodicity of $\lambda/2$ to which the silicon convolver plate is extremely insensitive. For the special case when $G(t)$ is a delta function—or at any rate, an rf pulse which is short as compared to the inverse bandwidth—the output will be a replica of the function $F(t)$. The storage time can range from the time delay of the device to several tens or even hundreds of milliseconds.

It is also possible to read the stored information from the opposed port 2. In this case, the output will take the form

$$S_0(t) = e^{i\omega t} \int F\left(-\frac{z}{v}\right) \cdot G\left(t + \frac{z}{v}\right) dz \tag{4.53}$$

where we have again assumed adequate length of the silicon sample. In this case, the required length is the spatial extent of G, or F, whichever is the greater. Equation (4.53) shows that in this case we obtain the convolution between $F(t)$ and $G(t)$ again without frequency or time transformation. For the particular case

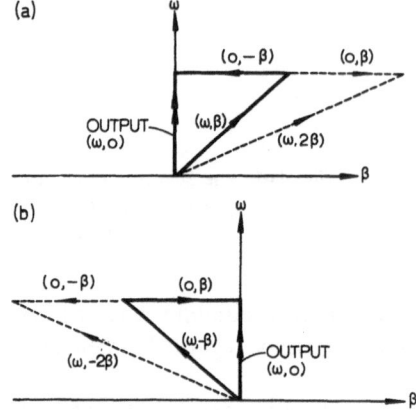

(a)

(b)

Fig. 4.42a and b. Interaction process in reading a memory correlator. Note that the output is at the signal frequency. (a) Correlation. (b) Convolution

when $G(t)$ is sufficiently short, the output will be the time-inverted version of the stored signal—available again at any time up to the storage limit. The nature of the interaction is indicated in Fig. 4.42b; it is seen that in this case it is the $(0, \beta)$ component which comes into play.

It is important to appreciate that the method of readout is nondestructive. Essentially one is measuring the width of the diode depletion region; no charge is transferred across it. The nondestructive nature of the mechanism has been confirmed by reading a distribution up to 10^5 times without observing any degradation of the information [4.123].

Rather similar results can be obtained with $P-N$ junction diodes [4.122, 124]. Using these it is possible to obtain even longer storage times— exceeding 10 s. However, the mechanisms that limit the speed of response of such diodes in normal circuits apply equally in this situation; the writing time is of the order of 0.5 μs—2 orders of magnitude longer than for the Schottky case. It has been shown [4.122] that this does not necessarily limit the bandwidth of the information which can be written; it does involve some considerable additional complexity in processing. It seems probable that Schottky diodes will be the preferred medium for all but the longest storage-time applications.

A further development has led to the capability of superimposing stored information on the array [4.125]. In the basic form of the diode system, one reads the local instantaneous electric field; the time constants of the diode are so fast that any knowledge of its previous state is forgotten. The key step in overcoming this problem was to deposit a layer of polysilicon onto the diode alloy contact. The polysilicon-alloy system acts effectively as another diode, in series with the Schottky diode. It is by all normal standards a very poor diode, with time constants in the reverse-biased state of the order of microseconds, perhaps 5 orders of magnitude shorter than for the Schottky diode itself. The effect in this situation is, however, altogether desirable. The charge transferred onto the diode alloy contact when reading a distribution F_1 leaks, in a matter of a microsecond, into the polysilicon layer. If now we introduce a second signal F_2, the amount of charge transferred across the depletion region will record the distribution of that signal. The reading period is too short to allow a significant portion of the charge from F_1, distributed within the polysilicon, to reach the diode alloy contact. After the second reading, the charge stored in the polysilicon will be a measure of the sum of F_1 and F_2. It is important to note that the presence of the $\cos \Phi$ term in (4.51) implies that the individual record, and hence also the addition, preserves phase information. The fact that one can thereby record and subsequently process the summation of a number of signals clearly enhances the potential processing power of the system.

Acoustic surface beams are typically several tens of wavelengths in width. Since, to avoid sampling errors, we must have at least two diodes per wavelength in the propagating direction, a typical design will involve over 1000 diodes for each sample to be stored. This fact is of great assistance in manufacture, as a small percentage of inoperative diodes in the array will have no measurable effect. Conversely, one is led to enquire as to whether one could not make additional use of the array by, in some manner, coding the signals to be stored. There are in fact a number of approaches: one can resort to the use of transducers which are phase coded normally to the direction of propagation. In this way it has been shown [4.126] that one can superimpose and separately read information corresponding to a number of distinct lateral codes. Alternatively, one can use a number of distinct acoustic beams inclined at small angles to each other [4.127]. The isolation which one can achieve by either of these two methods is analogous to that in an optical holographic store. It has therefore been suggested that such acoustic holographic stores might be used as a high density analog serial store.

These developments are in their earliest stages; however, the competition from established semiconductor storage techniques—notably CCD—is undoubtedly fierce. On the other hand, in the field of signal processing the ability separately to write, and subsequently to process, the recorded signals is likely to lead to an important new class of applicable signal processing techniques.

4.7 Multiport Acoustic Devices

The simplest surface wave devices, such as the basic delay line structure, utilize a single linear acoustic propagation path. The transducers have one electrical and two acoustic ports. However, the delay line can be thought of as a simple two-port, as in Fig. 4.43a, in that two of the acoustic ports arise only from the fact that the transducers are bidirectional; if we utilize one of the techniques for making them undirectional, they effectively disappear. In other cases, one is likely to minimize their effect by absorbing the acoustic energy radiated. If we progress to a transversal filter, or to a convolver with transmitting transducers on either end, the filter structure forms one additional electrical port, coupled to the acoustic system, as in Fig. 4.43b.

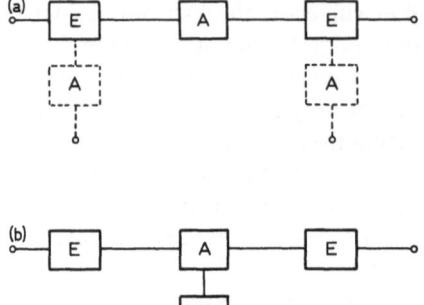

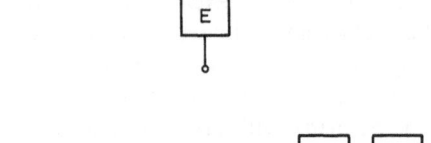

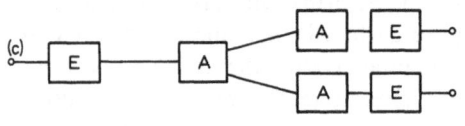

Fig. 4.43a–c. Structure of acoustic surface wave devices. (a) Simple delay. The transducers have two acoustic ports but only one is used. (b) Filter structure. The acoustic system has one additional electric port. (c) Acoustic multiplexer. Additional acoustic ports form integral parts of system

Pulse compression filters of the reflective array type to involve acoustic paths in which some of the energy is reflected through two right angles at a position depending on the frequency of the wave, while the remainder proceeds without reflection. In this sense, we could identify several acoustic ports in such a device. However, the energy which is not subjected to the double reflection does not serve any useful purpose—strenuous efforts are made to absorb it. Operationally, therefore, the system is still described by the configuration of Fig. 4.1a, though the acoustic path has been made to appear dispersive. Clearly, it is altogether possible selectively to reflect acoustic waves using grating structures, so as to realize a multiport acoustic structure, as in Fig. 4.43c. Power dividers based on this principle have been described [4.128]. It is at once clear that structures in this class can perform a richer variety of functions, such as, for example, channel separation.

One way of bifurcating an acoustic path is by the use of a partially reflecting grating formed either by metallization or by groove fabrication. There is, however, another way in which we could generate additional acoustic paths, indicated in Fig. 4.44. The interdigital transducer R_1 which is electrically connected to T_2 will transform some of the acoustic energy into electrical form and thus invoke the re-radiation of acoustic energy from T_2. The fact that the electrical signals travel with negligible delay implies that we can regard the combination of R_1 with T_2 as a means for the direct transfer of acoustic energy from one channel to another. The basis for adopting an interdigital transducer structure, with two fingers per wavelength, rests on the simplification in the driving circuitry when only 0 and π phases are required. Three phase transducers are nevertheless used [4.3] to achieve a unidirectionality. However, in the particular situation of Fig. 4.44, there is *no* external electrical circuit attached to

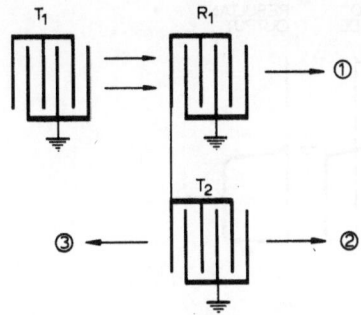

Fig. 4.44. Formation of additional acoustic ports by interconnection of transducers

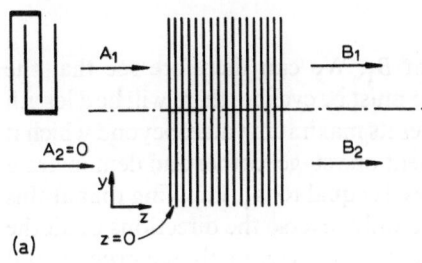

(a)

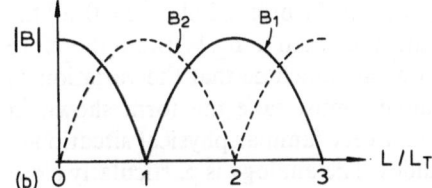

(b)

Fig. 4.45a and b. Multistrip coupler. (a) Basic configuration. (b) Magnitude of output as function of coupler length

either R_1 or T_2. The phase of the voltages induced in the fingers of R_1 is governed simply by their position. There is no reason why we should not be able to deposit more than two fingers per wavelength; the phase change between fingers will depend simply on their spacing. We arrive then at the remarkably simple structure shown in Fig. 4.45a. The array of conductors will still lead to a lateral transfer of acoustic energy. If we have more than two fingers per wavelength, the radiation will no longer be bidirectional. Indeed, one can at once see that, since the sequence of phases in the lower channel corresponds exactly to that produced by the acoustic wave travelling in the $+z$ direction in the upper channel, the new wave will also be launched preferentially in the $+z$ direction. The structure was named a multistrip coupler (msc) by its inventors, *Marshall* and *Paige* [4.129].

4.7.1 Multistrip Coupler—Basis of Operation

It is clear that if the incident surface wave amplitude is A_1, the output amplitude in the top channel B_1 will be less than A_1, since some of the energy has been used to launch B_2. As the wave B_2 grows with distance, it will react back to the first

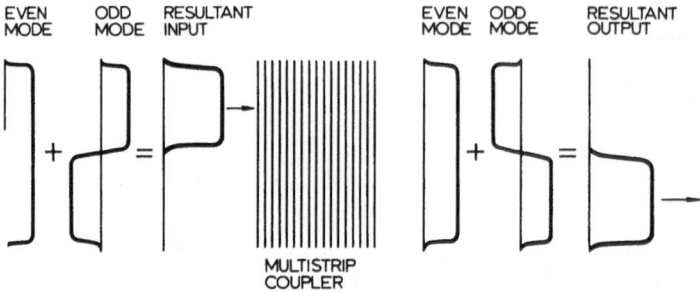

EVEN   ODD   RESULTANT   EVEN   ODD   RESULTANT
MODE   MODE   INPUT   MODE   MODE   OUTPUT

MULTISTRIP
COUPLER

Fig. 4.46. Operation of a total transfer multistrip coupler

channel and re-radiate so as to augment B_1. We can therefore see that the transfer of energy from channel one to two must be cyclic—there will be a length of multistrip coupler L_T at which B_2 reaches its maximum value, beyond which it will start to decay. We can take the argument one stage further and demonstrate that the peak value of B_2 (neglecting losses) is equal to A_1, implying that at this point B_1 must be zero. To see this we need only reverse the directions of all the waves, and think of B_2 as incident on the msc. We can apply reciprocity, and conclude that A_2 will again be zero. Therefore, at the original plane $z = 0$, all the energy has to be in the upper channel: A_1 must be equal to B_2. Without therefore having to write down a single equation, we can conclude that the variation of amplitude with distance in the two channels must take the form shown in Fig. 4.45b. We can recognize in this system a very familiar physical situation— the coupling of two synchronous waveguides. The analogy is particularly close to microwave directional couplers realized by the uniform coupling of two identical waveguides.

To calculate L_T we must obtain some estimate of the strength of the coupling between the guides. Here one can adopt an approach—also widely used in microwave systems—of describing the interaction in terms of the normal modes of the system. It is of course far from obvious just what form the normal modes of such a structure will take [4.130, 131], or even whether regarding the msc as a single waveguide, we can express the fields entirely in terms of nonleaky waveguide modes. However, in SAW technology it is usual to work with relatively wide beams, so that the amplitude distribution is approximately rectangular. In this situation, the two lowest normal modes will approximate the form in Fig. 4.46: a symmetrical mode and an asymmetrical mode. If, as in Fig. 4.45a, we have initially an amplitude A_1 incident on the top channel and $A_2 = 0$ on the bottom channel, we can see this distribution as resulting from the superposition of the two normal modes, each having an amplitude of $A_1/2$. The two modes will have a slightly different velocity, so that after a distance L_T there will be phase slippage of π. The total distribution is now formed by the subtraction of the two normal modes, resulting in the total transfer of power to B_2, which we have already anticipated. The length L_T is therefore governed by the difference between the phase velocity v_e, v_0 of the even and the odd normal modes. Specifically,

$$\left| \frac{\omega}{v_0} - \frac{\omega}{v_e} \right| L_T = \pi$$

so that

$$L_T = \frac{1}{2f} \frac{v_0 v_e}{v_e - v_0}. \tag{4.54}$$

If $v_e - v_0 \equiv \Delta v$ and $\Delta v / v_e \ll 1$

$$L_T = \lambda \left(\frac{v}{2 \Delta v} \right). \tag{4.55}$$

We can arrive at a very simple estimate for Δv, if we take the limiting case when the period of the grating is very much less than λ. The grating will then act as an inhomogeneously conducting sheet, with zero conductivity in the z direction and infinite conductivity in the y direction. Now, we know that the effect of a homogeneously conducting layer is to reduce the velocity of a surface wave by an amount which is proportional to K^2, the effective electromechanical coupling constant. The effect is due to the short circuiting of the longitudinal electric field associated with the surface wave. It is helpful to look on the shorting action as arising from a charge distribution induced in the film which is just such that it cancels the electric field which would obtain in its absence. The charge distribution is maintained by longitudinal currents which at any instant change direction every half-length.

The even normal mode will have electric fields in the z direction and fields normal to the propagating surface (i.e., the x direction); it will not, however, have a significant field in the y direction, except near the edges. To a good approximation, therefore, we can expect the even mode to propagate, in ignorance of the presence of the msc.

The situation is, however, quite different for the asymmetrical mode. The charges induced in the fingers in the top channel are equal in magnitude and opposite in sign to those induced in the lower channel. Since, however, the two channels are joined by a perfect conductor, transverse currents can flow between them. Though the situation is not identical to that arising in the case of a uniform metallic film, where the currents flow in the longitudinal direction, the effect is very nearly identical. The asymmetrical wave velocity will therefore be very nearly that associated with a metallized surface [4.132], and

$$\frac{\Delta v}{v} \simeq \frac{1}{2} K^2 \tag{4.56}$$

so that

$$L_T \simeq \frac{\lambda}{K^2}. \tag{4.57}$$

For LiNbO$_3$, the value of K^2 is approximately 0.05, which would suggest that L_T is of the order of 20λ. In practice, for reasons which we shall briefly examine, the coupling length is typically twice as great. For a low coupling-constant material such as quartz, (4.57) would predict a coupling length of $500\,\lambda$; it is clear that the msc technique is practical only for strongly piezoelectric materials.

The estimate of the coupling length which we have obtained is based on the model of an inhomogeneously conducting sheet, while in practice we have a finite grating period d. If, as is normally the case, the finger width is $d/2$, we see at once that we are coupling from one channel to the other only half of the time—suggesting that the coupling length will be doubled. In addition, there is, however, another important effect. An individual finger will transmit a voltage which will be an average of that which would obtain over its longitudinal extent $d/2$. If the amplitude at the leading edge of the strip is $A\exp(i\theta/2)$, that at the trailing edge will be $A\exp(-i\theta/2)$, where $\theta = \omega d/v$.

The amplitude coupled per unit angle is then

$$A(\theta) = \left| \frac{A(e^{i\theta/2} - e^{-i\theta/2})}{\theta} \right| = A\frac{\sin\theta/2}{\theta/2}. \tag{4.58}$$

The reduction factor clearly applies in the conversion from an acoustic to an electric signal in the top channel, and again when the electric signal is converted to an acoustic signal in the lower channel. The total length L_T is therefore increased by the square of this factor, as well as the factor of 2 arising from the partial coverage of the surface

$$L_T = \frac{2\lambda}{K^2} \left(\frac{\theta/2}{\sin\theta/2} \right)^2. \tag{4.59}$$

In our discussions so far we have tacitly assumed that the msc does not give rise to appreciable reflections. This will normally be true, except when the grating period is very close to $\lambda/2$. At this point, we encounter a stop band, which is of course precisely the phenomenon exploited in the realization of acoustic resonators. In multistrip couplers, it is however a region which is avoided; normally, msc devices are operated at frequencies ranging from perhaps $0.3f_0$ to $0.9f_0$, where f_0 is the stop band frequency. Within this range, detailed experiments [4.132] have indicated that (4.59) provides an excellent guide to the design of the correct coupling length. More detailed theories based on equivalent circuits [4.132] or on perturbation theory [4.133–135] have been obtained and confirm the essential correctness of this basic result.

Multistrip couplers are useful in applications where only part of the energy is transferred from one channel to the other. Furthermore, we shall be concerned with the phase of the coupled wave as well as with the amplitude. We must therefore obtain specific expressions for the distribution of the amplitudes of the waves. It is simplest to express these in terms of the normal mode amplitudes P_O and P_E for the odd and even modes, respectively, and their propagation constants β_O and β_E. If the modes take the form shown in Fig. 4.46, the amplitude in the upper channel is A_1 and that in the lower channel A_2, where

$$A_1(x) = P_E e^{i\beta_E x} + P_O e^{i\beta_O x}$$
$$A_2(x) = P_E e^{i\beta_E x} - P_O e^{i\beta_O x} \qquad (4.60)$$

or

$$A_1(x) = e^{i\frac{\beta_E + \beta_O}{2}x}\left(P_E e^{i\frac{\beta_E - \beta_O}{2}x} + P_O e^{-i\frac{\beta_E - \beta_O}{2}x}\right)$$
$$A_2(x) = e^{i\frac{\beta_E + \beta_O}{2}x}\left(P_E e^{i\frac{\beta_E - \beta_O}{2}x} - P_O e^{-i\frac{\beta_E - \beta_O}{2}x}\right). \qquad (4.61)$$

We are primarily interested in the amplitudes and relative phases of the waves in the two channels emerging after traversing a length L of the multistrip coupler. We are not usually concerned about the total delay which is described by the $\exp[i(\beta_E + \beta_O)x/2]$ factor, which we shall therefore suppress. For the case $P_E = P_O$, (4.61) shows that power transfer will be complete when

$$\frac{\beta_E - \beta_O}{2} L_T = \frac{\pi}{2}. \qquad (4.62)$$

It is useful to express the relationships in terms of the phase angle ϕ

$$\phi \equiv \frac{\beta_E - \beta_O}{2} x = \frac{\pi x}{2L_T}. \qquad (4.63)$$

With this notation, and writing the incident amplitudes $A_{1,2}(O)$ as $A_{1,2}$, and the emerging amplitudes $A_{1,2}(x)$ as $B_{1,2}$, (4.61) can be written in the form

$$\begin{vmatrix} B_1 \\ B_2 \end{vmatrix} = \begin{vmatrix} \cos\phi & -i\sin\phi \\ -i\sin\phi & \cos\phi \end{vmatrix} \begin{vmatrix} A_1 \\ A_2 \end{vmatrix}. \qquad (4.64)$$

The particular case already considered of an input in channel one only ($A_2 = 0$) then becomes

$$B_1 = A_1 \cos\phi$$
$$B_2 = -iA_1 \sin\phi. \qquad (4.65)$$

4.7.2 Full Transfer Multistrip Coupler

If we make the coupler length L_T, $\phi = \pi/2$, we achieve the total transfer from one channel to the other. There is an immediate application for this basic device in the suppression of bulk waves. The extent to which a transducer generates bulk waves depends on the material, the cut used, and also, very strongly on the bandwidth. For transducers having very few finger pairs so as to achieve a large bandwidth, the bulk-wave generation can easily be responsible for the largest spurious components in the output. If a full transfer msc is disposed in front of the transducer, it is only the surface wave component which is transferred to the second channel; the bulk waves will continue to propagate without being appreciably affected by the presence of the strips. It has been shown that in this way one can obtain an almost complete separation of the bulk waves [4.132].

The synthesis of filter structures requires some means of apodization of the transducers, which is normally effected by finger length weighting. In a typical design there will then be some finger pairs which overlap only a few wavelengths. The diffraction from these narrow sources is very much larger than from the wider sources where the overlap is greater. The variation of diffraction can be taken into account, but, for a number of reasons, it is preferable to minimize this effect. The use of a full transfer coupler can help in this respect [4.136]; if the filter structure is placed in one channel, the acoustic wave transferred to the second channel will extend at least very nearly over the full width of the second channel. The unequal diffraction from different sources is therefore confined only to the distance between the sources and the msc. It must be admitted that to characterize this application in a quantitative manner would require a more rigorous modal analysis. However, experimental results do confirm that the effect is, at least qualitatively, correctly rendered by the simple picture we have presented.

4.7.3 Partial Transfer Multistrip Coupler

It is clear from (4.65) that by choosing the length of the msc we can couple a predetermined portion of the acoustic energy into the second channel. Of particular importance in a number of applications is the case of the even power splitter, $\phi = \pi/4$,

$$B_1 = A_1/2^{1/2}$$
$$B_2 = -iA_1/2^{1/2}. \tag{4.66}$$

Such a device can be used, for example, to split the acoustic path from the transmitting transducer, and, by placing the receiving transducer at different distances from the msc, generate two distinct delays. The process can of course be repeated so as to obtain any even number of delays.

A particularly important device, based on a 3 dB power-splitting msc, is the reflector. It consists essentially of such an msc; folded in the manner shown in Fig. 4.47a. If the curved portions of the coupler are short as compared with the straight sections, the device is equivalent to the developed straight msc shown in Fig. 4.47b. The four ports of the coupler have been numbered to aid identification. From this it is seen that the inputs to the various ports arising from the folded geometry are as indicated. We assume that we have an input on port 1, A_1, and no input on port 3. The output from port 2 is $A_1/2^{1/2}$ and, by

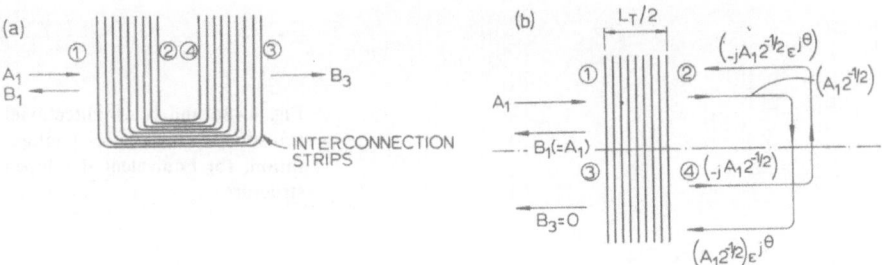

Fig. 4.47a and b. Multistrip coupler coupler reflector. (a) The interconnection links are normally semicircular and involve a relatively small portion of the total strip length. (b) Equivalent developed structure, showing the signal paths

virtue of the folded geometry, after a phase delay θ is incident on port 4. Similarly, the wave emerging from port 4 is $-iA_i/2^{1/2}$, and it suffers an identical delay before becoming incident onto port 2. If then we apply (4.66) again to the final outputs from ports 1 and 3 in response to these secondary inputs, we find,

$$B_1 = \frac{-i}{2^{1/2}}\left(e^{i\theta}\frac{A_1}{2^{1/2}}\right) + \frac{1}{2^{1/2}}\left(-ie^{i\theta}\frac{A_1}{2^{1/2}}\right)$$

$$= -ie^{i\theta}A_1$$

$$B_3 = \frac{1}{2^{1/2}}\left(e^{i\theta}\frac{A_1}{2^{1/2}}\right) - \frac{i}{2^{1/2}}\left(-i\frac{A_1e^{i\theta}}{2^{1/2}}\right) = 0. \qquad (4.67)$$

We see that the device acts as a perfect reflector. In practice, there are of course losses, associated with the ohmic resistance of the strips and further with the generation of acoustic waves by the curved portions of the folded transducer. Moreover, the operation of the device can be perfect at a single frequency only, the frequency dependence entering through the dependence of L_T on frequency [(4.59)]. Nevertheless, it is possible to achieve a reflection loss of less than 2 dB over a 60 % bandwidth.

The existence of such a surface wave reflector might suggest its use in the realization of a unidirectional transducer. If we place such a reflector on one side of a bidirectional transducer, we convert the transduction system from a three-port (two acoustic, one electrical) to a two-port (one acoustic, one electrical). In principle, it must then be possible to find matching conditions which would lead to total power transfer from the electrical to the acoustic system. The disadvantage of this approach is that since the msc is several tens of wavelengths thick, the system would inevitably turn out to be narrow-band. In fact, a far more elegant solution to the problem has been presented [4.137]; it is to place the transducer within the two arms of the reflecting msc, as shown in Fig. 4.48a. If the amplitude of the transducer-generated wave is A, the waves incident on ports 2 and 4 will be equal in amplitude but subject to phase delays which depend on the

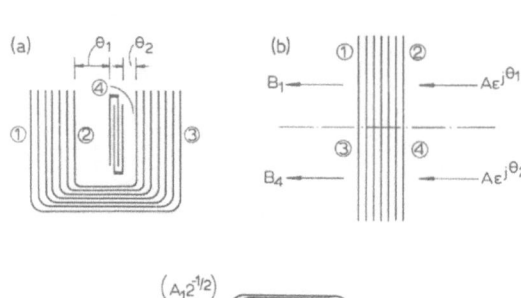

Fig. 4.48a and b. Unidirectional transducer. (a) Basic configuration. (b) Equivalent developed structure

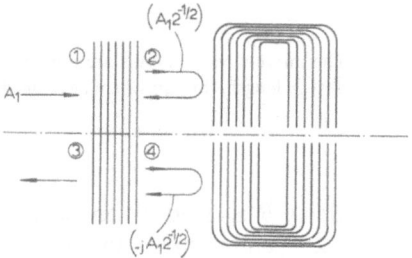

Fig. 4.49. Reflecting track changer

position of the transducer within the coupler. Using the developed version of the coupler as shown in Fig. 4.48b, the waves emerging from ports 1 and 3 are given by

$$
\left.
\begin{aligned}
B_1 &= \frac{A}{2^{1/2}} (e^{i\theta_1} - i\, e^{i\theta_2}) \\
B_3 &= \frac{A}{2^{1/2}} (e^{i\theta_2} - i\, e^{i\theta_1})
\end{aligned}
\right\}
\tag{4.68}
$$

If we make

$$
\left.
\begin{aligned}
\theta_1 - \theta_2 &= \pi/2, \\
B_1 &= e^{i\theta_1} 2^{1/2} A \\
B_3 &= 0
\end{aligned}
\right\}
\tag{4.69}
$$

so that we achieve perfectly unidirectional transduction. The fact that the emerging wave has an amplitude $2^{1/2}$ times larger than that produced by the interdigital transducer is a result of the addition of the two waves initially travelling in opposite directions. The total power is proportional to $(A2^{1/2})^2$, i.e., the $2A^2$ which one obtains from the sum of the two waves produced by a simple transducer. Using such transducers, it has been possible to realize delay lines with just over 2 dB of total insertion loss at the center of the band. The technique is particularly applicable to transducers having relatively few finger pairs. The attempt to apply it to long apodized transducer structures would involve a number of difficulties, including the large portion of the total length of the msc which would be used in the curved part of the structure.

A single msc reflector will redirect an acoustic wave back towards its source. However, a combination of an equal power split msc, used in conjunction with two msc reflectors, can fold an acoustic path [4.138] as shown in Fig. 4.49. Here

$$B_2 = A_1/2^{1/2}, \qquad B_4 = -iA_1/2^{1/2}. \tag{4.70}$$

This time these waves are totally reflected, so that, other than for a common factor, $A_{2,4} = B_{2,4}$. Applying the coupler equations (4.66) once more, we find $B_3 = A_1$, $B_1 = 0$. The wave has been totally reflected, but propagates is an adjacent channel. This procedure can be repeated many times to realize a long delay line, as already discussed in Section 4.2 (Fig. 4.5f). In practice, such reflecting track changers can have less than 3 dB loss over a 50 % bandwidth [4.139].

4.7.4 Asymmetrical Multistrip Coupler for Beam Compression

The width of acoustic surface wave beams used in practical systems is determined not only by the need to avoid excessive diffraction, but also by the requirement that the transducer input impedance can be readily matched to a 50 Ω system. If we resort to the use of a waveguide, this latter consideration alone remains. An example is given by the piezoelectric waveguide convolver [4.105] described in Section 4.6. To provide a low loss feed to the thin film waveguide, it was necessary to compress the width of the beam emerging from the transducer by a factor of 15. An asymmetrical version of the multistrip coupler has been shown [4.140] to be capable of performing this task. The concept is illustrated in Fig. 4.50. The wide channel structure with stripe period d_A is connected to the narrow channel, in which the period of the stripes d_B is somewhat smaller. As we shall see, the difference is in fact very small, and has been greatly exaggerated in the figure. A detailed theory of propagation under a metallized grating shows that the velocity depends, though weakly, on the widths of the channels w_A and w_B. To achieve total power transfer, it is necessary to ensure that the two channels are synchronous. The elegant solution to this problem was to adjust the period in each channel to compensate for the small velocity deviation produced by the difference in width. In addition to the small change in period, which can be accurately predicted, there is also an increase in the total number of stripes required. If N_1 is the number of stripes for a symmetrical total transfer coupler, it can be shown that, for a coupler with a compression ratio of ϱ, the required number is N_ϱ where,

$$N_\varrho = N_1(\varrho^{1/2} + \varrho^{-1/2})2. \tag{4.71}$$

Measurements have shown that for compression ratios of the order of 10, the inherent loss of the msc is no larger than for symmetrical couplers. Moreover, a 1 dB bandwidth of one octave has been demonstrated. The compressor used in conjunction with the waveguide convolver already discussed [4.105] had a

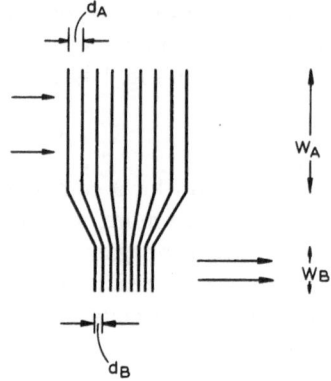

Fig. 4.50. Asymmetrical multistrip coupler for beam compression. (The difference between d_A and d_B is typically of the order of 1 %, and has been greatly exaggerated on the diagram)

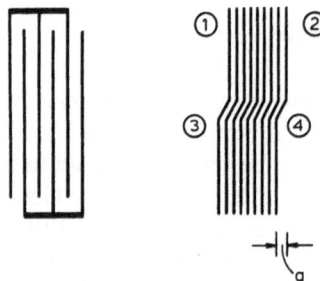

Fig. 4.51. Offset multistrip coupler frequency multiplexer

bandwidth of 50 MHz centered on 150 MHz. The compression ratio of 15:1 required 235 stripes. The period of the stripes in the wide channel was 1.1 % larger than that in the final compressed channel. This figure gives some indication of the degree of precision in the photolithography required in the implementation of such compressors.

4.7.5 Acoustic Surface Wave Multiplexers

We have so far considered a number of devices in which the existence of multiport acoustic elements plays a significant role in the operation of the device. However, in most cases we have ended up with a single useful acoustic output port. Thus, in the unidirectional transducer of Fig. 4.48, while we are undoubtedly using an msc having four identifiable ports, the aim of the exercise is to produce a transducer which will have but a single acoustic port. There is, however, a class of applications where the nature of the function to be performed requires a series of distinct acoustic output channels. The most important is that of frequency multiplexers. The multistrip coupler can be adapted to performing a multiplexing function [4.141] by a small modification to the structure—the provision of an offset between the two channels in an equal-split coupler (Fig.

4.51). If the structure is illuminated by a uniform transducer, there will be phase difference between the inputs at ports 1 and 3

$$A_1 = A, \qquad A_3 = A \cdot e^{-i\theta} \tag{4.72}$$

where

$$\theta = (2\pi f/v)g.$$

Using the basic coupler equations

$$B_2 = \frac{A}{2^{1/2}}(1 - ie^{-i\theta})$$

so that

$$|B_2|^2 = A^2(1 - \sin\theta) \tag{4.73}$$

and similarly

$$|B_4|^2 = A^2(1 + \sin\theta).$$

Thus there is a zero in the output from the top channel for $\theta = \pi/2$ and a zero in the output of the bottom channel for $\theta = 3\pi/2$. The output will therefore switch from one channel to the other whenever θ changes by π, or whenever the frequency changes by δf, where

$$\delta f = \frac{1}{2}\frac{v}{g} \equiv \frac{1}{2\tau} \tag{4.74}$$

where τ is the delay time associated with the offset g. The device can therefore be used to separate two frequency bands separated by δf. As it stands, it suffers from the multiplicity of responses, as well as the lack of control over the pass-band shape. Improvements in both respects can be engendered by using a series of such couplers, with progressively smaller offsets. Figure 4.52 shows the computed response [4.141] for a two-stage multiplexer having four output frequencies. Using this system, but with yet one further stage of separation, it has been possible to build an eight-channel multiplexer spanning the range 190–232 MHz, with a channel separation of 6 MHz. The insertion loss for each channel was 10 dB, and the sidelobe level was 13 dB below the peak response. These results, as well as the computed curves of Fig. 4.52, were obtained for broad band interdigital transducers in the receiving channels. It is, of course, entirely possible for these transducers to be realized in the form of high-performance band-pass filters. The task of the multiplexer is only to achieve an initial channel separation.

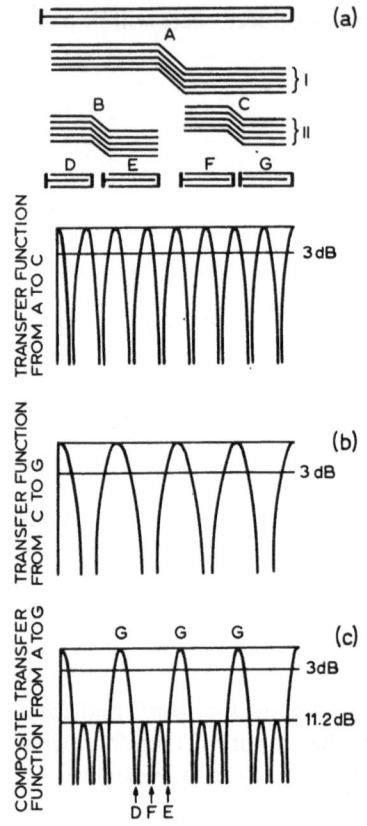

Fig. 4.52a–c. Schematic of a four-channel multiplexer together with predicted transfer function (after [4.141])

An approach entirely different from the realization of a multiplexer has been described [4.142–144]. In this, a msc is used as a reflecting array, see Fig. 4.53. A normal msc will have a stop band, i.e., a reflection peak, when the period of the fingers, d_A, is equal to $\lambda/2$. For couplers having a large number of fingers, this reflection peak is narrow, extending over a frequency range of the order of $1/N$. Analysis has shown that in the structure of Fig. 4.53, a very similar phenomenon takes place, with the reflected beam emerging not in the same channel but from a section of the structure where the period is d_B, where

$$d_A + d_B = \lambda. \tag{4.75}$$

If the period of the coupler spans the range from d_{B1} to d_{Bn}, the device will cover a frequency range Δf

$$\Delta f = \frac{d_{Bn} - d_{B1}}{(d_A + d_{B1})(d_A + d_{Bn})}. \tag{4.76}$$

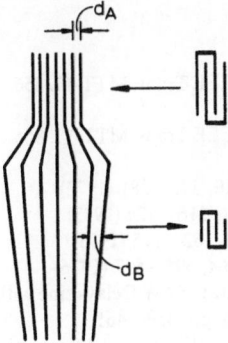

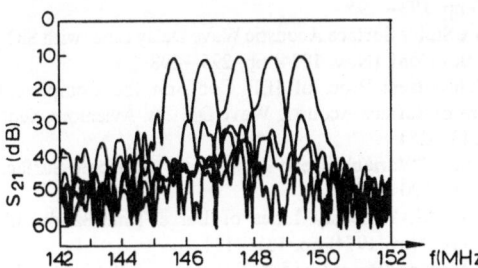

Fig. 4.53. Reflective multistrip coupler frequency multiplexer. Only a single one of the n output transducers is shown

Fig. 4.54. Frequency response of a five-port multistrip reflective coupler having 450 strips (after [4.143])

Since the bandwidth of each output channel is $1/N$, the total number of available channels n is given by

$$n = N \frac{\Delta f}{f_0}.$$ (4.77)

If all of these n potentially available output channels are used, the response will be approximately $\sin x/x$ in each. While this may be adequate, especially when used in conjunction with an output transducer which provides a further filtering action, it may be preferable to use some of the available bandwidth to provide a better shaped response for a smaller number of channels. An example of an experimental result [4.143] using this approach is shown in Fig. 4.54. This relates to a device with a center frequency of 150 MHz having 450 strips, and a range of d_B which would permit up to 15 output channels. The device was in fact designed to provide five output channels separated by 1 MHz. The total insertion loss was 12 dB, but of this only approximately 2 dB was associated with the multistrip coupler itself.

There would appear to be every prospect that multiplexers of the kind shown in this section will find application in frequency division multiplex communication systems.

References

4.1   W. R. Smith, H. M. Gerard, J. H. Collins, T. M. Reeder, H. J. Shaw: IEEE Trans. MTT-**17**, 856—864 (1969)

4.2   W. R. Smith, H. M. Gerard, J. H. Collins, T. M. Reeder, H. J. Shaw: IEEE Trans. MTT-**17**, 865—873 (1969)

4.3   C. S. Hartmann, W. Stanley Jones, H. Vollers: IEEE Trans. SU-**19**, 378—381 (1972)

4.4   F. G. Marshall, C. O. Newton, E. G. S. Paige: IEEE Trans. MTT-**21**, 216—224 (1973)

4.5   C. S. Hartmann, D. T. Bell, R. C. Rosenfeld: IEEE Trans. MTT-**21**, 162—175 (1973)

4.6   K. Shibayama, K. Yamanouchi, H. Sato, T. Meguro: Proc. IEEE **64**, 595—597 (1976)

4.7   P. H. Carr, T. E. Fenstermacher, J. H. Silva, W. J. Kearns, M. R. Stiglitz: "SAW Delay-Line with all Spurious 70 dB Down", Proc. IEEE Ultrasonics Symp. (1976), pp. 459—461

4.8   E. A. Ash: "Surface Wave Grating Reflectors and Resonators". IEE Symp. Microwave Theory and Techniques (Newport Beach, May 11—14, 1970)

4.9   F. Y. Cho, B. J. Hunsinger, L. L. Lee: "Programmable Planar Folded Path SAW Delay Line". Proc. IEEE Ultrasonics Symp. (1974) pp. 193—195

4.10 T. E. Parker, M. B. Schulz: "Temperature Stable Surface Acoustic Wave Delay Lines with SiO_2 Film Overlays". Proc. IEEE, Ultrasonics Conf. (Nov. 1974) pp. 295—298

4.11 R. L. Zimmerman, G. R. Nudd, B. P. Schweitzer: Proc. of IEE Conf. on the Component Performance and Systems Applications of Surface Acoustic Wave Devices. Aviemore, Sept. 1973 (IEE Conf. Publ. No. 109) pp. 243—254

4.12 J. D. Ross, S. J. Kapuscienski, K. B. Daniels: "Variable Delay Line using Ultrasonic Surface Waves". IRE Nat. Conf. Rec. pt. 2, 118—120 (1958)

4.13 C. M. Fortunko, S. L. Quilici, H. J. Shaw: "SAW Delay Lines of Large Time-Bandwidth Products". Proc. IEEE Ultrasonics Conf. (Nov. 1974) pp. 181—184

4.14 J. Chambers, E. Papadofrangakis, I. M. Mason: "Beam Guided, Temperature Stabilized Disc Long Delay Lines". Proc. IEEE Ultrasonics Conf. (1974) pp. 760—762

4.15 L. A. Coldren, H. J. Shaw: Proc. IEEE **64**, 598—609 (1976)

4.16 L. A. Coldren: Wave Electron. **1**, 177—189 (1975)

4.17 A. J. Slobodnik Jr., W. J. Kearns, J. H. Silva, T. E. Fenstermacher: "A Comparison of SAW Wrap-around Delay Line Geometries". Proc. IEEE Ultrasonics Conf. (Nov. 1974) pp. 185—188

4.18 I. M. Mason, E. Papadofrangakis, J. Chambers: Electron Lett. **10**, 63—65 (1974)

4.19 I. M. Mason, E. Papadofrangakis, J. Chambers, R. Ulrich: "A Chain Matrix Approach to the Design of Anisotropic Rayleigh Wave Delay Lines". Proc. IEEE Ultrasonics Symp. (Sept. 1975) pp. 523—525

4.20 L. R. Adkins, A. R. Hughes: J. Appl. Phys. **42**, 1819—1822 (1971)

4.21 P. Hartemann, M. Morizot: "Variation of Surface Acoustic Wave Velocity produced by Ion Implantation". Proc. IEEE Ultrasonics Conf. (Nov. 1974) pp. 307—310

4.22 R. V. Schmidt: Appl. Phys. Lett. **27**, 1 (1975)

4.23 R. Schmidt, L. A. Coldren: Appl. Phys. Lett. **22**, 482—483 (1973)

4.24 T. I. Browning, F. G. Marshall: "Compact 130 µs SAW Delay Line using Improved MSC Reflecting Track Changers". Proc. IEEE Ultrasonics Symp. (1974) pp. 189—192

4.25 R. L. Rosenberg, R. V. Schmidt, L. A. Coldren: Appl. Phys. Lett. **25**, 324—326 (1974)

4.26 G. D. Boyd, L. A. Coldren, R. N. Thurston: Appl. Phys. Lett. **26**, 31—34 (1975)

4.27 A. A. Oliner, K. H. Yen: "A New Class of Components which do not Require Piezoelectric Substrates". Proc. Ultrasonics Symp. (1974) pp. 108—113

4.28 T. Reeder: Private communication (1970)

4.29 E. J. Staples: "UHF Surface Acoustic Wave Resonators". Proc. of Symp. on Frequency Control (Fort Monmouth, 1974)

4.30 E. J. Staples, J. S. Schoenwald, R. C. Rosenfeld, C. S. Hartman: "UHF Acoustic Wave Resonators". Proc. IEEE Ultrasonics Symp. (Nov. 1974) pp. 245—252

4.31 De Lamar, T. Bell, R. C. M. Li: Proc. IEEE **64**, 711—721 (1976)

4.32 S. C. C. Tseng, G. W. Lynch: SAW Planar Network, loc. cit. pp. 282—285

4.33   G.D.Boyd, H.Kogelnik: B.S.T.J. **41**, 1347 (1962)

4.34   R.E.Collin: *Foundations of Microwave Engineering* (McGraw-Hill, New York 1966) Chap. 8

4.35   P.Hartemann, M.Morizot: "Variation of Surface Acoustic Wave Velocity Produced by Ion Implantation". Proc. IEEE Ultrasonics Symp. (1974) pp. 307—310

4.36   R.Schmidt: Appl. Phys. Lett. **27**, 8—10 (1975)

4.37   F.A.Jenkins, H.E.White: *Fundamentals of Optics* (McGraw-Hill, New York 1957)

4.38   W.R.Smith, H.M.Gerard, J.H.Collins, T.M.Reeder: IEEE Trans. MTT-**17**, 856—864 (1969)

4.39   I.M.Mason, J.Chambers, P.Lagasse: Electron Lett. V. **11**, 288—290 (1975)

4.40   K.M.Lakin, T.R.Joseph: "Surface Wave Resonators". Proc. IEEE Ultrasonics Symp. (Sept. 1975) pp. 269—278

4.41   W.P.Mason: *Electromechanical Transducers and Wave Filters*, 2nd ed. (Van Nostrand, New York 1948)

4.42   W.R.Shreve: "Surface Wave Resonators and Their Use in Narrowband Filters". Proc. IEEE Ultrasonics Symp. (1976) pp. 706—713

4.43   J.D.Maines, E.G.S.Paige, A.F.Saunders, A.S.Young: Electron Lett. **5**, 678 (1969)

4.44   W.S.Jones, C.S.Hartmann, T.D.Sturdivant: IEEE Trans. SU-**19**, 368—377 (1972)

4.45   J.Crabb, M.F.Lewis, J.D.Maines: Electron.Lett. **9**, 195 (1973)

4.46   T.R.Joseph: "Phase Locked SAW Delay Line and Resonator Oscillators". Proc. IEEE, Ultrasonics Symp. (1976) pp. 234—239

4.47   M.Lewis: "The Design, Performance and Limitations of SAW Oscillators", Aviemore 1973 (IEE Conf. on SAW Devices Conf. Publ. 109) pp. 63—72

4.48   M.Lewis: "The Surface Acoustic Wave Oscillator—A Natural and Timely Development of the Quartz Crystal Oscillator". Proc. 28th Annual Frequency Control Symp., Atlantic City (May 1974) pp. 304—314

4.49   I.Bale, M.F.Lewis: "Improvements to the SAW oscillator". Proc. IEEE Ultrasonics Symp. (1974) pp. 272—275

4.50   P.Hartemann, O.Menager: Electron Lett. **8**, 214 (1972)

4.51   P.Hartemann: "Surface Acoustic Wave Phase and Frequency Discriminator", Aviemore 1973 (IEE Conf. on SAW Devices Conf. Publ. 109) pp. 152—166

4.52   R.C.Williamson: "Problems Encountered in High Frequency Surface Wave Devices". Proc. IEEE Ultrasonics Symp. (1974) pp. 321—328

4.53   I.Browning, J.Crabb, M.F.Lewis: "A SAW Frequency Synthesizer". Proc. IEEE Ultrasonics Symp. (Sept. 1975) pp. 245—247

4.54   H.Gilden, J.M.Reeder, A.J.DeMaria: "The Mode-locked SAW Oscillator". Proc. IEEE Ultrasonics Symp. (Sept. 1975) pp. 251—254

4.55   J.D.Maines, J.N.Johnston: Ultrasonics **11**, 211—217 (1973)

4.56   P.Hartemann, E.Dieulesaint: Electron Lett. **5**, 299—320 (1969)

4.57   H.M.Gerard, W.R.Smith, W.R.Jones, J.B.Harrington: IEEE Trans. SU-**20**, 94—104 (1973)

4.58   J.D.Maines, G.L.Moule, N.R.Ogg: Electron Lett. **8**, 431—433 (1972)

4.59   H.Skeie: "Mechanical and Electrical Reflections in Interdigital Transducers". Proc. IEEE Ultrasonics Symp. (1972) pp. 408—412

4.60   E.G.S.Paige: "Dispersive Filters: Their Design and Application to Pulse Compression and Temporal Transformations". Component Performance and Systems Applications of Surface Acoustic Wave Devices (IEE Conf. Publ. No. 109 Sept. 1973) pp. 167—180

4.61   R.C.Williamson, H.I.Smith: Electron. Lett. **8**, 401—402 (1972)

4.62   R.C.Williamson: Proc. IEE **64**, 702—710 (1976)

4.63   H.I.Smith: Proc. IEE **62**, 1361—1387 (1974)

4.64   L.P.Solie: "A SAW Filter Using a Reflective Dot Array (RDA)". Proc. IEEE Ultrasonics Symp. (1976) pp. 309—312

4.65   J.Melngailis, R.C.Williamson, J.Holtham, R.C.M.Li: Wave Electron. **2**, 177—198 (1976)

4.66   T.A.Martin: IEEE Trans. SU-**20**, 104—112 (1973)

4.67   H.Skeie: J. Appl. Phys. **48**, 1098—1109 (1970)

4.68   R.C.Williamson, H.I.Smith: IEEE Trans. SU-**20**, 113—123 (1973)

4.69   M.G.Holland, L.T.Claiborn: Proc. IEEE **62**, 582—610 (1974)

184   E. A. Ash

4.70   J.H.Collins, P.M.Grant, B.J.Darby: Wave Electron. **1**, 311—342 (1976)
4.71   H.M.Gerard, O.W.Otto, R.D.Weglein: "Development of a Broadband Reflective Array 10000:1 Pulse Compression Filter". Proc. IEEE Ultrasonics Symp. (1974) pp. 197—201
4.72   S.deKlerk: Phys. Acoust. 11, 213—243 (1975)
4.73   T.W.Bristol, W.R.Jones, P.B.Snow, W.R.Smith: "Applications of Double Electrodes in Acoustic Surface Wave Device Design". Proc. IEEE Ultrasonics Symp. (1972) pp. 343—345
4.74   B.J.Darby: IEEE Trans. SU-**20**, 382—384 (1973)
4.75   D.T.Bell, J.D.Holmes, R.V.Ridings: Trans. IEEE MTT-**21**, 263—271 (1973)
4.76   C.F.Vasile, R.La Rosa: Electron. Lett. **8**, 479—480 (1972)
4.77   E.J.Staples, L.T.Claiborn: IEEE Trans. MTT-**21**, 195—205 (1973)
4.78   P.J.Hagon: "Programmable Analogue Matched Filters". Symp. on "Component Performance and Systems Applications of Surface Acoustic Wave Devices" (IEE Conf. Publ. No. 109) pp. 92—101
4.79   C.L.Grasse, D.A.Gandolfo: "Acoustic Surface Wave Dispersive Delay Lines as High Resolution Frequency Discriminators". Proc. IEEE Ultrasonics Symp. (1972) pp. 233—236
4.80   G.R.Nudd, O.W.Otto: "Chirp Single Processing Using Acoustic Surface Wave Filters". Proc. IEEE Ultrasonics Symp. (1975) pp. 350—354
4.81   M.A.Jack, P.M.Grant, J.H.Collins: "Real Time Network Analyzer Employing Acoustic Wave Chirp Filters". Proc. IEEE Ultrasonics Symp. (1975) pp. 359—362
4.82   M.A.Jack, G.F.Manes, P.M.Grant, C.Atzen, L.Masott, J.H.Collins: "Real Time Network Analysers Based on SAW Chirp Transform Processors". Proc. IEEE Ultrasonics Symp. (1976) pp. 376—381
4.83   J.D.Maines, G.L.Moule, C.O.Newton, E.G.S.Paige: "A Novel Saw Variable Frequency Filter". Proc. IEEE Ultrasonics Symp. (1975) pp. 355—358
4.84   R.M.Hays, W.M.Shreve, R.T.Bell, L.T.Claiborne: "Surface Wave Transform Adaptable Processing System". Proc. IEEE Ultrasonics Symp. (1975) pp. 363—370
4.85   R.M.Hays, C.S.Hartmann: "Surface Acoustic Wave Devices for Communications". Proc. IEEE (May 1976) pp. 652—671
4.86   W.S.Mortley: "Improvements in or Relating to Variable Electrical Delay Arrangements" (British Patent 994842)
4.87   A.F.Podell, R.F.Lee, A.S.Bahr: IEEE-GMTT Int. Microwave Symp. May, 1972 (IEE Conf. Publ. No. 72) pp. 92—94
4.88   V.S.Dolat, R.C.Williamson: "A Continuously Variable Delay-Line System". Proc. IEEE Ultrasonics Symp. (1976) pp. 419—423
4.89   V.Shtykov, I.Mason, M.Motz: IEEE Trans. SU-**22**, 131—136 (1975)
4.90   H.Louisell: *Coupled Modes and Parametric Electronics* (John Wiley, New York 1960)
4.91   A.A.Chaban: Sov. Phys. Sld. St. **9**, 2622—2623 (1968)
4.92   L.Svaasand: Appl. Phys. Lett. **15**, 300—302 (1969)
4.93   N.Ansteg, W.E.Lerwill: Proc. Roy. Soc. **290**, 430—477 (1963)
4.94   W.C.Wang: *A convolution integration*, presented at the Joint Services Technical Advisory Committee (Polytechnic Institute of Brooklyn, 1966)
4.95   C.F.Quate, R.B.Thompson: Appl. Phys. Lett. **16**, 494 (1970)
4.96   M.V.Lukkala, G.S.Kino: Appl. Phys. Lett. **18**, 393—394 (1971)
4.97   G.S.Kino, S.Ludvik, H.Shan, W.R.Shreve, S.M.White, D.K.Winslow: IEEE Trans. MTT-**21**, 244—255 (1973)
4.98   M.Motz, J.Chambers, I.M.Mason: "Suppression of Spurious Signals in a Degenerate SAW Convolver". Proc. IEEE Ultrasonics Symp. (1973) pp. 152—154
4.99   D.Morgan: IEEE Trans. SU-**22**, 274—277 (1975)
4.100   D.P.Morgan, J.M.Hannah, J.H.Collins: Proc. IEEE **64**, 751—753 (1976)
4.101   P.E.Lagasse, I.M.Mason, E.A.Ash: IEEE Trans. MT-**21**, 225—236 (1973)
4.102   P.Nanayakkara, E.A.Ash: Wave Electron. **1**, 247—263 (1976)
4.103   A.A.Oliner: Proc. IEEE **64**, 615—635 (1976)
4.104   R.A.Zakarevicius, E.A.Ash, I.M.Mason: Electron. Lett. **9**, 363—364 (1973)
4.105   P.Deframould, C.Maerfeld: Proc. IEEE **64**, 748—751 (1976)

4.106 O.Baiocchi, I.M.Mason: IEEE Trans. SU-**22**, 347—354 (1975)
4.107 W.R.Shreve, G.S.Kino, M.Lukkala: Electron. Lett. **7**, 764—766 (1971)
4.108 M.Lukkala, J.Surakka: J. Appl. Phys. **43**, 2510—2518 (1972)
4.109 G.S.Kino: Proc. IEEE **64**, 724—748 (1976)
4.110 T.M.Reeder, M.Gilden: Appl. Phys. Lett. **22**, 254—256 (1973)
4.111 T.M.Reeder: Electron. Lett. **9**, 254—256 (1973)
4.112 T.M.Reeder: Proc. of IEE Symp. on the Component Performance and Systems Applications of Surface Acoustic Wave Devices, Aviemore, Sept. 1973 (IEE Conf. Publ. No. 109) pp. 73—84
4.113 O.Menager, B.Desormiere: Appl. Phys. Lett. **27**, 1—2 (1975)
4.114 T.M.Reeder, T.W.Grdkouski: "Real Time Fourier Transform Experiments Using a 32-Tap Diode Convolver Module". Proc. IEEE Ultrasonics Symp. (1975) pp. 336—339
4.115 J.M.Smith, E.Stern, A.Bers, J.Cafarella: "Surface Acousto Electric Convolvers". Proc. IEEE Ultrasonics Symp. (1973) pp. 142—144
4.116 A.G.Bert, B.Epsztei, G.Kantorowicz: IEEE Trans. MTT-**21**, 255—263 (1973)
4.117 J.H.Cafarella, W.M.Brown, E.Stern, J.A.Alusow: Proc. IEEE **64**, 756—759 (1974)
4.118 G.Kantorowicz: French Patent No. 7345234 (December 1973)
4.119 A.Bers, J.H.Cafarella: Appl. Phys. Lett. **25**, 133—135 (1974)
4.120 H.Hayakawa, G.S.Kino: Appl. Phys. Lett. **25**, 178—180 (1974)
4.121 K.A.Ingebritsen, R.A.Cohen, R.W.Mountain: Appl. Phys. Lett. **26**, 596—598 (1975)
4.122 C.Maerfeld, P.Defranould, P.Tournois: Appl. Phys. Lett. **27**, 577—578 (1975)
4.123 K.A.Ingebritsen: Proc. IEEE **64**, 764—769 (1976)
4.124 Ph.Defranould, H.Gautier, C.Maerfeld, P.Tournois: "*P–N* Diode Memory Correlator". Proc. IEEE Ultrasonic Symp. (1976) pp. 336—347
4.125 K.A.Ingebritsen, E.Stern: Appl. Phys. Lett. **27**, 170—172 (1975)
4.126 K.A.Ingebritsen: Electron. Lett. **11**, 585—586 (1975)
4.127 K.A.Ingebritsen, E.Stern: "Holographic Storage of Acoustic Surface Waves with Schottky Diode Arrays". Proc. IEEE Ultrasonics Symp. (1975) pp. 212—216
4.128 P.V.H.Sabine: Electron. Lett. **9**, 136—138 (1973)
4.129 F.G.Marshall, E.G.S.Paige: Electron. Lett. **7**, 460—462 (1971)
4.130 I.M.Mason, J.Chambers, P.E.Lagasse: "Spatial Harmonic Analysis of the Multistrip Coupler". Proc. IEEE Ultrasonics Symp. (1973) pp. 159—162
4.131 K.A.Ingebritsen: "A Normal Mode Representation of Surface Wave Multistrip Coupler". Proc. IEEE Ultrasonics Symp. (1973) pp. 163—167
4.132 F.G.Marshall, C.O.Newton, E.G.S.Paige: IEEE Trans. MTT-**21**, 206—211 (1973)
4.133 C.Maerfeld, P.Tournois: Electron. Lett. **9**, 115—116 (1973)
4.134 C.Maerfeld: Wave Electron. **2**, 82—109 (1976)
4.135 G.W.Farnell, C.Maerfeld: "Modes in Multistrip Couplers", Proc. IEEE Ultrasonics Symp. (1976) pp. 480—485
4.136 C.S.Hartmann, D.T.Bell, R.C.Rosenfeld: IEEE Trans. MTT-**21**, 162—175 (1973)
4.137 F.G.Marshall, E.G.S.Paige, A.S.Young: Electron. Lett. **7**, 638—640 (1970)
4.138 F.G.Marshall: Electron. Lett. **8**, 311—312 (1972)
4.139 F.G.Marshall, C.O.Newton, E.G.S.Paige: IEEE Trans. MTT-**21**, 216—225 (1973)
4.140 C.Maerfeld, G.W.Farnell: Electron. Lett. **9**, 432—434 (1973)
4.141 H.Van de Vaart, L.P.Solie: Proc. IEEE **64**, 688—691 (1976)
4.142 M.Feldman, J.Henaff, M.Carel: Electron. Lett. **12**, 118—119 (1976)
4.143 M.Feldmann, J.Henaff: "The Reflective ASW Multistrip Array—A Frequency Sensitive Device for Filter and Channel Bank". Proc. European Microwave Conf. (1976) pp. 239—241
4.144 M.Feldmann, J.Henaff: "An ASW-filter Using Two Fan-Shaped Multistrip Reflective Arrays". Proc. IEEE Ultrasonics Symp. (1976) pp. 397—400

5. Waveguides for Surface Waves

A. A. Oliner

With 15 Figures

5.1 Background and General Considerations

This chapter discusses the basic features relating to waveguides for acoustic surface waves. The principal stress is on the *types* of waveguide which have been conceived, measured, and studied theoretically, and on their *properties*. Those discussions also include simple physical viewpoints to explain why specific waveguiding geometries possess particular properties.

Other important considerations relating to waveguides are treated more briefly. In the remainder of this section, we first review what is meant by a waveguide and why it should be of interest, touching briefly on some applications and some problems, and then summarize various waveguide types and indicate which properties are of importance. The various waveguides themselves, and their properties, are treated next, in Sections 5.2–5.4. In Section 5.5, Summary and Conclusions, we first summarize in tabular form the main properties of the various waveguides discussed herein, adding some special remarks on the question of loss, then discuss various applications for waveguides, both actual and potential, and speculate on the future.

5.1.1 Why Waveguides?

Almost all current acoustic surface wave devices utilize wide-beam surface waves. These surface wave beams possess certain limitations: beam spreading, inefficient use of the substrate area, and awkwardness in bending their paths. These limitations are all overcome by the use of waveguides for such surface waves, where the term waveguide implies a geometrical structure which confines the lateral extent of the surface wave and binds the wave to itself.

The most important difficulty associated with wide surface wave beams, which is beam spreading (causing cross talk between neighboring beams), is thereby automatically overcome since the field cross section associated with the waveguide remains constant as the wave progresses. The wave bound to the waveguide will also readily follow any waveguide bends, provided that the radius of curvature of that bend is not too small. Also, for certain waveguides the guide cross section can be *less* than a wavelength wide, and the acoustic wave field well confined, so that the substrate area required is very much less than that needed for wide-beam surface waves, which are typically 40–100 wavelengths wide.

Waveguides have so far not been widely used, for two principal reasons. One of these reasons relates to two problems customarily attributed to waveguides: loss and inefficient excitation. These problems are to a large extent more imagined than real; they are discussed briefly in Section 5.5. It is pointed out there that for some waveguides, in some applications, loss can be an important concern, whereas some other waveguides are not measurably lossier than wide-beam surface waves on the same material. With regard to methods of excitation, much progress has been made recently, and efficient excitation for most waveguides can be achieved in practice. It is believed that waveguides will be more widely used when it is recognized that these two former problems can in large measure be overcome.

The second reason for the comparatively sparse use to date of waveguides relates to the question of whether or not they are needed to satisfy current device requirements. Some waveguides are currently being considered seriously for use in long delay lines, principally in order to overcome the beam-spreading problem. However, this application is a very narrow one, with specific requirements, such as low loss and negligible dispersion. For most other device needs, waveguides are not essential but, with ingenuity, they could be useful in improving device performance. There do exist a number of applications for which waveguides would serve well, and some of these are described briefly in Section 5.5.2. One class of such applications which might be mentioned now involves nonlinear interactions; the efficiency of such interactions should be greatly increased by the use of waveguides because of the lateral confinement of the fields.

The most intriguing potential application for waveguides is that of a highly compact sophisticated circuit technology. This potential application furnished some of the early motivation for work in acoustic surface waves and was often referred to as "microsound" technology [5.1–3]; it still remains the most exciting possibility for waveguides. But, the acoustic wave device field has not moved in the direction of circuitry which performs several functions simultaneously. SAW devices generally (but not always) perform a single function, and consist of a simple circuit placed between input and output transducers. When the philosophy changes, and when the performance of multiple functions on the same substrate using wide surface wave beams is seen to require excessive substrate surface area, waveguides will be called upon. However, they will then not be ready because components for waveguides have not yet been developed. In turn, work on such components has been very sparse because the demand for them was absent.

In summary, waveguides overcome some of the limitations of wide surface wave beams, notably the problem of beam spreading, and indeed certain applications for waveguides have already presented themselves. These applications at present are few, but the potential for future use could be very great.

5.1.2 Types of Waveguide

A waveguide for acoustic surface waves is a geometrical structure which confines the lateral extent of the surface wave on the substrate surface and binds the wave to itself. Such waveguides are of four types:

1) Overlay waveguides, in which a strip of one material is placed on a substrate of another material,

2) topographic waveguides, which consist of a local deformation of the substrate surface itself (that is, a change in the local topography of the surface),

3) waveguides in which a local change has been produced in the properties of the substrate material, and

4) circular fiber waveguides, which do not employ a planar substrate.

Typical waveguiding structures in each of the four categories listed above are shown in Fig. 5.1.

Interest in these waveguides was given enormous stimulus by two independent symposium talks presented in the same year, 1967. The first of these, by *Ash* [5.5], advanced the concept of topographic waveguides, and presented a double-scratch arrangement as a possible embodiment. The topographic rectangular ridge waveguide [Fig. 5.1d], which was a simple modification of the double-scratch structure, emerged soon afterwards [5.6, 7] and was later seen to be identical with a ridge that was used earlier [5.8] to form a delay line by spiraling on the outside of a duralumin cylinder. The second talk was by *White* [5.9], who presented measurements on the propagation behavior of a strip waveguide [Fig. 5.1a], formed by a gold strip on a fused quartz substrate. The strip waveguide was first described in a pair of patents [5.10] by *Seidel* and *White*.

The structures shown in Fig. 5.1a–c are the most important examples of *flat overlay waveguides*; the strip (or ribbon) guide of Fig. 5.1a is the best known of these. All have been analyzed theoretically and measured. The central region in each case is "slower" than the outside regions (corresponding to a higher refractive index in optics), so that the acoustic surface wave field is pulled in laterally toward the central region. In the *strip* waveguide of Fig. 5.1a, the plated strip itself consists of material which is slower than that of the substrate. In the *slot* (or gap) waveguide of Fig. 5.1c, the plated material is "faster" than that of the substrate, so that the slot formed by the unplated central region is slower and acts as the waveguiding region. The structure in Fig. 5.1b consists of a metallic strip placed on a piezoelectric substrate; the strip short circuits the electric field associated with the piezoelectric surface wave and produces a slight slowing of the wave under the strip. This waveguiding structure is herein called the *shorting-strip* waveguide; it is usually referred to as the $\Delta v/v$ guide because the velocity is changed by the shorting strip. This notation is bad, however, because *all* waveguides are $\Delta v/v$ guides in the sense that some modification has been introduced to change the velocity in some local region. Further details relating to these flat overlay waveguides are presented in Section 5.2.

The two most important *topographic* waveguides are the rectangular ridge and the wedge, shown respectively in Fig. 5.1d and e. The guiding effect in these

FLAT OVERLAY WAVEGUIDES

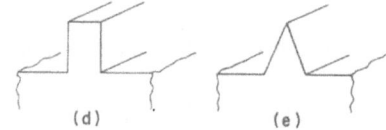

(a)   (b)   (c)

TOPOGRAPHIC WAVEGUIDES

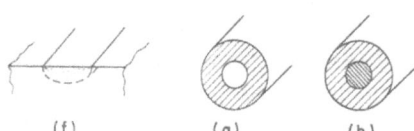

(d)   (e)

OTHER TYPES

Fig. 5.1a–h. Various types of waveguide for acoustic surface waves. Flat overlay waveguides: (a) strip, (b) shorting-strip or $\Delta v/v$ (metal strip on a piezoelectric substrate), (c) slot. Topographic waveguides: (d) rectangular ridge, (e) wedge. Other types: (f) in-diffused or ion-implanted, (g) capillary fiber, (h) cladded-core fiber (from *Oliner* [5.4])

(f)   (g)   (h)

topographic waveguides results from a reduction of restraining forces acting on the material [5.2]; such topographic structures have no counterpart in electromagnetic waveguides. The *rectangular ridge* waveguide possesses two dominant modes, one symmetric and the other antisymmetric. The antisymmetric, or flexural, mode is the better known of the two, and is characterized by a strong dispersive behavior; the symmetric mode has almost no dispersion and is referred to as the pseudo-Rayleigh mode because its field resembles that of a slice out of a Rayleigh surface wave. The *wedge* waveguide is of interest because it possesses no dispersion above a certain frequency (an infinite ideal wedge has no dispersion whatever). The slow nature of the antisymmetric mode on narrow-angle wedges is also of interest. Further information on these two waveguiding structures is contained in Section 5.3.

The structure in Fig. 5.1f is an example of a waveguide belonging to Type 3. The substrate surface remains geometrically flat, but the properties of the material in the region shown dotted have been modified. This change may be produced in various ways: by in-diffusion to produce loss or a velocity change, by ion implantation, by depoling in the case of a ferroelectric, etc. A difficulty associated with overlay waveguides (Type 1) is the fact that a layered substrate has been found to be lossier than an unlayered one. Recent work by *Schmidt* [5.11] on certain metals in-diffused into $LiNbO_3$ has shown that the diffusion process produces a velocity increase but a negligible change in loss. Such in-diffused waveguides should therefore possess very *low loss*, and this waveguide type could become important for certain applications. Further details concerning this class of waveguides appear in Section 5.4.1.

The *circular fiber* waveguides, indicated above as type 4, are being investigated for their potential use in long delay-line applications. In different ways they are analogous to optical fibers; the structure in Fig. 5.1g is a capillary structure which guides a surface wave, resembling a compressed Rayleigh wave, on its inner surface, while that in Fig. 5.1h is a direct analog of a cladded optical fiber, in which the inner core is composed of slower material than the outer cladding.

Each of these waveguides in the categories above possesses its own characteristic properties and therefore has a different range of potential applications. Of these properties, two may be regarded as the most important: the acoustic field confinement and the dispersion behavior.

The *field confinement* is measured by the rate at which the acoustic fields decay away from the guide itself along the substrate surface. A rapid decay, implying strong confinement, isolates the waveguide from neighboring circuitry, and also permits the guide to undergo a bend with a smaller radius of curvature before measurable radiation leakage occurs. For some applications, however, a weaker confinement is preferred; for example, weaker confinement permits easier design of a directional coupler composed of two waveguides which lie parallel to each other for a specified length. For the waveguides in Fig. 5.1a–c and f, the degree of confinement is directly related to the slowness of the wave; if v_R and v_z represent respectively the velocity of the Rayleigh wave and that of the guided wave along the waveguide (in the z direction), and k_R and $|k_t| = \alpha_t$ denote the wave number of the Rayleigh wave and the decay constant along the substrate surface transversely away from the waveguide, the decay constant $|k_t|$ is given simply by

$$\frac{|k_t|}{k_R} = \left[\left(\frac{v_R}{v_z} \right)^2 - 1 \right]^{\frac{1}{2}}. \tag{5.1}$$

For topographic waveguides, the field confinement is also governed by how much of the field is contained in the ridge or wedge itself, in Fig. 5.1d and e; if essentially all the field is contained in the guide itself, then the mechanical isolation is very strong even though the guide velocity is near to the Rayleigh velocity. As a general rule, the topographic waveguides offer strong field confinement or isolation, whereas the flat overlay waveguides possess weak field confinement unless the overlays produce strong loading effects.

The *dispersion behavior,* by which we mean the variation of propagation velocity with frequency, is quite different for each of the waveguides in Fig. 5.1. This behavior is described in detail in Sections 5.2–5.4, but we can mention now which waveguides are essentially dispersionless over some frequency range. This feature is important for application to long delay lines. The wedge waveguide (Fig. 5.1e) is dispersionless above a certain minimum frequency, the symmetric mode of the ridge waveguide (Fig. 5.1d) has very small dispersion over the entire frequency range, and the slot waveguide (Fig. 5.1c) possesses a fairly flat maximum in the plot of velocity vs frequency. The reasons for these behaviors are made clear in the discussions below.

A summary of these properties for each of the waveguides is presented in tabular form in Section 5.5.1 to serve as a guide.

The propagation characteristics of the most important waveguiding structures treated to date are presented next in separate sections. The discussions include simple physical viewpoints which explain the nature of the guiding process, and contain information on the field-confinement properties and the dispersion behavior for each of the waveguides considered.

5.2 Flat Overlay Waveguides

In this class of waveguide, the wave velocity of the substrate surface is altered by placing a strip of one material on a substrate of another material. The three most important structures in this class are shown as a, b, and c of Fig. 5.1, and their properties are discussed separately in the three subsections below. Because these structures are similar to each other in various ways, the basic features common to them are presented in more detail under Section 5.2.1, and then simply referred to in the other two subsections.

5.2.1 The Strip Waveguide

Let us consider more closely the best known of these, the *strip waveguide*, shown as Fig. 5.1a and repeated in Fig. 5.2a. The structure consists of a strip or ribbon of "slow" material on a "fast" substrate. The term slow means that a bulk acoustic wave in that material would travel more slowly than such a wave in the fast material. The strip itself is wide and flat, with customary aspect ratios ranging from 5 to 50.

The guiding mechanism of the strip waveguide is basically similar to that of an electromagnetic or optical waveguide comprised of a dielectric strip on a thin film placed on a substrate. The slow strip, relative to the fast substrate, corresponds in the optical case to a strip with a higher index of refraction. The field is bound to the strip, decaying away from it in all transverse directions.

One of the analytical methods used to theoretically treat these waveguiding structures employs microwave network procedures [5.12, 13]. Let us utilize one feature of that approach here because of its pedagogical value in this context. In that approach, it is necessary first to derive a transverse equivalent network for the structure under consideration. An incomplete such network for the dominant mode of the strip waveguide is shown in Fig. 5.2b. The central transmission line represents the modified Rayleigh wave supportable by the strip and substrate region, and the outer transmission lines represent the Rayleigh waves on the unplated substrate surfaces on each side of the strip. The geometric discontinuities between the strip and the unplated substrate are accounted for by the boxes shown.

The transverse equivalent network appearing in Fig. 5.2b permits us to view the propagation of the guided mode in terms of two modified Rayleigh waves in

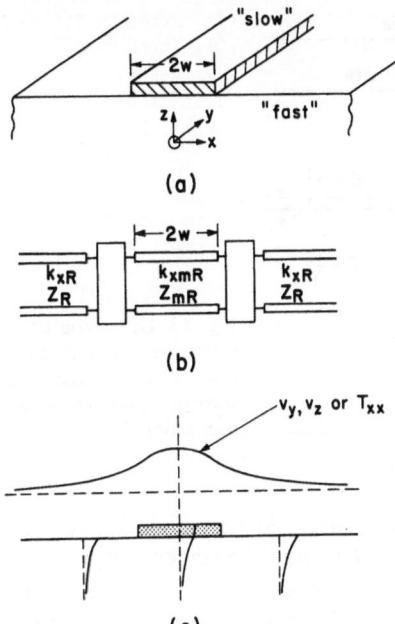

(a)

(b)

(c)

Fig. 5.2a–c. The strip waveguide, consisting of an overlay of slow material on a fast substrate. (a) Geometrical structure. (b) Transverse equivalent network for dominant mode; k_{xR} and k_{xmR} are the transverse wave numbers in the x direction. (c) Transverse acoustic field behavior; v_y and v_z are particle velocity components and T_{xx} is a stress tensor component

the strip region which bounce back and forth from the strip sides at angles which are greater than the critical angle, so that total reflection occurs within the strip and the wave becomes completely guided. We also conclude that the decays of the vertical field into the substrate in the strip and in the unplated regions correspond, respectively, to those for the modified Rayleigh and the Rayleigh waves. The transverse field behavior for the dominant mode is summarized in Fig. 5.2c.

The transverse decay in the x direction, along the substrate surface, is that of a simple exponential, consistent with the requirement in the network of Fig. 5.2b that the Rayleigh wave transmission lines be below cutoff. This exponential decay will be a slow one for most flat overlay guides, depending of course on the amount by which the guided wave is slowed down [see (5.1)]. In general for these waveguides the acoustic field extends out substantially so that the wave is weakly bound. Such behavior discourages sharp bends but may be useful, for example, when one wishes to obtain only mild guidance, or to construct directional couplers between adjacent lines.

The dispersion properties of the strip waveguide are summarized in Fig. 5.3. It is seen that the guided wave is slower than the Rayleigh surface wave on the substrate alone, and that the wave is slower for waveguides with higher aspect ratios (larger values of G). Furthermore, the wave is quite dispersive.

At low frequencies, the field spreads far out; therefore, almost none of the field is in the strip and it is almost all on the substrate in the form of a Rayleigh wave. Hence, at low frequencies the velocity approximates that of the Rayleigh wave on the substrate. As the frequency increases, more of the field is contained

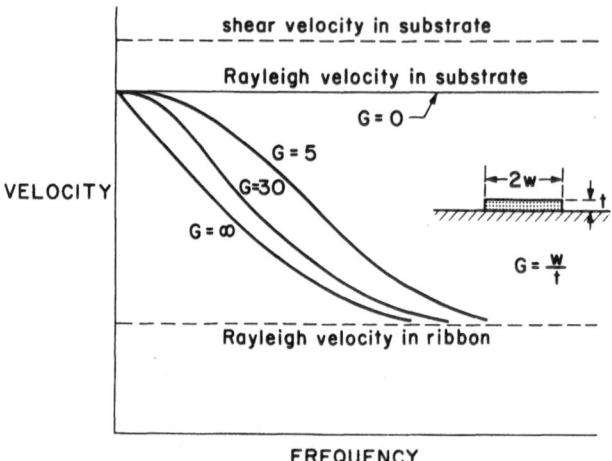

Fig. 5.3. Dispersion characteristics for the dominant mode on the strip waveguide, for various aspect ratios G

in the strip (slow) region, and the velocity decreases. At very high frequencies, almost all of the field in concentrated into the strip, and the velocity approaches that of the Rayleigh velocity on the strip material alone. Thus, both the transverse field distribution and the velocity of the guided wave are strong functions of frequency.

The strip waveguide was first proposed by *Seidel* and *White* in a series of two patents [5.10]. At that time (1967), *White* performed measurements [5.9] of the propagation characteristics of waveguides composed of a gold ribbon on a fused-quartz substrate, but these measurements were made only on strips which were *wide* and *flat*. Subsequently, *Tiersten* [5.14] published an approximate but accurate theory, and his calculations agreed very well with White's measurements. Tiersten's theory was completely analytical, but it utilized the equivalent of the transverse equivalent network of Fig. 5.2b in that he broke up his solution into the same separate regions in the same way, and introduced an approximate boundary condition to represent the equivalent of the boxes in Fig. 5.2b. Soon thereafter, a simpler theory was published by *Adkins* and *Hughes* [5.15] in 1969. Their theory also utilized the equivalent of the inner and outer transmission lines in Fig. 5.2b, but they dispensed entirely with the boxes shown; their solution was equivalent to a direct connection between the respective transmission lines. They also performed measurements on waveguides composed of gold ribbons on fused-quartz substrates, but again only for ribbons of large aspect ratio. Shortly afterward, a theory based on the microwave network approach was developed by *Oliner* et al. [5.12, 13], which approximated the boxes in Fig. 5.2b by simple transformers obtained by the use of overlap integrals.

For very wide and flat strips ($G > 30$), the results from all three theories were practically indistinguishable from each other and all agreed very well with the measurements. For $G = 15.6$, the theoretical values were still close to each other but distinguishable from each other, and they all agreed well with White's measurements. For smaller aspect ratios, such as $G = 5.0$, the theoretical values

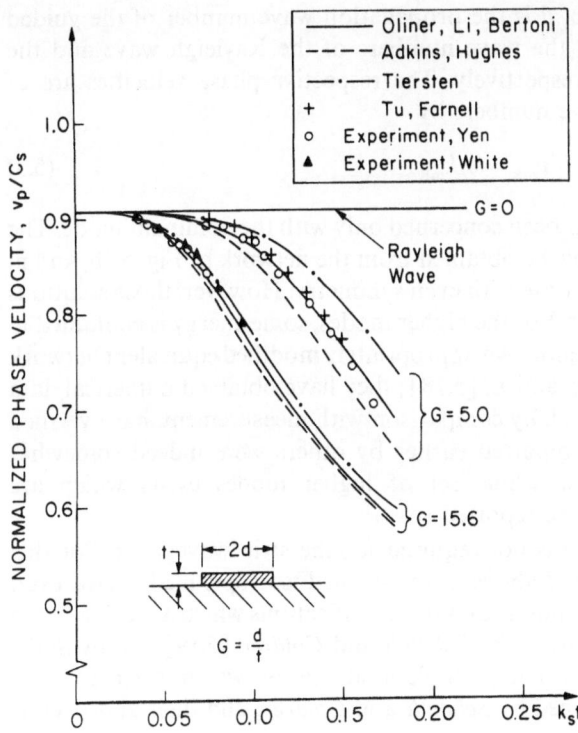

Fig. 5.4. Comparison with measurements of various theoretical solutions for the dominant mode on the acoustic strip waveguide (gold strip on a fused-quartz substrate). The theory of *Oliner* et al. is seen to agree best with the measurements (from *Yen* and *Oliner* [5.16])

were *no longer* close to each other. Recently, measurements were taken by *Yen* [5.16] at $G = 5.0$ to ascertain which of the theories gave the best correlation with measurement. It was found that his measurements agreed best with the theory of *Oliner* et al.; Tiersten's values were too high and those of *Adkins* and *Hughes* too low. Independent theoretical values obtained numerically by *Tu* and *Farnell* [5.17] agreed closely with the theory of *Oliner* et al. and with the measurements of *Yen*. These statements are summarized by the curves in Fig. 5.4. We may conclude that only for strips which are relatively narrow ($G < 15$ or so) is it necessary to distinguish among the theories; under those conditions, for accurate calculations, the theory of *Oliner* et al. [5.12, 13] is best.

Most of the flat overlay waveguides which have so far been considered for application (generally as delay lines in large capacity circulating memories) possess aspect ratios G greater than 15 or so. Since all of the available theories would yield essentially the same values in that range, the simplest formulation should be used. That formulation is the one by *Adkins* and *Hughes* [5.15], or equivalently, in terms of Fig. 5.2b, the transverse equivalent network without the connecting boxes. The dispersion relation for the dominant guided mode is given by the lowest resonance of this network, and it is obtained simply [5.13] as

$$\tan\left[(k_{mR}^2 - \beta^2)^{\frac{1}{2}} w\right] = \left(\frac{\beta^2 - k_R^2}{k_{mR}^2 - \beta^2}\right)^{\frac{1}{2}}, \tag{5.2}$$

where $2w$ is the strip width, β is the propagation wave number of the guided mode, and k_R and k_{mR} are the wave numbers of the Rayleigh wave and the modified Rayleigh wave, respectively. The respective phase velocities are of course related to these wave numbers by

$$v_p = \omega/\beta, \qquad v_R = \omega/k_R, \qquad v_{mR} = \omega/k_{mR}. \tag{5.3}$$

The discussion has so far been concerned only with the dominant mode. The *higher* modes can in principle be obtained from the network of Fig. 5.2b, and in fact (5.2) should represent those with even symmetry. However, those solutions would be somewhat in error. For the higher modes, some energy is contained in the Love wave in the strip region. An appropriately modified equivalent network has been derived by *Shimizu* and *Li* [5.18]; they have obtained numerical data for various higher modes and, by comparison with measurement, have verified that the simpler solutions reported earlier by others were indeed somewhat in error, and that another whole set of higher modes exists which are interspersed with those of the reported set.

A piezoelectric substrate is not required for the strip waveguide. For this reason, the analyses described above have assumed isotropic media. However, *anisotropic* substrates are common, and the modifications which arise due to the anisotropy have been considered by *Schmidt* and *Coldren* [5.19] and by *Sinha* and *Tiersten* [5.20]. A particularly simple modification which is accurate for very thin strips was introduced by *Schmidt* and *Coldren*, and their approach is summarized below.

They view the wave propagation along the strip guide in terms of modified Rayleigh wave rays which are totally reflected successively from the side walls of the strip, as mentioned above in this section. The angle ξ made by the ray with respect to the guide axis is given by

$$\beta = k_{mR}(\xi) \cos \xi, \tag{5.4}$$

where β is the propagation wave number of the guided wave, and the modified Rayleigh wave wave number is now a function of the angle ξ. The dependence of k_{mR} on ξ may be calculated numerically for specific materials, but *Schmidt* and *Coldren* use the simpler parabolic approximation [5.21] and express the wave number as

$$k_{mR}(\xi) = k_{mR}(\xi=0)(1 + \alpha\xi^2), \tag{5.5}$$

where α is the parabolic anisotropy factor. This approximation is valid for small deviations from the guide axis, which would be the case for very thin strips. The value for $k_{mR}(\xi)$ obtained from (5.5) is then substituted into dispersion relation (5.2) wherever k_{mR} appears. For k_R in (5.2), they use the value corresponding to $\xi=0$. When this simple approach was applied to other flat overlay waveguides by *Schmidt* and *Coldren*, they found very good agreement with their measurements.

5.2.2 The Shorting-Strip (or $\Delta v/v$) Waveguide

The shorting-strip waveguide consists of a very thin strip of metal plated on a substrate which is piezoelectric, as shown in Fig. 5.1b. The metal strip short circuits the electric field associated with the piezoelectric surface wave and, as a result, produces a slight reduction Δv in the velocity of that wave under the strip. Since the velocity v of the wave on the unplated substrate is thus somewhat higher than that of the wave under the strip, the strip portion is slow relative to the remainder of the substrate, and the structure is able to guide an acoustic wave in a fashion directly analogous to that discussed under Section 5.2.1. The strip there produced slowing by a mass-loading effect; here, the slowing is produced by a short-circuiting effect.

Because the effect of the short circuiting was originally referred to in terms of $\Delta v/v$, where the terms are defined above, this waveguide is usually called a $\Delta v/v$ waveguide. The notation is unfortunate, however, since *all* waveguides operate on the basis of some $\Delta v/v$ effect; it is therefore appropriate to employ a more descriptive name, such as the "shorting-strip" waveguide, as opposed to the "mass-loaded-strip" waveguide, or simply the "strip" waveguide.

Engan [5.22] was the first to recognize that the shorting strip would indeed guide surface waves. The theory presented by *Hughes* [5.23] appreciated the fact that piezoelectric, and therefore anisotropic, substrates are required, but his analysis is basically an isotropic one. His dispersion relation is similar to that in (5.2), but the wave numbers corresponding to plane waves in the plated and unplated regions are those along the guide *axis* in the anisotropic medium. Careful measurements by *Coldren* and *Schmidt* [5.24] showed deviations from the values predicted by Hughes' approximate theory, with the theoretical values for velocity coming out higher than the experimental data. When the additional simple modification [5.19] discussed under Section 5.2.1 was employed by *Coldren* and *Schmidt*, they obtained excellent agreement with their measurements. This modification was just to employ, in the strip region, the wave number corresponding to the zig-zag successive bounce direction rather than the guide axis direction.

The change Δv produced by the shorting strip is very small, so that the velocity of the guided wave is only 1 or 2 % less than the velocity of the Rayleigh wave on the substrate. As a result, the guided wave is only very weakly bound, and the acoustic field extends transversely for some distance. Nevertheless, the transverse dimensions of the guided wave remain the same as the wave progresses down the guide, thus avoiding the beam spreading associated with unguided surface wave beams. The shorted-strip waveguide is thus under consideration for long-line applications where the group velocity, rather than the phase velocity, must be maintained constant over some frequency range. Such behavior has been explored by *Adams* and *Shaw* [5.25] and by *Schmidt* and *Coldren* [5.19].

Typical results obtained by *Schmidt* and *Coldren* for an aluminum strip on a LiNbO$_3$ substrate are shown in Fig. 5.5. The dashed curves for both the phase

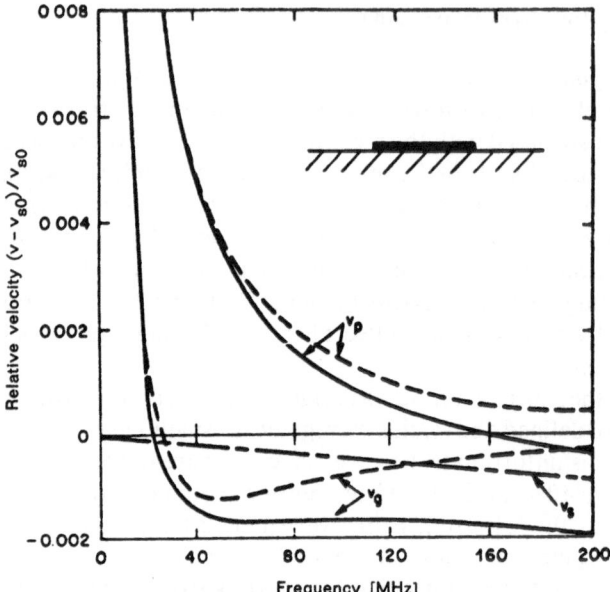

Fig. 5.5. Dispersion curves, showing group velocity dispersion compensation, for the shorting-strip (or $\Delta v/v$) waveguide consisting of an aluminum strip on a LiNbO$_3$ substrate. Dashed curves (for both phase velocity v_p and group velocity v_g) represent the shorting effect only. Solid curves include slight mass loading (represented by the v_s curve) added to produce dispersion compensation (from *Schmidt* and *Coldren* [5.19])

velocity v_p and the group velocity v_g are obtained by means of the approximate theory described above, and they take into account only the shorting effect of the thin metal strip. When the strip thickness is increased so as to provide a small amount of mass loading in addition (the amount indicated by the line with long and short dashes), the curves shown solid are produced. It is seen that dispersion in the curve for group velocity is nearly eliminated over a wide frequency range, from 45 to about 125 MHz. The aluminum strip was 180 microns wide and 500 Å thick. *Dispersion compensation* in the group velocity by means of added mass loading is therefore not only feasible, but also easily designed.

5.2.3 The Slot Waveguide

The slot waveguide, first proposed and analyzed by *Tiersten* [5.14], is in a sense the *dual* of the strip waveguide. The geometry of the slot waveguide is shown in Fig. 5.1c. It is seen that the plating now covers the substrate surface except for the region of the slot, within which the guiding occurs. Guiding is achieved because the material of the plating, which is stiff and light, is now faster than that of the substrate, a situation just opposite to that with the strip waveguide. As a result, the Rayleigh wave in the free-space region of the slot is slower than the modified

Rayleigh wave on the exterior plated region, and field confinement to the slot region is produced.

A transverse equivalent network similar to the one shown in Fig. 5.2b for the strip waveguide can be drawn by inspection since the plated and free-space regions are simply interchanged for the slot waveguide. The basic mechanism for guiding is similar to that for the strip waveguide except that now a Rayleigh wave is multiply reflected from the step discontinuities at the edges of the slot guide; similarly, the transverse field distribution strongly resembles that shown in Fig. 5.2c.

Since the basic mechanism for guiding is the same as that for the strip waveguide, the dispersion relation (5.2) also holds for the slot waveguide when appropriate duality replacements are made, namely, when the terms k_R and k_{mR} are interchanged. Furthermore, the modification introduced by *Schmidt* and *Coldren* [5.19] to account for substrate anisotropy, and discussed under Section 5.2.1, also holds here.

Despite the similarity to the strip waveguide in both the basic guiding mechanism and the transverse field behavior, the dispersion behavior for the slot waveguide differs strongly from that for the strip waveguide. For the slot waveguide, the curve of phase velocity vs. frequency possesses a maximum.

For a slot waveguide composed of an aluminum plating on a T-40 glass substrate, dispersion curves [5.13] in the form of normalized velocity vs normalized frequency are presented in Fig. 5.6 for the dominant mode, for several different aspect ratios G. At low frequencies, the acoustic field spreads far out laterally and also downward into the substrate, so that the plating exerts a negligible influence. Hence, at low frequencies the velocity approximates that of the Rayleigh wave on the substrate alone. As the frequency increases, the presence of the plating is noted by the field, and the velocity increases because the plating is fast. At very high frequencies, the field becomes almost completely confined to the slot region, which is unplated; the velocity must therefore again approach the Rayleigh wave velocity at the high frequency limit. The curve thus must possess a maximum, which turns out in some cases to be rather broad in frequency.

Because of the maximum exhibited in these curves, the phase velocity is essentially constant over a substantial frequency range. However, the deviation from the Rayleigh wave is small altogether, so that the wave is actually very loosely bound. Tighter binding can be achieved by a different choice of materials; *Sinha* and *Tiersten* [5.20] show that about twice the deviation in Fig. 5.6 can be achieved with sapphire on T-40 glass.

Theoretical calculations for isotropic materials were also made by *Li* et al. [5.13] employing the microwave network approach; their results agreed very well with those of *Tiersten* [5.14]. No accurate measurements are available for the isotropic slot waveguide, but some approximate experimental data by *Knox* and *Owen* [5.26] follow the shape of the theoretical curves very well over a 60% frequency range.

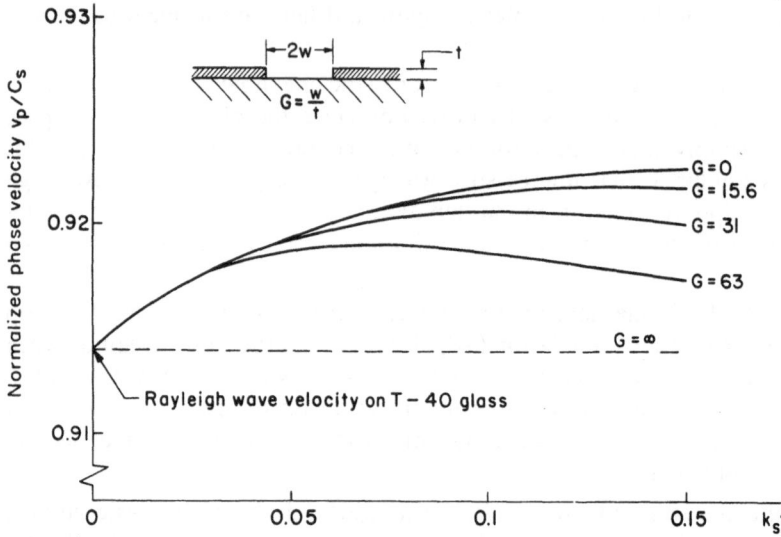

Fig. 5.6. Dispersion characteristics for the dominant mode of the slot waveguide shown in Fig. 5.1c, consisting of aluminum on a T-40 glass substrate, for various aspect ratios G. C_s and k_s are the substrate shear wave velocity and wave number (linearly proportional to frequency). Note the expanded ordinate scale (from Li et al. [5.13])

Accurate measurements on slot waveguides employing an anisotropic substrate were made by *Coldren* and *Schmidt* [5.27]. They employed an SiO film on $Bi_{12}GeO_{20}$, and they obtained excellent agreement with theory for both the phase velocity and the group velocity. The phase velocity data indeed showed the expected maximum in the curve of velocity vs frequency, and the group velocity curve flattened out above a certain frequency. The slot waveguide might therefore be considered a candidate for long-line applications which require a group velocity with small or negligible dispersion; the difficulty, however, is that higher modes are also present in that frequency range.

Coldren and *Smithgall* [5.28] have investigated the effect of tapering or otherwise varying the junction between the plated and unplated regions. This junction ordinarily consists of a rectangular shape, i.e., the walls are perpendicular to the substrate surface. A transition region between the plated and unplated regions is thereby introduced, and this added flexibility can be employed to improve the guiding characteristics. Measurements, which agree well with theory, were made only on slot guides with linearly tapered walls.

5.3 Topographic Waveguides

Topographic waveguides are guiding structures which are produced by a local deformation in the surface of the substrate, i.e., in the substrate topography. The guiding results from a reduction of the restraining forces acting on the material;

the deformation must be such as to create a protrusion rather than a depression, however.

The topographic waveguides are distinctly different from the flat overlay structures described in Section 5.2; the mechanism of guiding must be viewed differently, their modes are very tightly bound whereas those of the flat overlay structures are loosely bound, the loss in these modes is probably substantially lower (not measured), and so on. It is also pertinent to note that in many ways the behavior of the flat overlay waveguides resembles that of certain optical structures, so that experience could be carried over from electromagnetics to acoustics. Such experience is much less useful in the case of topographic structures since there are no counterparts to such waveguides in electromagnetics.

The two most important topographic waveguides are the rectangular ridge and the wedge; their properties are discussed in the subsections following. The rectangular ridge waveguide possesses two equally important and distinctly different dominant modes, however, and these modes are discussed independently. When the rectangular ridge becomes very tall and thin (such structures have been fabricated [5.29]), it resembles the edge of a semi-infinite plate, and the modes on such a structure take on special forms that are of interest in themselves. Such modes are treated in Section 5.3.3.

5.3.1 The Antisymmetric, or Flexural, Mode of the Rectangular Ridge Waveguide

This mode is the better known of the two dominant modes; it is strongly dispersive with strong field confinement. The topographic rectangular ridge structure is shown in Fig. 5.1d.

It was advantageous for the flat overlay structures to view the propagation in terms of constituent waves which bounce back and forth horizontally. In the topographic ridge, where the structure is tall and thin rather than flat and wide, it is best to think in terms of constituent waves which go vertically instead of horizontally. We shall return to this viewpoint shortly.

The mode possessing odd, or antisymmetric, symmetry is also known as the *flexural mode* because the ridge rocks back and forth as the wave progresses down it. Furthermore, the constituent waves in the plate structure comprising the ridge, which bounce up and down between the top and bottom of the ridge and form a standing wave there, are waves corresponding to the lowest flexural Lamb mode in the plate structure. The fields are strongest at the top of the ridge and they decrease to the bottom of the ridge in the form of an approximate quarter of a standing wave. The remaining energy decays exponentially into the substrate. These features are summarized in Fig. 5.7a.

Basically, two distinct approaches have been used to determine theoretically the properties of this mode: a finite-element procedure refined by *Lagasse* et al. [5.30–32] which produced excellent agreement with experimental data, and a *microwave network* approach [5.33, 34] which was less accurate but neverthe-

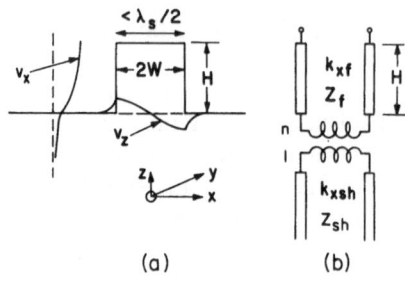

Fig. 5.7a and b. The dominant antisymmetric (flexural) mode on the topographic rectangular ridge waveguide. (a) Transverse field behavior; v_x and v_z are particle velocity components and λ_s is the shear wave wavelength. (b) Approximate transverse equivalent network; the vertical transverse wave numbers k_{xf} and k_{xsh} are respectively the x components of the lowest flexural Lamb wave in the plate and the horizontal shear wave in the substrate

less yielded good agreement with measurement. The network approach had the added virture of providing simple physical insight, and for that reason the transverse equivalent network obtained there is reproduced here as Fig. 5.7b.

In this particularly simple network, in which the transmission line direction is seen to be vertical, the transmission line with length equal to the ridge height represents rigorously the lowest flexural Lamb mode, the lower transmission line represents only approximately the influence of the substrate, and the junction between the ridge and substrate is approximated by a simple transformer. The resonances of this network yield the approximate dispersion relation [5.33, 34]

$$\cot[(k_f^2 - \beta^2)^{\frac{1}{2}}H] = \frac{k_s^2}{k_f^2}\left(\frac{k_f^2 - \beta^2}{\beta^2 - k_s^2}\right)^{\frac{1}{2}} \tag{5.6}$$

which is valid for the dominant and higher order flexural modes within the frequency range $\lambda_s/2 > 2W$. In (5.6), H is the height of the ridge, β is the propagation wave number of the guided wave, k_s is the shear wave number, and k_f is the wave number of the lowest flexural Lamb wave of the plate.

The description of the wave behavior summarized by Fig. 5.7a follows directly from the network in Fig. 5.7b. Furthermore, despite the very simple form of this network, numerical results computed from dispersion relation (5.6) agree quite well with measured data, especially for tall ridges.

The dispersion behavior for the lowest flexural mode on a duralumin ridge waveguide is shown in Fig. 5.8 for several different ridge aspect ratios. These data were computed by means of the finite-element method [5.30], and they agreed very well with measurements taken by *Mason* et al. [5.7]. It is seen that the behavior is very strongly dispersive, and that the curves always lie above the one for the lowest flexural Lamb mode on an infinite plate, shown dashed in Fig. 5.8. At lower frequencies, the curves seem to approach a low-frequency cutoff; this result is contradicted by dispersion relation (5.6), which predicts that propagation continues down to zero frequency, without a low-frequency cutoff, and in fact terminates at the shear wave velocity. Unfortunately, this discrepancy is not resolved by either the measurements or the finite-element calculations because both become inaccurate in that frequency range.

For a large portion of the frequency range it is seen that the guided wave is substantially slower than the Rayleigh wave. As a result [see (5.1)], the field decay on the substrate surface transversely away from the ridge is moderately

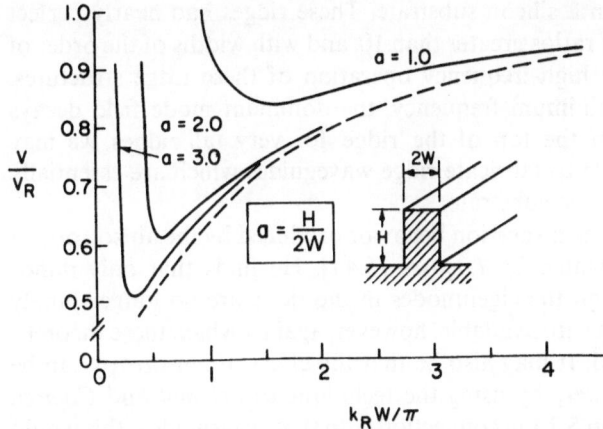

Fig. 5.8. Dispersion characteristics of the dominant antisymmetric (flexural) mode on the topographic rectangular ridge waveguide, for different aspect ratios, $a = H/2W$. v_R is the Rayleigh wave velocity (after *Mason* et al. [5.7])

strong. However, most of the acoustic wave energy resides in the ridge itself, and that feature is the one which contributes primarily to the strong confinement for this mode.

It has also been shown [5.30] that if the sides of the ridge are sloped (rather than remaining vertical), with the top of the ridge becoming narrower than its base, the dispersion characteristic in Fig. 5.8 becomes flattened. Increasing the slope still further would result in a wedge located on a substrate, as shown in Fig. 5.1e. Such a structure would become *nondispersive* above a certain frequency, as predicted by *Lagasse* [5.35].

The network of Fig. 5.7b also yields information on the *higher* flexural modes. Such modes have a low-frequency cutoff, but they become leaky at frequencies below cutoff. *Markman* [5.36] has studied these higher modes in some detail. Other studies on the behavior of higher modes have been reported by *Lagasse* et al. [5.37]; they present both dispersion data and displacement patterns.

Recent analyses [5.38–43] have been made of a flexural mode guided along the top edge of a ridge of semi-infinite height. Some inconsistencies appear in these papers, as has been observed [5.42, 43], but all authors agree that the mode decays exponentially from the top edge, rather than possessing the standing wave behavior shown in Fig. 5.7a. It has also been noted [5.42] that at higher frequencies the dominant flexural mode field distribution on a ridge of finite height *crosses over* at some frequency from a standing wave form to the decaying form exhibited by the wave on a ridge of semi-infinite height, and maintains this decaying form at all higher frequencies. With respect to Fig. 5.8, it means crossing the dashed line by going below it. Furthermore, at very high frequencies, the velocity for the dominant mode approaches that for a 90° top-angle wedge. Further details are presented in Section 5.3.3.

The above results are significant when taken in conjunction with the fabrication technique reported by *Rosenfeld* and *Bean* [5.29]. They demonstrated that by using selective etching techniques on silicon they could fabricate very tall and narrow ridges on a silicon substrate. These ridges had nearly perfect vertical walls, with aspect ratios greater than 10, and with widths of the order of microns, as required for high-frequency operation of these ridge structures. Since, above a certain minimum frequency, the dominant mode field decays exponentially down from the top of the ridge for very tall ridges, we may conclude that it is possible to fabricate ridge waveguides which are essentially completely isolated from the substrate.

The modifications in the dispersion behavior produced by the anisotropy of silicon have been investigated by *Lagasse* [5.44]. He finds that only minor changes result even though the eigenmodes in the ridge are no longer purely flexural. No measurements are available, however, against which these theoretical results can be checked. It may also be that the effect of anisotropy can be accounted for approximately by using the technique of *Schmidt* and *Coldren* [5.19], described in Section 5.2.1 in connection with strip waveguides; this would require replacing k_f in dispersion relation (5.6) by the value k_f would have in the zig-zagging ray direction in the particular anisotropic crystal.

5.3.2 The Symmetric, or Pseudo-Rayleigh, Mode of the Rectangular Ridge Waveguide

This mode propagates down to zero frequency, and is characterized by very strong field confinement, particularly for tall ridges, and by almost no dispersion over a very wide frequency range. It is sometimes called the pseudo-Rayleigh mode because the field distribution of the mode in the ridge resembles a slice out of a Rayleigh wave, particularly at the lower frequencies [5.45].

Whereas the field in the ridge in the case of the flexural mode was essentially that of a single mode, the lowest flexural Lamb mode, here that field consists of two modes, the lowest symmetric Lamb mode and the lowest SH mode in the plate comprising the ridge. A transverse equivalent network similar to that of Fig. 5.7b for the flexural mode has been derived [5.34, 45], but it is not as simple in form. Nevertheless, a relatively simple dispersion relation follows from it, namely,

$$(k_s^2 - 2\beta^2)^2 \frac{\sqrt{\beta^2 - k_L^2} + \sqrt{\beta^2 - k_p^2} \tanh \sqrt{\beta^2 - k_L^2} H}{\sqrt{\beta^2 - k_p^2} + \sqrt{\beta^2 - k_L^2} \tanh \sqrt{\beta^2 - k_L^2} H} = 4\beta^2 \sqrt{\beta^2 - k_s^2} \sqrt{\beta^2 - k_L^2},$$

$$(5.7)$$

where H is the ridge height, β is the propagation wave number of the guided wave, k_s and k_p are the shear and compressional wave wave numbers, and k_L is the wave number of the lowest symmetric Lamb wave in the plate comprising the ridge.

The two waves in the ridge couple together at the top of the ridge in such a way as to resemble the coupling of the p and s bulk waves which combine to form a Rayleigh wave at a free planar surface. An examination of the acoustic field

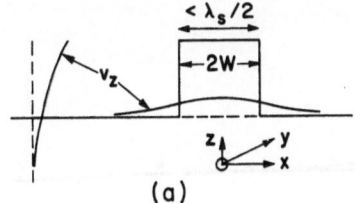

(a)

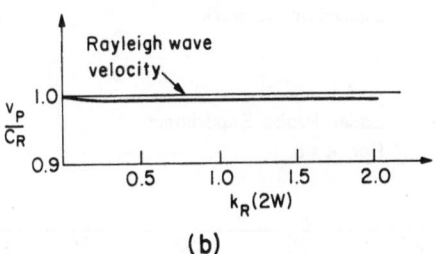

(b)

Fig. 5.9a and b. The dominant symmetric (pseudo-Rayleigh) mode on the topographic rectangular ridge waveguide. (a) Transverse field behavior; v_z is a particle velocity component and λ_s is the shear wave wavelength. (b) Dispersion behavior; C_R and k_R are the velocity and wave number of the Rayleigh wave

components of the ridge guided mode brings out further its resemblance to a slice of a Rayleigh wave. The field configuration at some low frequency is summarized by Fig. 5.9a; at high frequencies the decay from the top of the ridge is so strong that the wave is essentially isolated from the substrate.

The dispersion behavior, as obtained from (5.7), is shown in Fig. 5.9b; the same data are repeated in Fig. 5.10 using a greatly expanded ordinate scale to show the effect of different aspect ratios. Measured results by *Yen* et al. [5.46] superimposed on Fig. 5.10 show very good agreement with the theory.

At very low frequencies, the field decay is small so that the presence of the substrate is felt. At zero frequency, all the energy must be in the substrate so that the curve must approach the Rayleigh wave velocity. Above a certain frequency, which is seen to be quite low, the behavior becomes identical with that of a symmetric mode guided along the top edge of a ridge of semi-infinite height. That mode, which is discussed further in Section 5.3.3, is known to possess very little dispersion over a very wide frequency range [5.46, 47]. Dispersion relation (5.7) is actually valid only for frequencies below about $k_R(2W) = 2$; above that range additional propagating modes in the plate would need to be taken into account.

As pointed out in Section 5.3.3, it is expected that the curve would approach the velocity for a 90° top-angle wedge at very high frequencies. The velocity would then be expected to vary only a little more than 1 % from the frequency for which the substrate can be ignored all the way to infinite frequency. Thus, the mode may be viewed as *essentially dispersionless* over some finite frequency range, but a more accurate quantitative statement cannot be made at this time.

As in the case of the flexural mode, the tall and narrow ridge waveguides obtainable by selective etching would permit the acoustic field to be confined near the top of the ridge and to be isolated from the substrate. In contrast to the flexural mode, however, which is highly dispersive, the symmetric mode

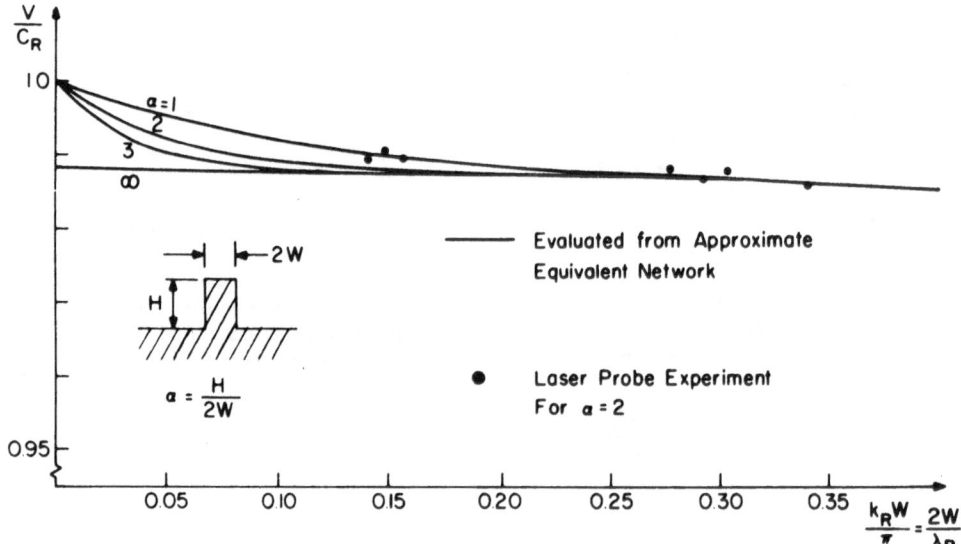

Fig. 5.10. Dispersion characteristics for the dominant symmetric (pseudo-Rayleigh) mode on the topographic rectangular ridge waveguide, using an expanded ordinate scale to show the effect at low frequencies of different aspect ratios. Curves represent theory using (5.7); points are measured (from *Yen* et al. [5.46])

possesses essentially no dispersion over a wide frequency range. Furthermore, it does not suffer as strong competition from higher modes. The symmetric mode should therefore be considered seriously for long delay-line applications.

The symmetric ridge guide mode is also as close as one might come to an acoustic coaxial line mode or strip line mode, in the sense that it is very tightly confined, it propagates down to zero frequency, and it is essentially dispersionless.

5.3.3 Waves Guided by a Plate Edge

Ridge waveguides of finite and semi-infinite height are shown in Fig. 5.11; the basic modes which can be supported by the top edge of the semi-infinite structure, also called a plate edge, are discussed now. As stated above, the ridge waveguide of Fig. 5.11a can support two basic propagating modes, an antisymmetric, or "flexural", mode and a symmetric, or "pseudo-Rayleigh", mode. The field of the flexural mode forms a *standing* wave in the ridge region, but it decays into the substrate away from the ridge-substrate interface. The pseudo-Rayleigh mode field, on the other hand, *decays down* the height of the ridge from the top edge, and then decays further into the substrate at a different rate.

The edge of the semi-infinite plate (shown as Fig. 5.11b) can also support two basic propagating modes. The field of the symmetric, or pseudo-Rayleigh, mode

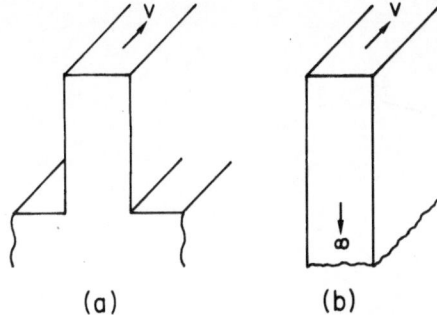

Fig. 5.11a and b. Ridge waveguides of finite and semi-infinite height. (a) Usual ridge waveguide, with ridge of finite height. (b) Edge of a semi-infinite plate, corresponding to a ridge of semi-infinite height (from *Oliner* and *Lagasse* [5.42])

(a)   (b)

decays down from the plate edge, and it therefore resembles the corresponding mode of the ridge waveguide. The field of the antisymmetric, or flexural, mode *also decays down* from the plate edge; it therefore *differs* in its behavior from the flexural mode of the ridge waveguide even though other features of the field structure may be similar in each.

Let us consider first the symmetric, or pseudo-Rayleigh, mode guided by the plate edge. Several different solutions are available in the literature for the behavior of this mode. The oldest and simplest is that employing generalized plane stress; *Oliver* et al. [5.48] utilized this approach in the analysis of thin plates for two-dimensional seismological models. This approach takes into account only the two lowest symmetrical modes of the infinite plate, and it produces the result that if Poisson's ratio v is replaced by $v/(1 + v)$, the solution for the Rayleigh wave on an elastic half-space yields the result for the wave guided by the free edge of a plate. This result is of course dispersionless.

A similar approximation was made by *Tamm* and *Weiss* [5.49], although the phraseology was slightly different. They described a "disc" approximation in which the plate was assumed to be acoustically thin with constant stresses across the plate cross section. Their analysis also involved the two lowest plate modes, and also led to a dispersionless solution in which Poisson's ratio v was replaced by $v/(1 + v)$.

McCoy and *Mindlin* [5.47] extended the generalized plane stress result by taking into account three additional modes of the infinite plate, while still employing second-order equations. These additional modes are higher modes which are below cutoff (two with complex propagation constants and one with an imaginary one) at the lower frequencies of interest. At really low frequencies, their solution reduces to the generalized plane stress result; however, it is also valid at the higher frequencies where it exhibits slight dispersion. Their approach also shows that a second edge wave appears above a certain frequency.

More recently, another result for the edge wave was obtained [5.46] which should be very accurate for low frequencies, and which exhibits slight dispersion. It reduces identically (in analytic form) to the solution presented by *Tamm* and *Weiss* [5.49] in the limit of zero frequency. This dispersion relation, which is valid for frequencies up to $k_R W/\pi (= 2W/\lambda_R) = 0.5$, where $2W$ is the width of the plate

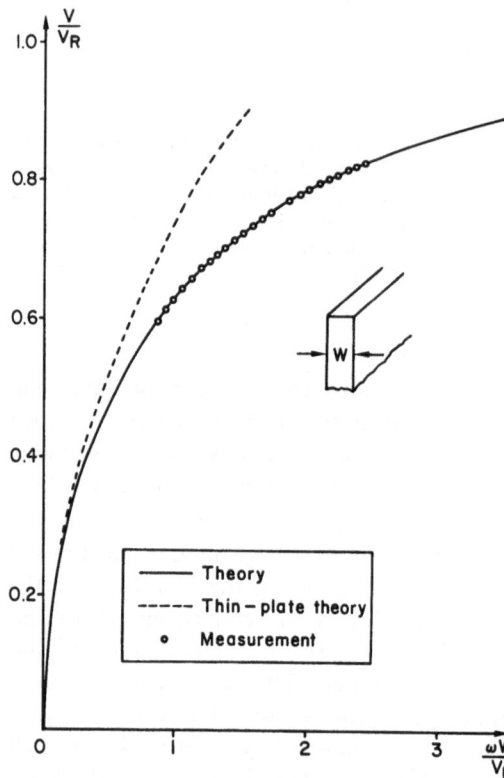

Fig. 5.12. Dispersion behavior for the flexural plate-edge mode. Comparison of the finite-element calculations with thin-plate theory and with measurements (from *Oliner* and *Lagasse* [5.42])

and $k_R = 2\pi/\lambda_R$ is the propagation wave number of a Rayleigh wave, has the following simple form:

$$(k_s^2 - 2k_z^2)^2 = 4k_z^2 \sqrt{k_z^2 - k_s^2} \sqrt{k_z^2 - k_L^2}. \tag{5.8}$$

Wave number k_z is the propagation wave number of the edge wave under discussion, and k_s and k_L are respectively the wave numbers of shear bulk waves and of the lowest symmetrical Lamb mode in the plate. For the remaining quantities, ϱ is the mass density, μ the rigidity, and ω the angular frequency.

The derivation of dispersion relation (5.8) also involves only the two lowest symmetric (propagating) plate modes, and it is this feature which limits the range of validity for relation (5.8). The method employed a microwave network approach, and the derivation is presented in [5.46].

Two sets of measurements have been made of the wave guided by the free edge of a plate: those reported by *Sinclair* and *Stephens* [5.50] on a brass plate, and more recent ones taken by *Yen* et al. [5.46] on a microscope slide (soda lime glass). The latter measurements were made with the aid of a laser probe and are believed accurate to within 0.2%.

The results of these measurements are summarized in [5.46], and comparisons are made with all the available theories. These comparisons show first that the *symmetric* edge wave does possess a very small amount of dispersion; the phase velocity is only a percent or two different from the Rayleigh wave velocity, and its variation is monotonic and is less than 1 percent over the frequency range covered. It is also concluded that for very low frequencies, that is, for $2W < 0.2\lambda_R$ or so, the generalized plane stress result should be used because of its relative simplicity. For intermediate frequencies, that is, $0.2\lambda_R < 2W < 2\lambda_R/3$, greatest accuracy is obtained by using (5.8); for still higher frequencies, that is, for $2W > 2\lambda_R/3$, the *McCoy* and *Mindlin* solution [5.47] is the best presently available.

With respect to the *antisymmetric* or *flexural* plate-edge mode, it was observed in Section 5.3.1 that some inconsistencies have appeared in several papers treating this mode theoretically. In [5.38, 39], the authors present solutions based on the thin-plate approximation, which are correct in the range of very low frequencies but not valid for higher frequencies. The limitations of those results and comments on other results are contained in [5.42, 43]. The latter references also present numerical results for this mode determined from a finite-element analysis, and measured results obtained by means of a ring resonator experiment. The measurement details and some aspects of the calculations are presented in *Oliner* and *Lagasse* [5.42], together with many results.

A comparison between the measured and calculated dispersion data is presented in Fig. 5.12; it is seen that the agreement is excellent. Also shown are the thin-plate theory results, which are seen to be correct only for very low frequencies. From the calculations, it was also determined that the phase velocity of this mode is only about one-half percent less than the phase velocity of the lowest flexural mode of the infinite plate. Furthermore, except very near to the plate edge, the vertical decay rate and the field dependence across the plate width correspond to those of the infinite-plate mode.

The mode behavior at high frequencies is particularly interesting. In the high-frequency limit, where the width of the plate is large compared to the wavelength, the flexural Lamb wave of an infinite plate is known to tend to a combination of Rayleigh waves, each associated with one face of the plate. At high frequencies, the mode can thus be viewed as the interaction between two Rayleigh waves. Similarly, one may speculate that at high frequencies the flexural plate-edge mode can be modeled as a combination of *two wedge modes* in antiphase propagating along the 90° wedges formed by the corners of the plate edge. This would mean that the high-frequency asymptote for the phase velocity of the plate-edge mode is given by the velocity of a 90° top-angle wedge mode.

This point of view has been verified by means of dispersion data and by displacement patterns obtained via finite-element computations. The dispersion behavior of the plate-edge mode over a very wide frequency range is shown in Fig. 5.13. The velocity data are plotted on an expanded scale, normalized to the velocity of the infinite-plate flexural mode, in order to clearly indicate the asymptotic behavior at both low and high frequencies. We see that the low-

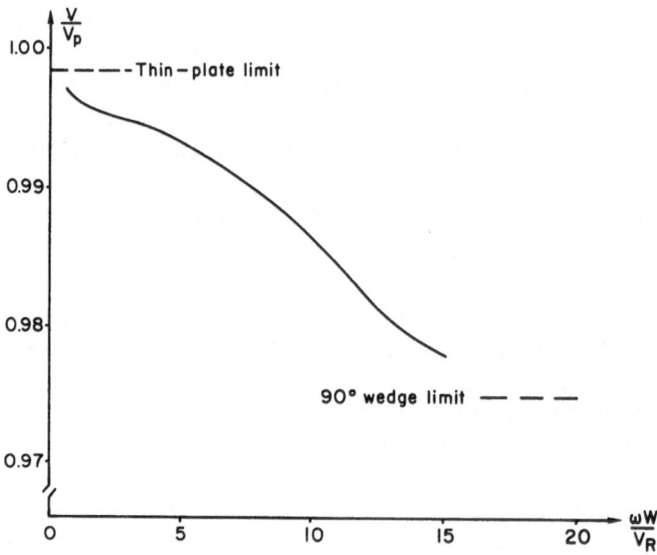

Fig. 5.13. Dispersion behavior for the flexural plate-edge mode over a very wide frequency range. Note the greatly expanded ordinate scale (from *Oliner* and *Lagasse* [5.42])

frequency asymptote is given by the thin-plate theory, and that the high-frequency asymptote corresponds to the velocity for a 90° wedge.

Oliner and *Lagasse* [5.42] also show how the flexural mode on the plate edge is related to the flexural mode on a ridge guide of finite height. Their conclusion is the following: On a ridge of finite height, there exists only one dominant flexural mode. At lower frequencies, its form is that of the usual ridge guide mode, with a standing-wave vertical dependence. At some sufficiently high frequency, depending on the ridge size and aspect ratio, this mode *changes over* into the plate-edge mode, with a vertical decay. The plate-edge mode then is maintained for all higher frequencies.

The flexural plate-edge mode, which is the only flexural mode on a semi-infinite plate, is thus seen to be the high-frequency form of the dominant flexural mode on a ridge of finite height. When real ridge structures are very tall and thin (and they have indeed been fabricated in that fashion [5.29]), the plate-edge solution applies over most of the usable frequency range.

It was shown above that in the high-frequency limit the flexural plate-edge mode is comprised of two 90° wedge modes in anti-phase. Similarly, the symmetric, or pseudo-Rayleigh, plate-edge mode reduces to two in-phase 90° wedge modes. Both plate-edge modes should thus possess the same asymptotic value of phase velocity—that appropriate to a 90° top-angle wedge. The dispersion behavior for both plate-edge modes on aluminum is summarized in Fig. 5.14, where it is seen that the antisymmetric, or flexural, mode is strongly dispersive, in contrast to the symmetric one. In line with observations above, the symmetric mode varies only about 1 % from zero frequency to infinite frequency,

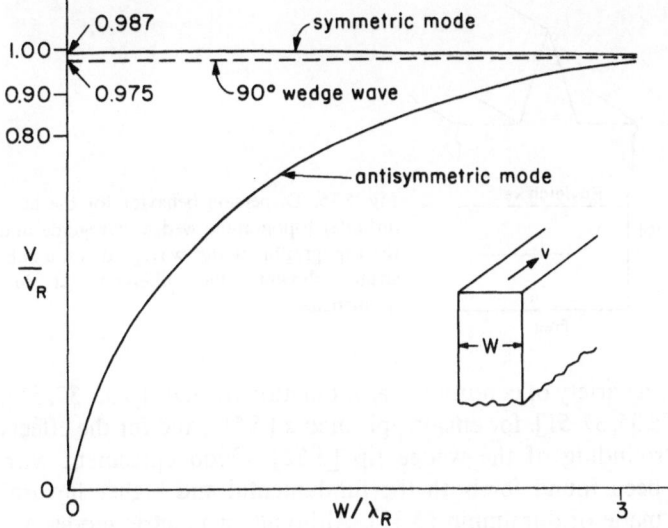

Fig. 5.14. Phase velocities, normalized to the Rayleigh velocity, of the symmetric and antisymmetric plate-edge modes over a very wide frequency range. Both modes are asymptotic at high frequencies to that on a 90° wedge

so that over some reasonable finite bandwidth the mode is *essentially dispersionless.*

The essentially dispersionless nature of this mode becomes of particular importance when it is recalled that tall and thin ridges can be fabricated [5.29].

5.3.4 The Wedge Waveguide

The topographic waveguide consisting of a wedge on a substrate is shown in Fig. 5.1e. An ideal wedge (substrate infinitely far away) is defined only by the angle of the wedge, without any characteristic lengths; as a result, one would expect the propagation velocity to be wavelength independent, i.e., dispersionless. Calculations made by *Lagasse* [5.35] show that this is indeed the case.

Typical dispersion behavior for an ideal infinite wedge and a wedge (with the same angle and material properties) on a substrate are shown in Fig. 5.15. The curve for the ideal wedge is a straight line, constant with frequency. For the wedge on a substrate, essentially all of the energy resides within the wedge portion at the higher frequencies, and the structure behaves as an infinite wedge. As the frequency is lowered, some of the energy spreads into the substrate and the velocity value rises, as shown in Fig. 5.15. This effect is borne out by both calculations and measurements [5.37]. It is not known, however, precisely how the curve approaches zero frequency.

No analytical theory is available for the wedge waveguide, but *Lagasse* and his colleagues have presented accurate finite-element calculations for wedge

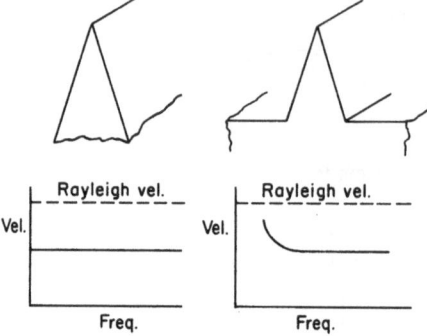

Fig. 5.15. Dispersion behavior for the ideal (infinite) topographic wedge waveguide and the topographic wedge waveguide on a substrate, showing the difference at low frequencies

performance under a variety of conditions: as a function of angle [5.35, 37, 51], for higher modes [5.35, 37, 51], for anisotropic media [5.51], and for the effects of truncation or rounding of the wedge tip [5.52]. Good agreement with measurement has been found for both the fundamental and higher flexural modes on wedges made of duralumin [5.37]. Although symmetric modes are possible under certain conditions on wedges, no systematic knowledge is available on the angles for which they exist or where they leak and where they are completely bound. The antisymmetric, or flexural, modes on wedges are better behaved and are capable of more varied performance; the considerations below apply only to such modes.

Depending on the wedge angle, more than one mode may be present simultaneously. For each mode, the acoustic wave field is tightly bound to the apex, with the particle displacement occurring primarily normal to the plane bisecting the wedge apex angle, and decaying exponentially in radial directions away from the apex. It was found empirically [5.37] that the guided wave velocity v is strongly dependent on the apex angle θ, according as

$$v \cong v_R \sin(m\theta), m = 1, 2, ..., m\theta < 90^\circ, \tag{5.9}$$

where v_R is the Rayleigh wave velocity and m is an integer denoting the mode order.

It is seen from (5.9) that for small wedge angles for the dominant mode one can obtain an essentially dispersionless mode which is quite *slow*. For example, a 15° wedge would support a guided wave with a velocity only 1/4 that of a Rayleigh wave. A narrow-angle wedge wave, because of its high acoustic field density, dispersionless feature, and low velocity, is an excellent candidate for devices which employ *nonlinear interactions*; this fact has been recognized by *Ash* and his colleagues [5.37, 53] and experiments have been performed.

Two difficulties mitigate against the practicality of narrow-angle wedges. One is that higher modes will be present simultaneously; however, if the wedge angle exceeds 45° the guide in effect will support only a single mode. The second difficulty involves fabrication and the integrity of the wedge apex, since most of the field is concentrated there. *Lagasse* et al. [5.52] have examined theoretically the effect of tip truncation or rounding and found that some dispersion (a few percent in velocity, depending on the truncation) is produced. However, if a discontinuity in truncation is present, as a result of damage, for instance, a serious reflection could be created. In an attempt to overcome this difficulty, *Wagers* and *Weirauch* [5.54] fabricated wedge structures in a novel supporting arrangement: a z-shaped structure with the horizontal portions of the z extended, coinciding with the substrate surfaces which are of course different on the two sides of the z. The overhanging wedge, with an angle about 60°, was selectively etched from quartz; the apex roughness is stated to be less than 0.1 micron. Preliminary measured results were encouraging. Unless fabrication techniques such as that above are successful, wedge waveguides may be useful only at lower frequencies.

5.4 Other Types of Waveguide

This section considers the properties of other types of waveguide, which are here placed into three groups: in-diffused, or similar, structures; rectangular (not flat) overlay structures; and circular fiber waveguides.

The in-diffused structures were proposed some time ago, but little work has been done on them; however, as a result of recent investigations on planar in-diffused layers, which showed them to possess very low loss, we may well expect a revival of interest. The rectangular, as opposed to flat, overlay waveguides comprise structures for which calculations are available, but which may not be practical. Several circular fiber configurations are being investigated, particularly lately, as candidates for long delay lines. These waveguide types are discussed in the subsections which follow.

5.4.1 In-Diffused Waveguides

The waveguides in this category, represented by Fig. 5.1f, are ones for which a local change has been produced in the properties of the substrate material. After the change, the substrate surface remains geometrically flat. This change may be produced in several different ways, such as depoling in the case of a ferroelectric [5.55], in-diffusion to produce loss [5.10] or a velocity change [5.2, 11], or ion implantation [5.56].

These structures were proposed some time ago, and were mentioned in the excellent review paper by *White* [5.57], but almost no work has been done on them since then despite the fact that in-diffused waveguides have been treated seriously in the context of integrated optics. In a sandwich form, which corresponds to a realistic model if the diffusion penetrates deeply and uniformly enough, the propagation characteristics of the resulting waveguide can be *determined using the dispersion relation* (5.2) presented in the context of flat overlay waveguides. This calculation is valid provided that the modified Rayleigh wave decay in the diffused region occurs completely within it, that the diffusion is uniform and the sides are steep, and that we can obtain the region's properties from other measurements.

It is more likely, however, that the in-diffusion process is not uniform with depth. In that case, other calculations are required based on the actual diffusion profile, which may, for example, be Gaussian.

The most interesting recent development in this connection relates to the question of *loss*. Waveguides were not fabricated in this study, but it was found recently by *Schmidt* [5.11] that a) the diffusion of certain metals into $LiNbO_3$ produced a velocity increase of about 1–2%, and b) there was no apparent increase in loss due to the in-diffusion process. The metals used were titanium, nickel, or chromium, and it is believed that the velocity change, which is roughly proportional to the percentage of diffusant atoms to Nb atoms, can be made as high as 5% by this process. The loss measurements, which were made over a 1 cm propagation length at frequencies as high as 400 MHz, indicate that the increase in loss due to the diffusion process must be less than 0.1 dB. At 400 MHz, the $LiNbO_3$ propagation loss itself is about 0.4 dB/cm, and the air loadening loss is about 0.2 dB/cm.

The biggest criticism leveled against flat overlay waveguides is that the layer itself contributes significant loss. If the velocity change can instead be produced by such an in-diffusion process, which introduces negligible added loss, the properties of some of the flat overlay guides could be achieved without the loss disadvantage. Since the process described by *Schmidt* results in a velocity increase, it could be used to create waveguides of the slot type.

Hartemann [5.56] has reported recently that a velocity *decrease* of the order of 1 or 2% can be produced in $LiNbO_3$ by ion implantation. He attributed the effect to a decrease which is produced in the electromechanical coupling coefficient, but he had not performed any loss measurements. His process could produce a strip waveguide, but we do not yet know what its loss properties would be.

5.4.2 Rectangular Overlay Waveguides

The class of structure considered here is formed of a rectangular ridge of one material placed on a substrate of another material. These structures differ from the flat overlay type in that the overlay is not thin, but is rather thick and may even resemble the topographic ridge guide in aspect ratio even though the ridge is now comprised of a different material. There is no indication at this time that these structures are of practical value; rather, calculations of their properties have been made, and such availability should be noted.

Calculations of the guided wave characteristics of such rectangular overlay structures have been carried out by *Waldron* [5.58], *Tu* and *Farnell* [5.17],

Lagasse et al. [5.37], and *Markman* et al. [5.34]. In the first three of these, the aspect ratio is effectively arbitrary, and the calculation is complicated by the fact that many constituent transverse modes couple together at the guide walls. In the work by *Markman* et al., the guide size and aspect ratio are similar to those of the topographic rectangular ridge waveguide. As a result, a simplified analysis is possible, and it has been accomplished using the microwave network approach. A possible advantage of this structure over its topographic counterpart is that it possesses stronger field confinement properties, but the nondispersive character, for the symmetric mode, will occur only above a certain frequency.

5.4.3 Circular Fiber Waveguides

This class of waveguide has been under recent investigation by personnel from the Bell Laboratories with the aim of obtaining a low loss, low dispersion waveguide suitable for long delay-line applications. Three structures of circular cross section have been analyzed and measured; two are shown in Fig. 5.1g and h, and the third is an anisotropic solid cylinder.

The earliest of these is the *capillary* structure shown in Fig. 5.1g, reported by *Rosenberg* et al. [5.59]. In this capillary tubing, the acoustic surface wave is guided along the interior surface of the tubing; the fundamental mode is basically a Rayleigh wave which has been curled up into a cylinder coaxial with the propagation direction. There is no azimuthal component of displacement, and the energy falls off approximately exponentially with radial distance from the inner surface. The mode has a low-frequency or low-radius cutoff, and its phase velocity possesses dispersion, whereas the group velocity is relatively flat with frequency over some range. Measurements were performed on a fused silica tubing which was acoustically lossier than the best silica grade, but "extremely clean" rf signals were transmitted through it nevertheless.

An advantage of this structure is that it can be drawn out of standard fused-quartz tubing by available thermal techniques without measurably raising the material attenuation above the bulk-wave level. If similar techniques would work with single-crystal materials with lower loss, the potential of this structure would be enhanced. In [5.59], considerable difficulty was mentioned in connection with efficient excitation of the desired mode; since that time, transduction has been greatly improved, with the total insertion loss through a tube excited at 16 MHz with a 15 % relative bandwidth being as low as 6 dB, and that amount being virtually all assignable to transducer conversion loss [5.60].

The second type of circular waveguide, shown in Fig. 5.1h, is analogous to the well-known *cladded-core* optical fiber. The experimental version, built and tested by *Boyd* et al. [5.61], used a core consisting of TiO_2-doped fused quartz, in which the doping did not significantly increase the loss, and a cladding of pure fused quartz; the bulk shear velocity difference between core and cladding was 3.8 %. Structures of this type, but with a cladding of infinite diameter, were first analyzed by *Waldron* [5.62]. The modes of interest must have most of the acoustic wave energy confined to the core region; with that restriction, so that negligible acoustic energy is present at the cladding outer boundary, the lowest radial mode and the lowest torsional mode, each above their "cutoff" frequencies and below those for the higher modes, are both suitable. Their behaviors are similar; the phase velocities are dispersive, but the group velocities possess a broad minimum around which the wave is reasonably nondispersive. The dispersion is in fact comparable to that found for shorting-strip (or $\Delta v/v$) waveguides.

Good results were obtained experimentally. With respect to loss, the bulk attenuation of the core and cladding, which varies as the square of the frequency, is about 1.8 dB/cm at 30 MHz, so that loss is not expected to be a problem. This waveguide has an advantage over planar structures for long delay-line applications because of the simple fabrication possible since no surfaces need be prepared except where the transducers are attached. Furthermore, the cladding provides a nonlossy support which permits the coiling of the guide for long delays. In addition, the air-loading loss associated with surface wave waveguides is avoided.

A novel type of waveguide which takes advantage of *anisotropy* was described recently by *Coldren* [5.63]. The structure which he analyzed and measured was a circular uniform rod, but that description is insufficient. What is needed is a small hump or depression, rounded in cross section and uniform in length, so as to form a waveguide; then the Rayleigh wave velocity as a function of the crystalline surface normal angle must experience a minimum at the center of this hump or depression. Even though this velocity reduction is due to a change in crystal cut, rather than a reduction in material restraining forces, *Coldren* refers to this guide as a topographic one. If the rounded hump is a part of a circular cylinder, then guiding occurs along only a small portion of the cylinder's periphery.

Measurements were performed on a $Bi_{12}GeO_{20}$ rod with a radius of 1.5 mm at a frequency of 90 MHz; good agreement was found with theory. The guide may be of practical interest because of the relatively small and noncritical amount of surface curvature required for good acoustic energy confinement, the lack of any critical dimensions, and the absence of any deposited films.

Sapphire has been identified as a good single-crystal material for the various waveguides of circular cross section described above, because it is both low loss and rugged. However, sapphire is anisotropic, and the information presently available for modes along circular rods, whether cladded or uncladded, holds for isotropic materials. A recent study aimed at investigating the effects of anisotropy in sapphire rods was conducted by *Wilson* [5.64]. Another development which may occur in the near future is the adaptation of the indiffusion process to the creation of graded-index circular rods for low loss, dispersionless acoustic guiding [5.65], in analogy to developments in fiber optics.

5.5 Summary and Conclusions

The most important properties of a variety of waveguiding structures for acoustic surface waves, and in some cases the implications of those properties, were presented in Sections 5.2–5.4. In this concluding section, we first summarize those properties in a comparative fashion and then, on the basis of those properties, describe certain applications for waveguides, both actual and potential.

The summary of properties is presented in tabular form, with special attention devoted to the category of loss. The elaboration with respect to loss is made because many people have the largely unfounded belief that waveguides are in general lossy, and have been discouraged from considering them for applications on that ground. The discussion on applications summarizes those areas in which waveguides are of particular value, presents illustrations to indicate that with ingenuity waveguides should be useful in improving device performance, and presents a speculation with respect to the future.

5.5.1 Summary of Waveguide Properties

Each of the waveguides discussed in Sections 5.2–5.4 possesses its own characteristic properties and therefore has a different range of potential applications. For a clear picture of how they compare with each other it is necessary to refer to the individual descriptions. Nevertheless, certain general characterizations can be offered as a guide, and with this aim some of the gross features are summarized in Table 5.1.

In some cases, the characterization is a matter of judgment, and could change with the advent of new information. Other categories could have been added also, such as ease of fabrication or ease of transduction. Those categories, particularly, could change at any time; moreover, they form subjects in their own right.

Some clarification of the terms used is appropriate. *Slowing* means relative to the Rayleigh wave; in that context, the term has less significance for the circular fibers. Slowing and *field confinement* are related directly [through (5.1)] for the first four waveguides listed because they are flat structures; for the remaining ones the categories are almost unrelated because of the differences in structure. The term *dispersion* refers principally to phase velocity behavior, but the characterization happens to be useful for group velocity as well.

Loss is a category that deserves some elaboration. In most cases careful measurements of loss have not been made for waveguides, so that the conclusions in Table 5.1 are in part speculative but based on sound considerations. The first three waveguides, which are flat overlay structures, employ a layer of some material to effect the guiding. It has long been assumed, from early measurements on layered media, that overlays introduce loss; for this reason, these waveguides were assumed to be relatively lossy. The slot waveguide

Table 5.1. Summary of Waveguide Properties

Type of waveguide	Section	Field confinement	Amount of slowing	Dispersion	Loss	Versatility
Strip	5.2.1	Moderate	Moderate	Moderate	Medium	Moderate
Shorting-strip or $\Delta v/v$	5.2.2	Weak	Small	Low	Medium	Low to moderate
Slot	5.2.3	Weak to moderate	Small to moderate	Low	Low to medium	Moderate
In-diffused	5.4.1	Weak	Small	Low	Low	Low to moderate
Topographic ridge flexural mode	5.3.1	Strong to very strong	Moderate to large	High	Low	High
Topographic ridge pseudo-Rayleigh mode	5.3.2	Very strong	Small	Very low	Low	High
Topographic wedge	5.3.4	Very strong	Moderate to very large	Very low to nil	Low	High
Rectangular overlays	5.4.2	Strong to very strong	Moderate to large	Variable	High	Moderate
Circular fibers	5.4.3	Self-contained	Small to moderate	Low to moderate	Low	Low

may possess less loss since in the usual range of operation most of the energy resides in the bare slot region. Nevertheless, when these guides were tried in long delay-line applications it was found that whatever loss was introduced was far less than that caused by beam spreading before the waveguide was introduced. In Table 5.1, the loss for these waveguides has been called "medium".

It was mentioned in Section 5.4.1 that recent investigations [5.11] have shown that the in-diffusion process seems, in certain cases at least, to introduce negligible additional loss. It is this feature that justifies the characterization of "low" loss for in-diffused waveguides, even though the measurements were not made on the guides themselves but simply on wide layers. This presumably negligible increase in loss is in itself a potentially exciting development since it could eliminate the "added loss" criticism often leveled at acoustic waveguides in general.

The topographic waveguides of Section 5.3 and two of the circular fiber waveguides discussed in Section 5.4.3 are essentially single material structures. The essential question here is whether the topographic alteration of the material has produced any measurable increase in the loss properties. The only structure in this group on which careful measurements of this type were made is the circular capillary waveguide described in Section 5.4.3. It was found for that waveguide [5.59] that within experimental error the measured material loss was equal to the bulk attenuation (after subtracting out the measured air-loading loss). This result is extrapolated in Table 5.1 to justify the "low" loss characterization for all waveguides of this type.

The final column, *versatility*, relates to the number of actual or potential applications for which the waveguide in question is suitable. If the waveguide can be used for long delay-line purposes but for almost nothing else, its versatilitity is deemed "low"; if it can also be utilized in a wide variety of potential circuit applications its versatility is "high".

5.5.2 Applications: Actual and Potential

It was pointed out in Section 5.1.1 that waveguides are not yet widely used in acoustic surface wave devices, and that there were two principal reasons for this. One of these reasons is that there exists a largely unjustified belief that waveguides in general are lossy, and that they cannot be excited efficiently. The discussion regarding *loss* in Section 5.5.1 summarized that question and showed that many waveguides should indeed have only small loss, in many cases negligibly greater than that for wide Rayleigh wave beams on the same material.

With respect to methods of *efficient excitation* of waveguides, substantial progress has been made recently. A thin film prism technique for exciting waveguides laterally was demonstrated by *Yen* and *Li* [5.66]; applied to a strip waveguide, it achieved a 65% efficiency of excitation without any cut and try. This technique has been extended [5.46] to topographic rectangular ridge waveguides, and efficiencies of 20–30% have been obtained without any attempt whatever at optimization; with proper design 70% should be achievable easily. *Nanayakkara* and *Ash* [5.67] investigate both theoretically and experimentally three different methods for the efficient excitation of topographic rectangular ridge and wedge waveguides. They conclude that the methods are effective; in their closing sentence they state, "they thereby remove what has been a major obstacle to their exploitation".

Another important reason is that many believe that waveguides are not needed to satisfy current device requirements. In this context, there is a lack of appreciation regarding the properties and potential of waveguides; better dissemination of information should help overcome that obstacle.

The principal application with which waveguides are generally associated is that of *long delay lines*. Such delay lines of course form the basis for the storage of either digital or analog signals. The class of circular fiber waveguide is in fact being developed with only this application in mind. A tough low-loss flexible fiber could be wound on a spool, and a long delay could be achieved in a compact form. Shorting-strip ($\Delta v/v$) and slot waveguides have also been considered seriously in this context. The waveguide would serve to confine the field laterally and prevent beam spreading; in addition, in applications where the beam traverses a helical path, the waveguide prevents cross talk between neighboring paths, a serious difficulty caused by beam spreading in the absence of any waveguide. Using flat plates of $Bi_{12}GeO_{20}$ with rounded ends, *Adams* and *Shaw* [5.25] obtained delays of several milliseconds at 50 MHz; they found explicitly that substantially lower overall loss was obtained when a shorting-strip waveguide was used. Waveguides with strong confinement properties, low loss, and negligible dispersion, such as the topographic waveguides, should be ideal candidates for compact long delay lines obtained either by a helical path on the outside of a cylinder or a spiral path on a planar substrate (both techniques have been suggested some time ago).

Devices employing *nonlinear interactions* can also benefit significantly by using waveguides. The principal advantage relates to the greatly increased acoustic energy density in the waveguide produced by the lateral field confinement; in strongly confining waveguides, such as the topographic structures, the energy density increase is pronounced since the guide widths are typically one or two orders of magnitude less than those of the Rayleigh wave beams. Lesser advantages include the single-mode nature of the waveguide, as contrasted with the (small) wave number spread associated with Rayleigh wave beams, which should offer sharper phase matching in the nonlinear processes, and the availability of lower velocities with some waveguides, since the interaction efficiency increases as the velocity decreases [5.68]. The added requirement, however, of negligible dispersion in convolution and correlation operations involving parametric interactions, limits the available waveguide candidates. Nevertheless, the topographic wedge waveguide and the symmetric mode on the topographic ridge waveguide both provide the strong confinement and negligible dispersion needed, and the wedge in addition provides lowered wave velocity. The advantages of wedges in this connection have been appreciated by *Ash* and his colleagues [5.37, 53] and they have conducted various experiments relating to convolver performance.

Two experiments, quite unrelated to each other, are now presented to illustrate the applicational potential of waveguides. One of these was conducted by *Coldren* and *Kino* [5.69] on a cw monolithic surface wave *amplifier*. They successfully operated an acoustic amplifier which intrinsically incorporated a narrow (2.5 wavelengths wide) shorting-strip waveguide of semiconducting InSb on a LiNbO$_3$ substrate (with a thin intervening layer of SiO), and they achieved a cw gain of 23 dB at 340 MHz. The waveguide configuration produced *less* dispersion than they measured earlier for a corresponding system employing wide unguided waves; the dispersion cancellation was achieved because the mass loading due to the film compensated for the dispersion produced by the shorting property of the strip (see discussion in Sect. 5.2.2). In a very different application, that of a *high Q* surface wave *filter*, *Tiersten* and *Smythe* [5.70] employ a pair of coupled strip waveguides which are placed side by side. The coupling produces symmetric and antisymmetric modes which resonate in the filter at slightly different frequencies and consequently produce a flatter topped resonance. In this application, loss is a critical parameter in limiting the Q that can be achieved. Since the application requires waveguides which are not strongly confined, but which possess very low loss, the most appropriate waveguide type would be the in-diffused structures which have exhibited negligible increase in loss after diffusion.

The above two examples are selected to show how, in disparate ways, device performance can be enhanced significantly by the clever use of waveguides, rather than wide Rayleigh wave beams. Many other examples could be cited, including Doppler velocity selectors, Fourier transformers, and various filter functions.

As was mentioned in Section 5.1.1, the most exciting potential application for waveguides in the opinion of the writer is that of a highly compact sophisticated circuit technology, often referred to as "microsound" [5.1–3]. In this technology, many functions would be performed simultaneously on a single substrate, in an integrated fashion [5.1]. If this were tried with wide surface wave beams it would require excessive substrate area; waveguides would then be needed. However, disappointments would soon follow because components for waveguides have not yet been developed; in turn, such development has been lacking because the demand was absent.

Some work on waveguide components has been attempted, with modest success, but the efforts have been quite sparse. Several groups have treated directional couplers comprised of parallel strip waveguides, power splitters in strip waveguide, and ring resonators in both strip waveguide and topographic rectangular ridge waveguide. But the work has not gone much further. Many questions remain unanswered. For example, we are not sure that large reflections can be produced, comparable to a short circuit or an open circuit; such large reflections do not seem possible with flat overlay guides, but they may be feasible with topographic structures which have strong field confinement. Additionally, we do not yet know how to produce the equivalent of lumped capacitances or inductances, but then again probably no one has tried.

Nevertheless, the prospects seem exciting. A glance backwards to the electromagnetic microwave field shows that sophisticated electromagnetic microwave circuitry did not appear until waveguides were well utilized. One is encouraged to speculate, therefore, that the full potential of acoustic waves will not be realized until acoustic waveguides are thoroughly exploited.

References

5.1 E. Stern: IEEE Trans. MTT-17, 835—844 (1969)
5.2 E. A. Ash, R. M. de la Rue, R. F. Humphryes: IEEE Trans. MTT-17, 882—892 (1969)
5.3 R. A. Waldron: IEEE Trans. SU-18, 219—230 (1971)
5.4 A. A. Oliner: Proc. IEEE 64, 615—627 (1976)
5.5 E. A. Ash: IEEE G-MTT International Microwave Symposium (Boston, Massachusetts, May 1967)
5.6 E. A. Ash, D. P. Morgan: Electron. Lett. 3, 462—463 (1967)
5.7 I. M. Mason, R. M. de la Rue, P. E. Lagasse, R. V. Schmidt, E. A. Ash: Electron. Lett. 7, 395—397 (1971)
5.8 J. D. Ross, S. J. Kapuscienski, K. B. Daniels: IRE Conv. Record 6 II, 118—120 (1958)
5.9 D. L. White: IEEE Symposium on Sonics and Ultrasonics (Vancouver, B.C., Canada, October 1967)

5.10 H.Seidel, D.L.White: Patent No. 3, 406, 358 (October 15, 1968), and Patent No. 3, 488, 602 (January 6, 1970)

5.11 R.V.Schmidt: Appl. Phys. Lett. **27**, 8—10 (1975)

5.12 A.A.Oliner, R.C.M. Li, H.L.Bertoni: Final Report, ECOM-0418-F, Polytechnic Institute of Brooklyn, Report No. PIBEP-71-092, Farmingdale, N. Y., August 1971

5.13 R.C.M.Li, A.A.Oliner, H.L.Bertoni: IEEE Trans. SU-**24**, 66—78 (1977)

5.14 H.F.Tiersten: J. Appl. Phys. **40**, 770—789 (1969)

5.15 L.R.Adkins, A.J.Hughes: IEEE Trans. MTT-**17**, 904—911 (1969)

5.16 K.H.Yen, A.A.Oliner: Appl. Phys. Lett. **28**, 368—370 (1976)

5.17 C.C.Tu, G.W.Farnell: IEEE Symposium on Sonics and Ultrasonics (Miami Beach, Florida, December 1971)

5.18 Y.Shimizu, R.C.M.Li: IEEE Ultrasonics Symposium (Monterey, Cal., Paper R-7, November 1973)

5.19 R.V.Schmidt, L.A.Coldren: IEEE Trans. SU-**22**, 115—122 (1975)

5.20 B.K.Sinha, H.F.Tiersten: J. Appl. Phys. **44**, 4831—4854 (1973)

5.21 T.L.Szabo, A.J.Slobodnik,Jr.: IEEE Trans. SU-**20**, 240—251 (1973)

5.22 H.Engan: Electronics Research Lab., Norwegian Institute of Technology (Trondheim, Norway, ELAB Report TE-128, July 1969)

5.23 A.J.Hughes: J. Appl. Phys. **43**, 2569—2586 (1972)

5.24 L.A.Coldren, R.V.Schmidt: Appl. Phys. Lett. **22**, 481—482 (1973)

5.25 P.L.Adams, H.J.Shaw: Ultrasonics Symposium (Boston, Mass., Paper M-5, October 1972)

5.26 R.M.Knox, D.B.Owen: Digest of Technical Papers, G-MTT International Microwave Symposium (Newport Beach, California, May 1970), pp. 370—374

5.27 L.A.Coldren, R.V.Schmidt: IEEE Trans. SU-**21**, 128—130 (1974)

5.28 L.A.Coldren, D.H.Smithgall: IEEE Trans. SU-**22**, 123—130 (1975)

5.29 R.C.Rosenfeld, K.E.Bean: Proc. IEEE Ultrasonics Symposium (Boston, Massachusetts, October 1972), pp. 186—189

5.30 P.E.Lagasse, I.M.Mason: Electron. Lett. **8**, 82—84 (1972)

5.31 P.E.Lagasse: IEEE Trans. SU-**20**, 354—359 (1973)

5.32 P.E.Lagasse: J. Acoust. Soc. Am., **53**,1116—1122 (1973)

5.33 R.C.M.Li, H.L.Bertoni, A.A.Oliner, S.Markman: Electron. Lett. **8**, 211—212 (1972)

5.34 S.Markman, R.C.M.Li, A.A.Oliner, H.L.Bertoni: IEEE Trans. SU-**24**, 79—87 (1977)

5.35 P.E.Lagasse: Electron. Lett. **8**, 372—373 (1972)

5.36 S.Markman: Ph. D. Dissertation, Polytechnic Institute of Brooklyn, June 1973

5.37 P.E.Lagasse, I.M.Mason, E.A.Ash: IEEE Trans. MTT-**21**, 225—236 (1973)

5.38 B.K.Sinha: J. Acoust. Soc. Am. **56**, 16—18 (1974)

5.39 R.N.Thurston, J.McKenna: IEEE Trans. SU-**21**, 296—297 (1974)

5.40 T.M.Sharon, A.A.Maradudin, S.L.Cunningham: Electron. Lett. **10**, 229—230 (1974)

5.41 T.M.Sharon, A.A.Maradudin, S.L.Cunningham: Letters in Appl. and Eng. Sciences **2**, 161—174 (1974)

5.42 A.A.Oliner, P.E.Lagasse: Proc. 1975 IEEE Ultrasonics Symposium (Los Angeles, California, September 1975), pp. 544—548

5.43 P.E.Lagasse, A.A.Oliner: Electron. Lett. **12**,11—13 (1976)

5.44 P.E.Lagasse: Proc. European Microwave Conference (Brussels, Belgium, September 1973)

5.45 R.C.M.Li, H.L.Bertoni, A.A.Oliner, S.Markman: Electron. Lett. **8**, 220—221 (1972)

5.46 K.H.Yen, H.L.Bertoni, A.A.Oliner, S.Markman: Proc. Symposium on Optical and Acoustical Micro-Electronics (Polytechnic Institute of New York, New York, N. Y., April 1974), pp. 297—309

5.47 J.J.McCoy, R.D.Mindlin: J. Appl. Mech., **30**, 75—78 (1963)

5.48 J.Oliver, F.Press, M.Ewing: Geophys. **19**, 202—219 (1954)

5.49 K.Tamm, O.Weiss: Acustica **11**, 8—17 (1961)

5.50 R.Sinclair, R.W.B.Stephens: Acustica **24**, 160—165 (1971)

5.51 P.E.Lagasse: IEEE Ultrasonics Symposium (Boston, Massachusetts, October 1972)

5.52 P.E.Lagasse, M.Cabus, M.Verplanken: Proc. IEEE Ultrasonics Symposium (Milwaukee, Wisconsin, November 1974)

5.53 I.M.Mason, M.D.Motz, J.Chambers: Proc. IEEE Ultrasonics Symposium (Boston, Massachusetts, October 1972), pp. 314—315

5.54 R.S.Wagers, D.F.Weirauch: IEEE Ultrasonics Symposium (Los Angeles, California, September 1975), pp. 539—543

5.55 M.Yamanashi, K.Yoshida: Japan J. Appl. Phys. **9**, 1276 (1970)

5.56 P.Hartemann: Appl. Phys. Lett. **27**, 263—265 (1975)

5.57 R.M.White: Proc. IEEE **58**, 1238—1276 (1970)

5.58 R.A.Waldron: IEEE Trans. SU-**18**, 8—16 (1971)

5.59 R.L.Rosenberg, R.V.Schmidt, L.A.Coldren: Appl. Phys. Lett. **25**, 324—326 (1974)

5.60 R.L.Rosenberg: Private communication (1974)

5.61 G.D.Boyd, L.A.Coldren, R.N.Thurston: Appl. Phys. Lett. **26**, 31—34 (1975)

5.62 R.A.Waldron: IEEE Trans. MTT-**17**, 893—904 (1969)

5.63 L.A.Coldren: Appl. Phys. Lett. **25**, 367—370 (1974)

5.64 L.O.Wilson: IEEE Ultrasonics Symposium (Los Angeles, California, September 1975), pp. 529—532

5.65 R.V.Schmidt: Private communication (1975)

5.66 K.H.Yen, R.C.M.Li: Appl. Phys. Lett. **20**, 284—286 (1972)

5.67 P.Nanayakkara, E.A.Ash: Wave Electronics **1**, 247—263 (1976)

5.68 M.Luukkala, J.Surakka: J. Appl. Phys. **43**, 2510—2518 (1972)

5.69 L.A.Coldren, G.S.Kino: Appl. Phys. Lett. **23**, 117—118 (1973)

5.70 H.F.Tiersten, R.C.Smythe: IEEE Ultrasonics Symposium (Los Angeles, California, September 1975), pp. 293—294

6. Materials and Their Influence on Performance

A. J. Slobodnik, Jr.

With 67 Figures

The proper choice of a substrate material for surface acoustic wave (SAW) delay lines, filters, and other signal processing devices can result in a substantial improvement in performance. This choice depends [6.1] principally on velocity, coupling constant [6.2, 3], temperature coefficient of delay [6.4, 5], propagation loss [6.6–8], beam steering [6.9], and diffraction [6.10–12]. The latter three properties become particularly important in the upper UHF and microwave frequency regions.

The purpose of this chapter is to present design information on each of the above quantities and assess their influence on performance. Major emphasis will be placed on presenting quantitive propagation loss data, including air and gas loading, as well as the effect of surface quality on attenuation, plus design curves illustrating the beam steering, diffraction trade-offs. The use of this information will be illustrated in several ways, including the design of a UHF delay line having minimum insertion loss in the presence of the material limitations discussed above as well as transducer conduction loss and parasitic elements [6.13–15]. In addition, the properties of one of the newly discovered [6.16, 17] minimal diffraction orientations of $Bi_{12}GeO_{20}$ will be described, as will a method of diffraction compensation [6.18] in the design of periodic apodized SAW filters. A section will also be devoted to the description of nonlinear effects in the propagation of acoustic surface waves. A unifying technique used for the observation and measurement of many of the topics outlined here will be found to be the laser probe [6.19]. This basic measurement tool is described in the first section of this chapter.

6.1 The Laser Probe as a Basic Measurement Tool

The subject of this section is the detailed description of a laser probe [6.20] developed for studying microwave acoustic surface wave phenomena. This probe is essential for obtaining design data for surface wave devices operating in the microwave frequency region (where small wavelengths make the use of mechanical probes quite difficult) [6.21].

The basic operation of this device is illustrated schematically in Fig. 6.1. Coherent light from a low power laser is directed onto a surface wave delay line. Due to the presence of the surface wave on the crystal acting as a diffraction grating, light is deflected into side lobes in addition to the specular direction. The

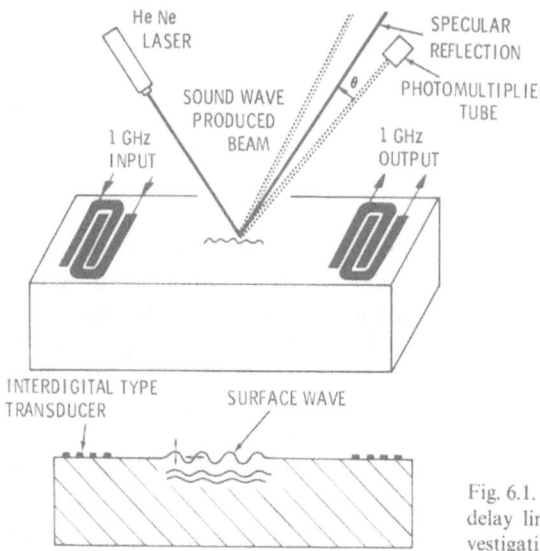

Fig. 6.1. Schematic diagram of a surface wave delay line and the laser probe used for investigating surface wave devices

intensities of these side lobes, which are directly related to the intensity of the surface wave, can then be monitored by a photomultiplier tube. This provides a convenient method of probing the behavior of the surface wave [6.22–24]. For example, by scanning the laser along the acoustic beam (that is, on a path between the two transducers) one can measure the acoustic propagation loss [6.6–8, 25]. At microwave frequencies this is an important quantity as it is a significant portion of the overall insertion loss of any device. By investigating various materials with the laser probe, one can choose the material having the lowest loss for a given device application. In addition, by scanning the laser across the acoustic beam, the surface wave profile or shape of the beam can be determined [6.9]. This information provides beam steering and diffraction data [6.12].

6.1.1 Description of the Laser Probe

It is well known in electromagnetic theory that a periodically modulated surface causes plane waves of electromagnetic energy to be deflected to side lobes in addition to the specular direction [6.26–27]. The angular directions of these side lobes are given by the grating equation

$$\sin \theta_m = \sin \theta_0 + \frac{m\lambda}{\Lambda_s} \tag{6.1}$$

where λ is the wavelength of the incident energy, and Λ_s is the surface wavelength. It can also be shown that the intensity of the deflected light is directly related to the intensity of the surface wave [6.26]. Thus using a low

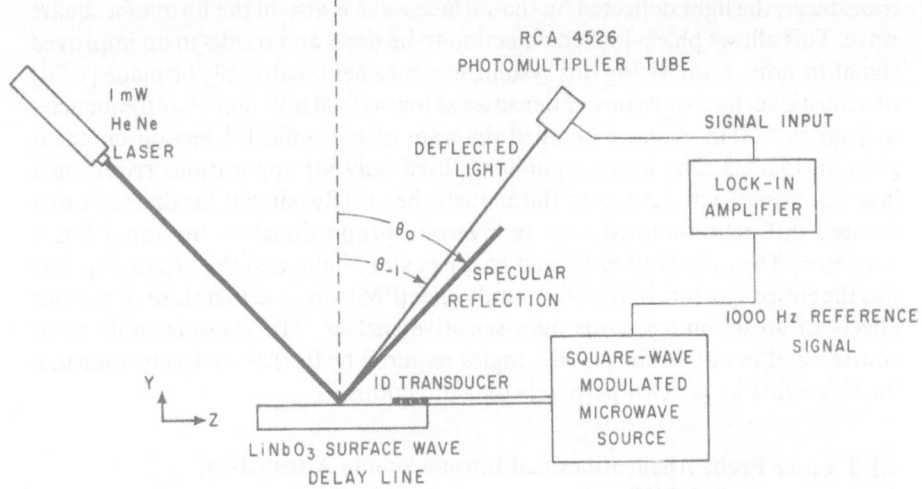

Fig. 6.2. Detailed block diagram of the laser probe system used for investigating microwave-acoustic surface waves. The angles shown in this figure are defined in (6.1)

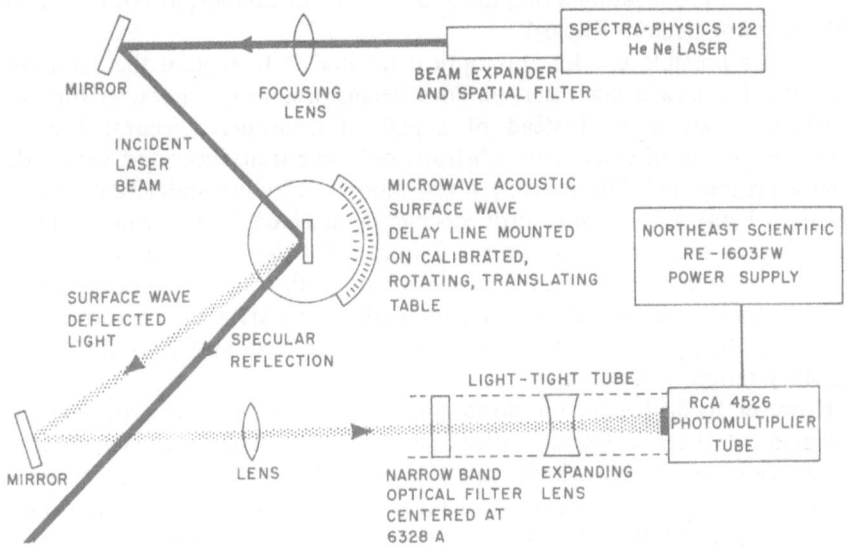

Fig. 6.3. Detailed schematic diagram of the optical circuit used with the laser probe

power HeNe laser ($\lambda = 6328\,\text{Å}$) as the incident electromagnetic energy, and placing a light detector at an angle corresponding to one of the side lobes, yields a direct method of monitoring surface wave disturbances.

A detailed block diagram of the experimental laser probe system is shown in Fig. 6.2. By square-wave modulating the electromagnetic input to the interdigital transducer, the light deflected by the surface wave is also in the form of a square wave. This allows phase-locked detection to be used, and results in an improved signal-to-noise ratio. Using this system, measurements can easily be made [6.20] of acoustic surface wave power densities as low as 0.08 mW mm^{-1} at frequencies as high as 4 GHz. A more detailed diagram of the optical detection circuit is given in Fig. 6.3. The beam expander is used only for applications requiring a laser spot small with respect to the acoustic beam. (Recall that the diameter of a focused diffraction-limited spot is inversely proportional to the input beam diameter.) The optical filter is used to reject stray light, and the expanding lens fills the entire aperture of the photomultiplier (PM) tube and therefore eliminates effects of an inhomogeneous light-sensitive surface. The rotating table is, of course, used to obtain the precise angles required by (6.1) as well as to translate the delay line in the two perpendicular directions.

6.1.2 Laser Probe Applications and Introduction to Attenuation, Beam Steering, and Diffraction

The validity of (6.1) can be experimentally confirmed by measuring the angle of deflection of the first side lobe for light incident at various angles. Results for surface waves propagating along the Z axis of Y cut LiNbO$_3$ at both 635 and 1950 MHz are given in Fig. 6.4.

Another possible use for this optical method is to replace the standard technique for measuring insertion loss versus frequency characteristics of interdigital transducers. Instead of a *pair* of transducers separated by a reasonable length of crystalline substrate, only one transducer and very little material are required. Alternatively, if one chooses to study standard delay lines, variations between the two transducers can be detected. Experimentally, both methods yield the same results versus frequency for the case illustrated in Fig. 6.5. Relative laser measurements were made on both transducers of the delay line, and their sum (in decibels), normalized at 905 MHz to the transducer measurements, agrees with the double transducer technique over the entire frequency range of the data.

However, by far the most important applications for the laser probe are the measurement of beam steering, diffraction, and attenuation. Each is illustrated schematically in Fig. 6.6 which depicts the launching and subsequent propagation of an acoustic surface wave on an anisotropic crystalline substrate [6.2]. The angle θ defines the direction of propagation (phase velocity direction) with respect to a reference crystalline axis, while the angle ϕ defines the deviation of the power flow (group velocity) from the phase velocity direction. Although practical devices are generally designed for surface wave propagation along pure-mode axes (specific directions for which $\phi = 0$), some unintentional misalignment always results thus causing beam steering [6.9]. The slope of the power flow angle $\partial\phi/\partial\theta$ can be used to provide a direct measure of the seriousness of a given misalignment in the amount of beam steering it causes. High values of $\partial\phi/\partial\theta$ obviously result in large amounts of beam steering.

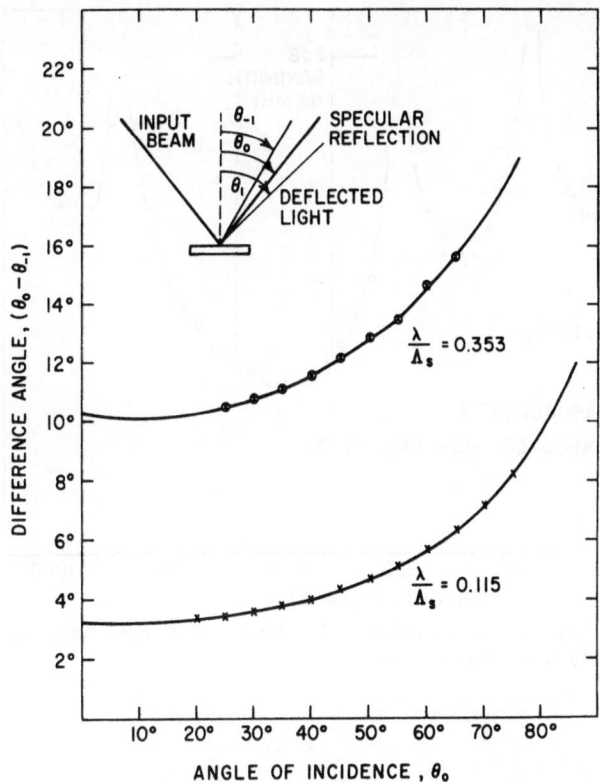

Fig. 6.4. Confirmation of the grating equation using deflected light from microwave-acoustic surface waves at 635 and 1950 MHz. Experimental data points are shown together with theoretical curves derived from (6.1)

Diffraction [6.12] is also illustrated in Fig. 6.6 by the changing acoustic beam profile and eventual beam spreading. In the anisotropic media with which we must deal to minimize the third major source of loss—attenuation (discussed in detail in Sect. 6.2)—the same parameter $\partial\phi/\partial\theta$ is also used to estimate the extent of beam spreading. Depending on the value of $\partial\phi/\partial\theta$, diffraction may be quite drastically increased or significantly retarded in contrast to the analogous case for light diffraction in an isotropic medium. Within limits, proper choice of surface wave orientation results in the ability to trade off beam steering against diffraction losses [6.1]. This will be discussed in detail in Section 6.3.

6.2 Propagation Loss

One of the major sources of overall device insertion loss at microwave frequencies is propagation loss or attenuation [6.6–8]. Not only is the magnitude of this phenomenon important for predicting absolute insertion loss and dynamic range, but its frequency dependence is equally important when, for example, designing filters having particular bandpass characteristics.

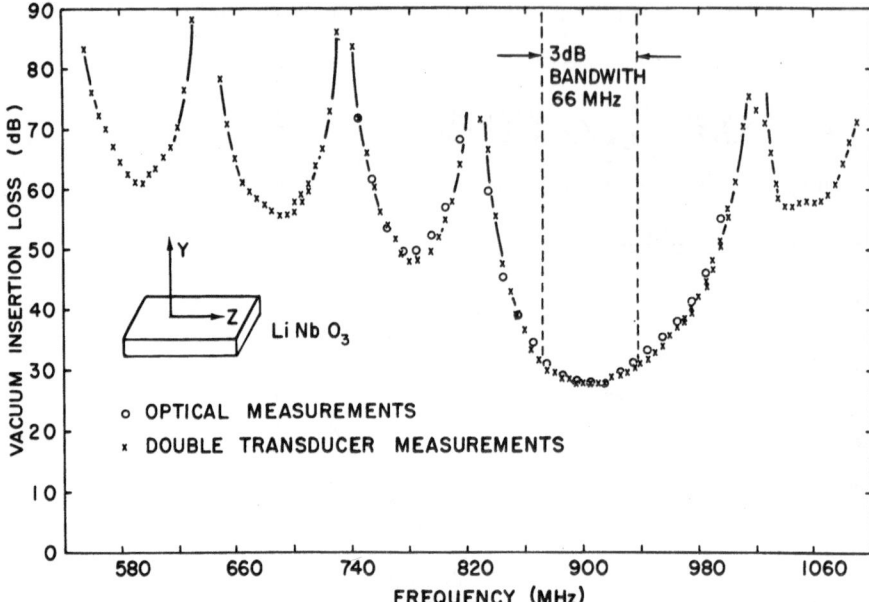

Fig. 6.5. Insertion loss versus frequency characteristics of a $LiNbO_3$ surface wave delay line measured by both double transducer and optical methods

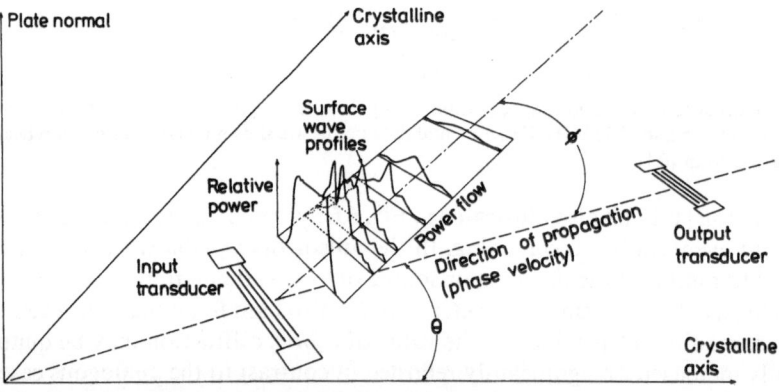

Fig. 6.6. Schematic representation of the launching and subsequent propagation of an acoustic surface wave on a crystalline substrate. Angle θ defines the direction of propagation with respect to the reference crystalline axis and angle ϕ defines the deviation of power flow from the phase velocity direction. A pure-mode axis is one for which $\phi = 0$

6.2.1 Room-Temperature Attenuation

Total room-temperature propagation loss is easily measured by scanning the laser probe described in Section 6.1 along the length of the acoustic column. The focusing lens of Fig. 6.3 is adjusted such that the light beam is slightly wider than the acoustic beam to provide an averaging mechanism over the changing surface

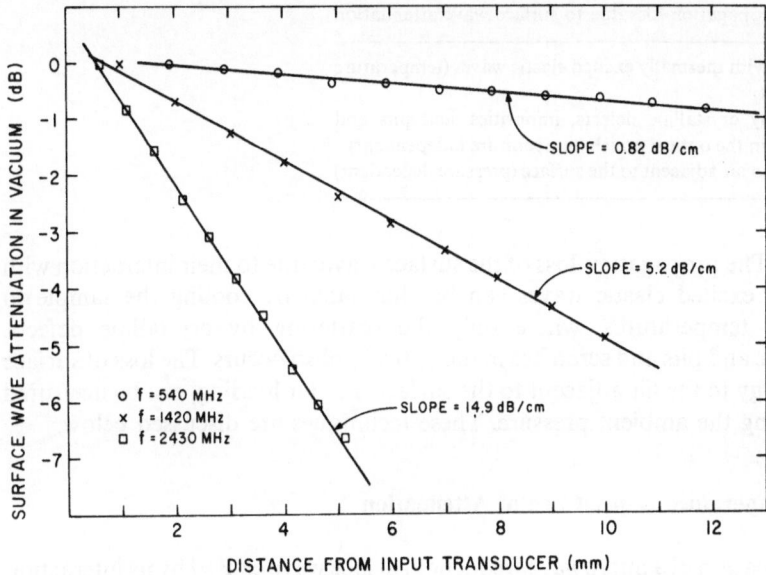

Fig. 6.7. Surface wave attenuation versus distance on Y cut Z propagating LiNbO$_3$. Data taken at room temperature

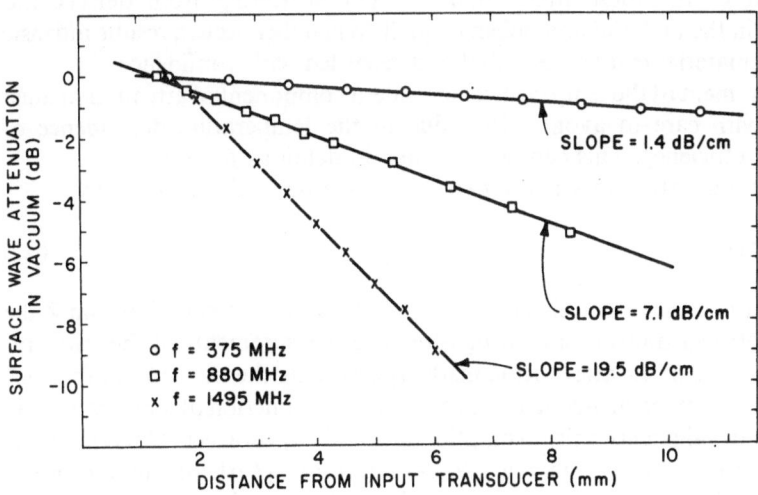

Fig. 6.8. Acoustic surface wave propagation loss vs distance for 001 cut 110 propagating Bi$_{12}$GeO$_{20}$ as measured with the laser probe at room temperature

wave profile. Examples [6.6, 8] of room-temperature propagation loss measurements for YZ LiNbO$_3$ and 001, 110 Bi$_{12}$GeO$_{20}$ are given in Figs. 6.7 and 6.8.

Neglecting beam steering and diffraction, total propagation loss arises from a combination of the three effects [6.6] listed in Table 6.1. These components can be separated by means of their dependence on temperature and ambient

Table 6.1 Propagation loss due to surface wave attenuation

Interaction with thermally excited elastic waves (temperature
  dependent)
Scattering by crystalline defects, impurities, and pits and
  scratches in the optical polish (temperature independent)
Energy lost to air adjacent to the surface (pressure dependent)

---

pressure. The propagation loss of the surface waves due to their interaction with
thermally excited elastic waves can be eliminated by cooling the sample to
cryogenic temperatures, where only the scattering by crystalline defects,
impurities, and pits and scratches in the optical polish occurs. The loss of surface
wave energy to the air adjacent to the surface, i.e., air loading, can be measured
by reducing the ambient pressure. These techniques are discussed below.

6.2.2 Temperature Dependence of Attenuation

The attenuation of a microwave frequency surface wave caused by its interaction
with thermal phonons represents the minimum propagation loss which can be
achieved with any given material. That is, if the temperature-*independent*
component of the total attenuation is small, scattering from defects and
impurities in the material and polish is small. When this occurs, results intrinsic
to a given material can be measured and recorded with confidence.

Measurement of the temperature-dependent component of attenuation must
be made with care to avoid effects due to the temperature dependence of
transducer efficiency. This can be explained in detail as follows.

The total insertion loss I of a two-transducer delay line is given by

$$I = E_1 E_2 \exp(-\alpha_T Z), \tag{6.2}$$

where E is the transducer efficiency, α_T the surface wave attenuation, and Z the
distance between transducers. Straightforward measurements of the temperature dependence of the attenuation made by assuming the transducer efficiency
to be temperature independent and measuring I as a function of temperature are
not generally valid since this assumption is not always correct. Thus the three-
transducer method shown in Fig. 6.9 was developed [6.6] for the purpose of
temperature-dependence measurements. The center transducer is excited with a
short pulse of electromagnetic energy which is converted into two equal surface
waves to be detected at transducers 1 and 2. One can thus measure the ratio of the
insertion losses of the two detection transducers

$$I_2/I_1 = (E_2/E_1) \exp[-\alpha_T(Z_2 - Z_1)]. \tag{6.3}$$

The requirement for measuring the desired temperature dependence of α_T is that
the temperature dependence of E_2/E_1 be small in comparison. This requirement

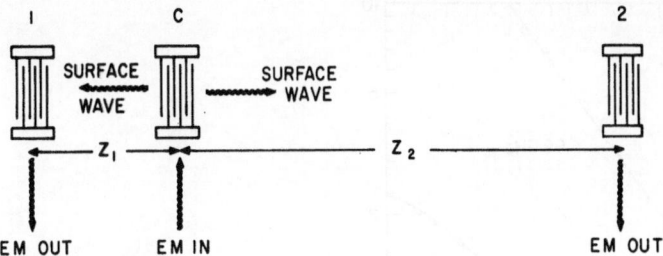

Fig. 6.9. Schematic diagram of three-transducer technique used for temperature-dependence measurements

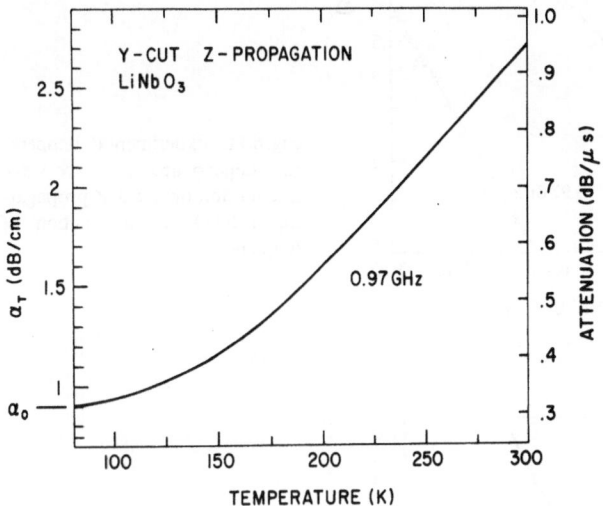

Fig. 6.10. Experimental temperature dependence of surface wave attenuation on Y cut Z propagating LiNbO$_3$ at 0.97 GHz. Room-temperature value normalized to laser measurements

is far easier to meet than that for the two-transducer method of (6.2) and was indeed satisfied in experiments where $Z_2 - Z_1$ was approximately 1 cm. However, this method cannot be used to reliably measure absolute room-temperature attenuation due to the possibility of the beamsteering and diffraction losses discussed in Section 6.3.

The attenuation on YZ LiNbO$_3$ measured [6.6] by this three-transducer method decreased rapidly below room temperature and approached a temperature-independent value α_0, as illustrated by the data [6.6] shown in Fig. 6.10. The room-temperature value was normalized to that obtained with the laser deflection technique. The value of α_0 is small compared to the room-temperature value α_T. Subtracting the residual value α_0 from the total observed attenuation α_T yields the results shown in Fig. 6.11. The temperature was measured with a calibrated platinum resistance thermometer accurate to 1 K and the samples were slowly cycled over the temperature range in a period of

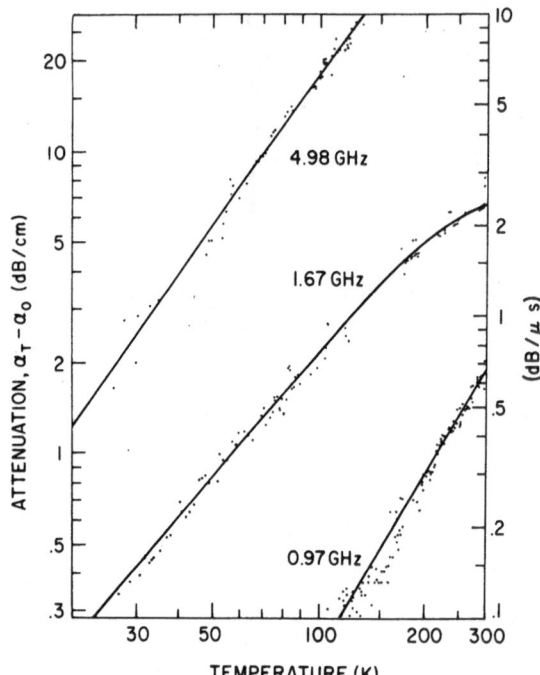

Fig. 6.11. Experimental temperature dependence of surface wave attenuation on Y cut Z propagating LiNbO$_3$ as a function of frequency

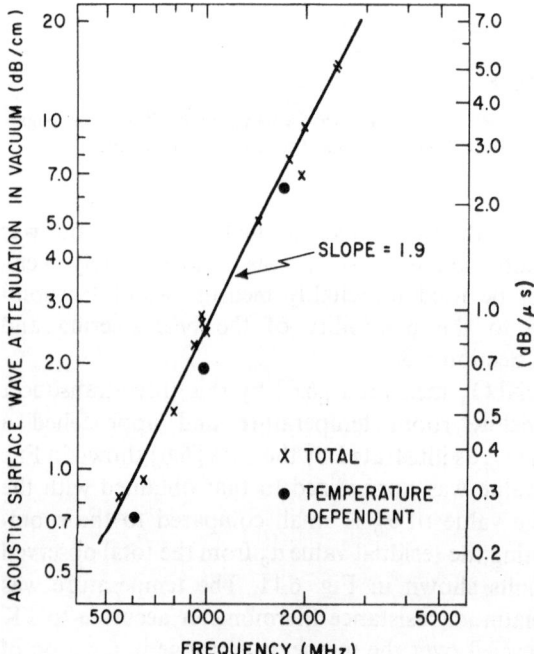

Fig. 6.12. Surface wave attenuation versus frequency on Y cut Z propagating LiNbO$_3$. The total room-temperature values are laser measurements while the temperature dependent values do not include α_0

4–8 h to minimize thermal stresses which can shatter $LiNbO_3$. There was no observable difference between the warm-up and cool-down runs [6.6]. The efficiency for the conversion of electromagnetic energy into elastic surface wave energy (in both directions) was $-2\,dB$ for the 0.97 GHz fundamental frequency interdigital transducers. These transducers had 10 electrode pairs and were 200 μ long with a linewidth of 0.85 μm and a center-to-center spacing of 1.7 μm. The 1.67 GHz fundamental frequency transducers were fabricated by electron beam exposure of resist [6.28] and had 20 electrode pairs of 200 μm long lines with a linewidth of 0.6 μm and a 1 μm center-to-center periodicity. Their efficiency was $-3\,dB$. It should be noted that these relatively high values minimize the temperature dependence of the transducer efficiencies. The 4.98 GHz measurements were made by operating the 1.67 GHz transducers at their third overtone.

Calculation of the thermal relaxation time τ from available data [6.29, 30] yields $\omega\tau = 0.1$ at 85 K for a frequency of 5 GHz. Since the condition $\omega\tau \gg 1$ is not satisfied, direct comparison of the results reviewed here with the theoretical temperature dependence of the attenuation ωT^4 computed by *Maradudin* and *Mills* [6.31] is not possible. The experimental data have a higher frequency dependence and a lower temperature dependence.

The total room-temperature attenuation, as measured by the laser deflection technique plus the temperature-dependent component, are plotted as a function of frequency in Fig. 6.12. These measurements were, of course, made [6.6] at power levels low enough to avoid any nonlinear effects [6.32, 33]. At 1 GHz, the temperature-dependent component is about $0.7\,dB\,\mu s^{-1}$ while the total attenuation in vacuum is $0.9\,dB\,\mu s^{-1}$. This low value of only $0.2\,dB\,\mu s^{-1}$ for the temperature-independent component was obtained [6.6] only for highest quality $LiNbO_3$ having scratch-free polishes. Further details on the effect of surface quality on attenuation will be found in Section 6.2.5. For measurements made in air, an additional loss of $0.2\,dB\,\mu s^{-1}$ occurs [6.6] as will be described in Section 6.2.3. The observed frequency dependence of the temperature-dependent component of the attenuation is equal, within experimental uncertainty, to the frequency-squared dependence observed for volume waves [6.30, 34, 35]. The temperature-dependent component for 1 GHz transverse volume waves propagating along the Z axis of $LiNbO_3$ is $0.9\,dB\,\mu s^{-1}$ [6.34], while the value for longitudinal waves is $0.3\,dB\,\mu s^{-1}$ [6.34, 35]. It is apparent that in this case the surface wave attenuation is dominated by the transverse bulk-wave attenuation.

Temperature-dependence measurements [6.8] on 001 cut 110 propagating $Bi_{12}GeO_{20}$ illustrate (Figs. 6.13 and 6.14) the loss peaks in the vicinity of 50–100 K which are characteristic [6.36, 37] of this material. This anomalous attenuation is due to a relaxation process closely connected to vacancies in the Ge sublattice [6.36]. The surface wave peaks shown in Figs. 6.13 and 6.14 were of such a large amplitude as to not allow any details of their structure to be observed. However, the lower temperature limit showed [6.8] negligible frequency dependence while the upper temperature increased with increasing frequency. This is in rough agreement with the results of *Spencer* et al. [6.37] for bulk shear waves propagating along the (100) axis. As in the case of $LiNbO_3$,

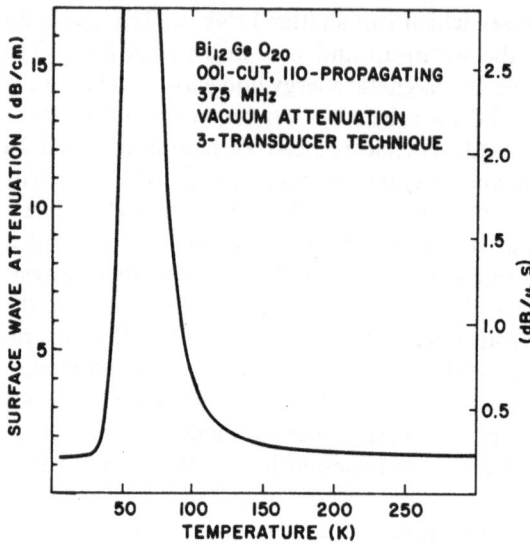

Fig. 6.13. Acoustic surface wave propagation loss vs temperature at 375 MHz

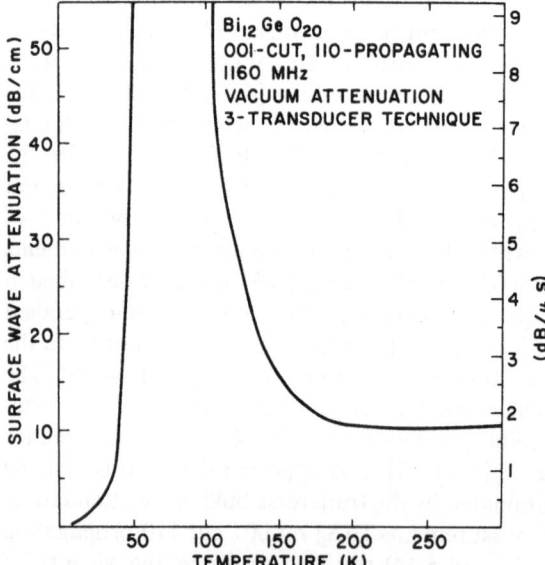

Fig. 6.14. Acoustic surface wave propagation loss vs temperature at 1160 MHz

excellent agreement could also be obtained [6.8] between the total room-temperature and temperature-dependent data; this again indicates that losses are due to the intrinsic interactions with thermally excited waves rather than to crystalline defects or pits and scratches in the polish.

When using the three-transducer method shown in Fig. 6.9, it is essential that the electromagnetic input power be directed to the center transducer. Use of the more conventional tapped delay-line mode, in which the electromagnetic input power went to end transducer number 1 with the delayed surface wave being observed at transducers C and 2, can result [6.6] in an apparent temperature-dependent component larger than the total room-temperature attenuation. Laser deflection measurements at room temperature revealed that the surface wave undergoes a loss in passing through the center transducer, thus causing the observed discrepancy to the temperature dependence of this loss.

6.2.3 Air and Gas Loading

Acoustic surface wave propagation along the boundary between a solid and a low-density fluid results in both attenuation along the direction of propagation and perturbation of the vacuum Rayleigh wave velocity [6.38–42]. At microwave frequencies, the attenuation of this leaky wave due to propagation in air can result in a significant increase in device insertion loss. This air loading phenomenon has been described in detail for microwave-acoustic surface wave delay lines by *Slobodnik* et al. [6.6].

The purpose of this subsection is to outline a study [6.42] of microwave-acoustic surface wave propagation in the monatomic gases, helium, neon, argon, krypton, and xenon. Gas loading will be illustrated as a function of frequency, pressure, atomic weight, and temperature and the results will be compared with two existing theories [6.39, 43, 44].

The two most likely causes of attenuation due to gas loading are emission of compressional waves and frictional loss [6.45]. Since *Dransfeld* and *Salzmann* [6.45] have shown the former to be the dominant loss mechanism, it will be the only one described here. Figure 6.15 is an illustration in the continuum limit (atomic mean-free path much smaller than the Rayleigh wavelength) of the generation of longitudinal acoustic waves by the propagation of a surface wave in a gas. Bulk waves are generated whenever the phase-matching condition,

$$\cos\zeta = \lambda_G/\Lambda_s, \tag{6.4}$$

is satisfied. Here ζ is the angle at which the bulk waves are launched, λ_G is the acoustic wavelength of the longitudinal wave in the gas, and Λ_s is the surface or leaky wave wavelength. Since (6.4) is satisfied for virtually all common solids and gases, it can be assumed that gas loading will be present.

The experimental procedures which can be followed for the quantitative measurement of gas loading are straightforward but require considerable care to obtain maximum accuracy. The block diagram of a typical apparatus is illustrated in Fig. 6.16. The delay line is placed in a vacuum system connected to a supply of gas. The total gas loading can then be measured as the difference in delay-line insertion loss between vacuum and a pressure of 1 atm of the gas under study. Use of the triple-transit echo provides a longer propagation path and yields 3 times the gas loading of the direct delayed output. Intermediate pressures can also be investigated using, for example, a precision pressure indicator.

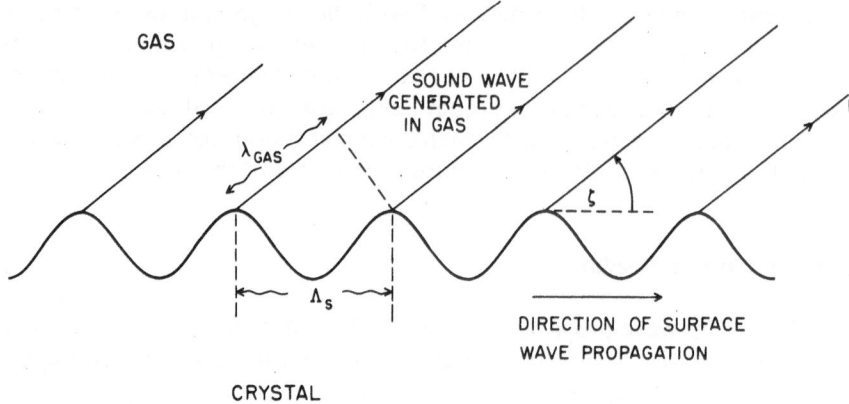

PHASE MATCHING CONDITION: $\cos \zeta = \dfrac{\lambda_{GAS}}{\Lambda_s}$

Fig. 6.15. Schematic diagram illustrating the generation of longitudinal bulk waves by a surface wave propagating along the boundary between a solid and a gas. Phase-matching condition: $\cos\zeta = \lambda_G/\Lambda_s$

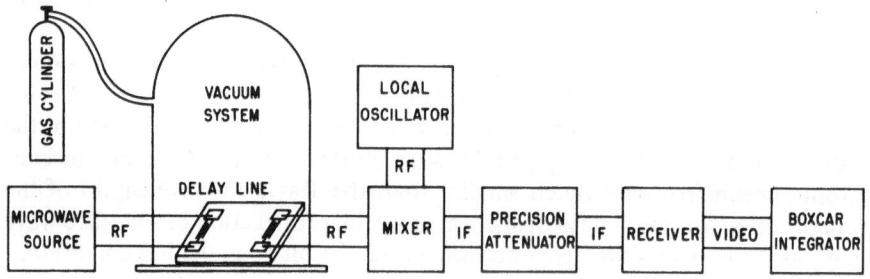

Fig. 6.16. Schematic and block diagram of the experimental apparatus used to measure microwave-acoustic surface wave attenuation due to gas loading

To attain maximum precision, the rf output of the delay line can be mixed down to an if of 30 MHz allowing use of a precision if attenuator accurate to 0.02 dB. Low power levels should always be used and careful checks made on the system to guarantee linearity. A boxcar integrator provides the final stage in the detection system, allowing the experimental measurements to be limited by the accuracy and ultimately the resetability of the attenuator (± 0.01 dB).

Measurements taken [6.42] according to the procedures outlined above were compared [6.42] to the predictions of two existing theories. The first is an approximate theory developed by *Arzt* and co-workers [6.39, 43], in which the attenuation α due to gas loading can be expressed in the following form [6.45]:

$$\alpha = \varrho_G V_G / \varrho_S V_S \Lambda_s. \qquad (6.5)$$

Here ϱ_G is the density of the gas, V_G is the longitudinal wave velocity in the gas, ϱ_S is the density of the surface wave substrate, and V_S is the Rayleigh wave velocity.

To facilitate the determination of the frequency, pressure, temperature, and atomic weight dependence of α, (6.5) will be rewritten in terms of these parameters. The first step utilizes the expression for the surface wave frequency $f = V_S/\Lambda_S$, and the equation relating the longitudinal wave velocity to the modulus of compression of the gas, λ, as given by

$$V_G = (\lambda/\varrho_G)^{1/2}. \tag{6.6}$$

Equation (6.6) is, of course, valid only under a continuum approximation. Equation (6.5) thus becomes

$$\alpha = \frac{f}{\varrho_S V_S^2}[\varrho_G(\lambda/\varrho_G)^{1/2}]. \tag{6.7}$$

For perfect gases below frequencies of approximately 1 GHz (which also corresponds to the continuum region), the following additional relations are valid [6.46]:

$$\varrho_G = MP/RT, \tag{6.8}$$

and

$$\lambda = (1/K_S)\gamma P, \tag{6.9}$$

where M is the molecular weight, P is the pressure, R is the universal gas constant, T is the temperature, K_S is the adiabatic compressibility, and γ is the ratio of the specific heat at constant pressure to the specific heat at constant volume.

Equation (6.7) can then be rewritten in its final form as

$$\alpha(\mathrm{Np\,m^{-1}}) = \frac{fP}{\varrho_S V_S^2}\left(\frac{\gamma M}{RT}\right)^{1/2}. \tag{6.10}$$

The second theory used to compare with experimental results was that originally described by *Campbell* and *Jones* [6.44]. This is basically a continuum mechanical approach in which arbitrary anisotropic piezoelectric substrates are treated explicitly. It is exact within the continuum limit and for ideal compressible fluids, for which the linearized equations of motion are valid.

The method of solution used in the continuum mechanical theory consists of solving simultaneously the equations of motion, the static forms of Maxwell's equations, and the constitutive relations for both the gas and the substrate. The boundary conditions at the interface between the two media must also be satisfied. Implementation of this technique requires an exceedingly complex

Table 6.2 Molecular weight M and ratio of the specific heat at constant pressure to specific heat at constant volume γ for various gases

Gas	M	γ
He	4.0	1.6667
Ne	20.2	1.6667
Ar	39.94	1.6667
Kr	83.8	1.6667
Xe	131.3	1.6667
N	28.0	1.400
Air	$\sim$ 29	~ 1.4

$R = 8.3 \times 10^3$

computer program, the input parameters for which were generated using (6.8) and (6.9), together with the data listed in Table 6.2.

The results [6.42] of surface wave attenuation due to argon loading vs frequency are given in Fig. 6.17. As with the results for air loading previously reported [6.6], these data are linear with frequency, in this case with a slope of 6.8×10^{-4} dB/MHz^{-1} cm. The experimental curve falls between the approximate and continuum mechanical theoretical curves. At higher frequencies, agreement appears to be better with the approximate curve, while at lower frequencies the continuum mechanical theory appears to be more accurate. This is actually to be expected since at the higher frequencies the atomic mean free path l is comparable to the wavelength, leading to a breakdown in the continuum approximation. (For example, for YZ LiNbO$_3$ and argon, $l \approx \Lambda_s \approx 0.1\,\mu$m at 3.4 GHz.) On the other hand, the basic form of the approximate theory as given in (6.5) (and from which Fig. 6.17, at least, is derived directly) has been shown by *Dransfeld* and *Salzmann* [6.45] to be equally valid in both the high- and low-frequency limits, $l \ll \Lambda_s$ and $l \gg \Lambda_s$.

Surface wave attenuation due to argon loading at 539, 1420, and 2430 MHz as a function of pressure is shown [6.42] in Fig. 6.18. Dependence is linear with pressure at all frequencies, which qualitatively agrees with both theories. As can be seen from Fig. 6.18 virtually 100% of the attenuation due to air loading occurred between 1 mm and atmospheric pressure. Going to very high vacuums produces no further discernible effect [6.42].

Surface wave attenuation due to gas loading as a function of molecular (and for monatomic gases, atomic) weight is illustrated [6.42] in Fig. 6.19. The approximate theory is in better agreement with the experimental results for small atomic weight than is the continuum theory. These roles are reversed for high molecular weight although neither theory appears to exhibit the correct functional form. It can thus be concluded that the ease in use of the approximate theory weighs heavily in its favor. Figure 6.19 illustrates an important practical fact. For operational systems, lower insertion loss (approaching that of vacuum) can be attained by encapsulating delay lines and other signal-processing devices in a light gas whenever vacuum encapsulation cannot be accomplished.

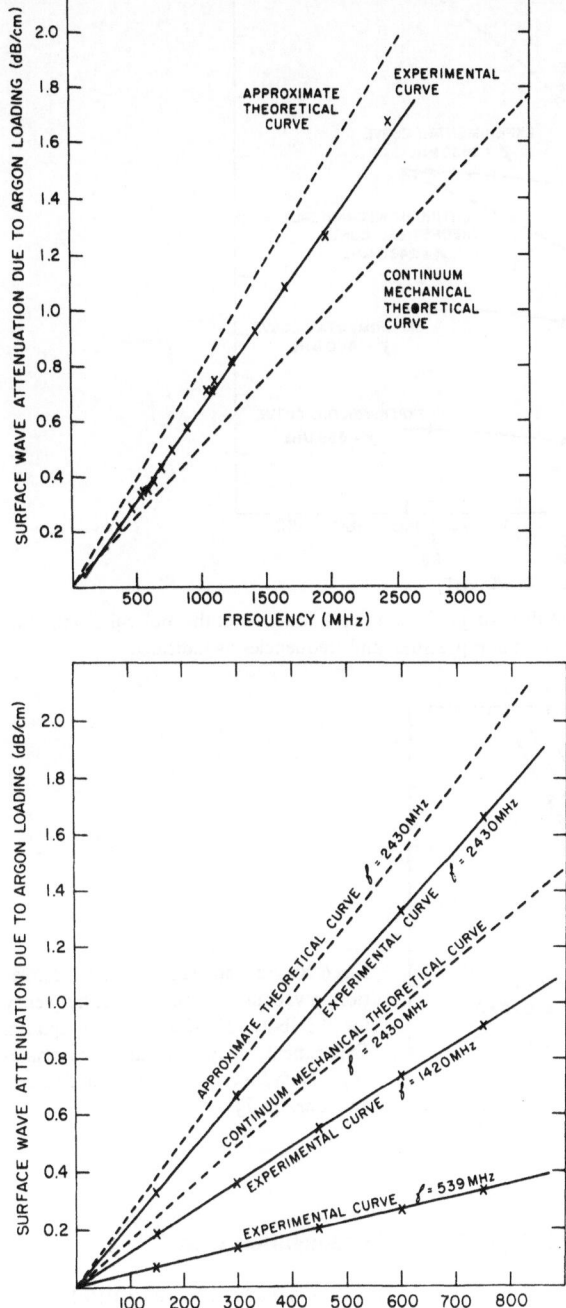

Fig. 6.17. Frequency dependence of surface wave attenuation due to argon loading. Data refer to room temperature $(298 \, \text{K} = 25° \, \text{C})$ and 1 atm pressure $(10^5 \, \text{N/m}^{-2})$

Fig. 6.18. Pressure dependence of surface wave attenuation due to argon loading. Room temperature and frequencies as indicated

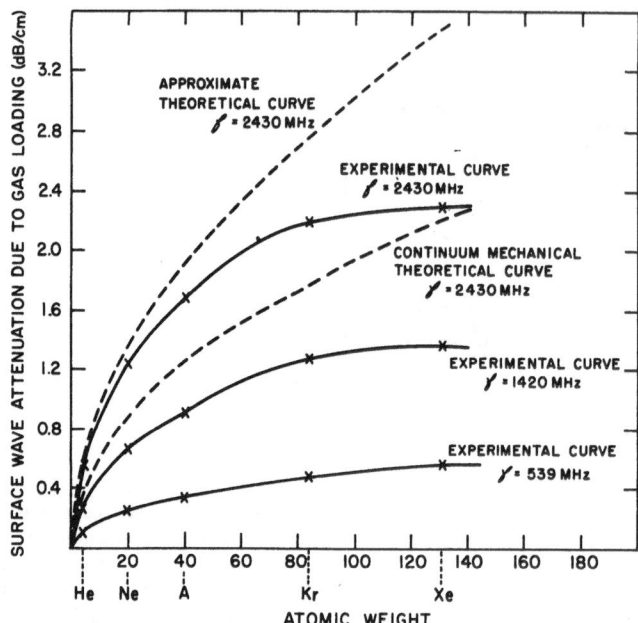

Fig. 6.19. Surface wave attenuation due to gas loading as a function of the molecular weight of monatomic gases. Room temperature, 1 atm pressure, and frequencies as indicated

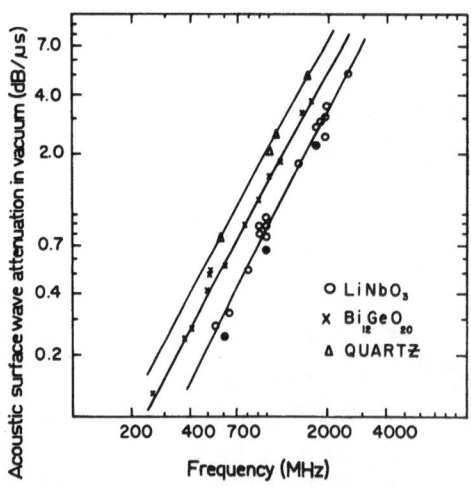

Fig. 6.20. Acoustic surface wave attenuation in vacuum as a function of frequency for $LiNbO_3$, $Bi_{12}GeO_{20}$, and quartz. Experimental slopes are all approximately f^2. Data for quartz courtesy of *Budreau* and *Carr* [6.7]

6.2.4 Frequency and Material Dependence of Propagation Loss

Using the techniques described in the subsections above, the magnitude and frequency dependence of loss for the three most important surface wave materials—lithium niobate, bismuth germanium oxide, and quartz—have been experimentally determined [6.6–8] and are illustrated in Figs. 6.20 and 6.21.

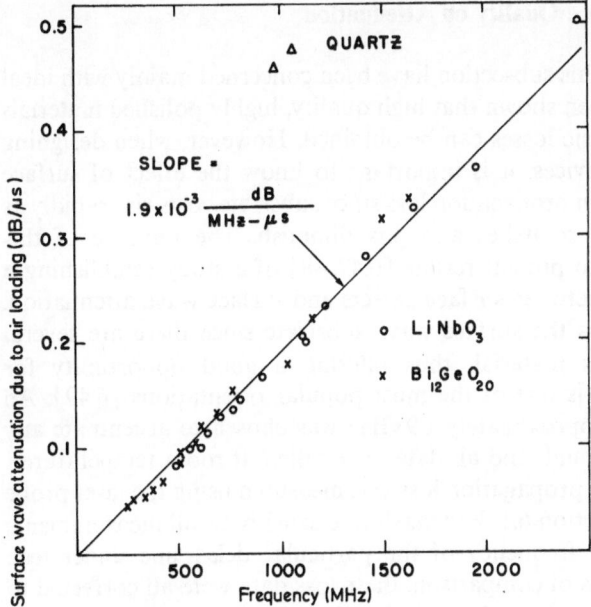

Fig. 6.21. Acoustic surface wave attenuation due to air loading as a function of frequency for LiNbO$_3$, Bi$_{12}$GeO$_{20}$, and quartz. It is interesting to note nearly identical results for LiNbO$_3$ and Bi$_{12}$GeO$_{20}$

Data for many other materials and orientations can be found in the summary in Section 6.8. Quantitative results are presented in (6.11)–(6.13), where F is in GHz.

$$YZ \, \text{LiNbO}_3 \, \text{Loss}[\text{dB} \, \mu s^{-1}] = 0.88F^{1.9} + 0.19F \qquad (6.11)$$

$$001, 110 \, \text{Bi}_{12}\text{GeO}_{20} \, \text{Loss}[\text{dB} \, \mu s^{-1}] = 1.45F^{1.9} + 0.19F \qquad (6.12)$$

$$YX \, \text{Quartz Loss}[\text{dB} \, \mu s^{-1}] = 2.15F^2 + 0.45F. \qquad (6.13)$$

The $F^{1.9}$ or F^2 term refers to surface wave propagation in vacuum, while for propagation in atmosphere the linear (with frequency) term due to air loading must be added. Note that propagation losses are negligible at low frequencies (below 200 MHz) unless exceedingly long time delays are being considered.

An example of the use of (6.11)–(6.13) is to compute the total attenuation loss on a 60 μs delay line on YX quartz. For a center frequency of 70 MHz, (6.13) becomes

$$\text{Loss}[\text{dB} \, \mu s^{-1}] = 2.15(0.07)^2 + (0.45)(0.07) \qquad (6.14a)$$

$$\text{Loss}[\text{dB} \, \mu s^{-1}] = 0.0105 + 0.0315 = 0.042 \qquad (6.14b)$$

$$\text{Total loss} = 2.52 \text{dB}. \qquad (6.14c)$$

It is seen that in this example air loading is the dominant source of propagation loss.

6.2.5 The Effect of Surface Quality on Attenuation

Data presented prior to this subsection have been concerned mainly with ideal crystals. That is, it has been shown that high quality, highly polished materials which exhibit only intrinsic losses can be obtained. However, when designing acoustic surface wave devices, it is important to know the effect of surface preparation and quality on propagation loss since substrate costs rise rapidly as the number and size of scratches and pits diminish. The purpose of this subsection is, therefore, to present results [6.47, 48] of a study establishing a quantitative correlation between surface defects and surface wave attenuation.

YZ LiNbO$_3$ is used as the surface wave substrate since there are several reliable suppliers of this material, thus offering a good opportunity for comparison, and since it is one of the most popular orientations [6.49]. An operating frequency of approximately 1.9 GHz was chosen to accentuate any differences which were found, and all data were taken at room temperature.

Acoustic surface wave propagation loss was measured using the laser probe technique described in Section 6.1. For maximum sensitivity, all measurements were made at the center frequency of the particular delay line under test. However, for the purposes of comparison, these loss data were all corrected to 1.866 GHz by subtracting the calculated air loading at the center frequency and scaling this vacuum attenuation to 1.866 GHz using an F-squared relation.

Loss data obtained using these procedures are directly compared to surface-finish and crystalline quality in Table 6.3. A definite correlation with surface finish, as determined using electron microscopy analysis [6.50, 51] at 15000X, is evident. In fact, surface roughness of the order of 0.25 wavelength causes an approximately 50% increase in surface wave attenuation. In addition to this increased propagation loss, substantial difficulties in device fabrication were encountered [6.52] with sample C50. These fabrication problems are probably equally as important as loss in establishing the lower limit to the surface quality which can be tolerated.

Figure 6.22 shows photographs [6.50, 51] of the surface of four of the samples listed in Table 6.3. The samples designated A were polished by a firm experienced in the preparation of lithium niobate surface wave substrates using 0.25 µm diamond followed by a chemical-mechanical polish. The specifications called for a 100% scratch- and pit-free surface (flat to 3 waves over 38 mm) under 80X magnification. The samples designated B were polished by a firm used to handling LiNbO$_3$ but not necessarily for critical surface wave applications. Specifications called for substrates as free as possible from scratches and pits. And, finally, the sample designated C was processed by an optical company more used to fused quartz than materials such as LiNbO$_3$. Again specifications called for plates as free as possible from scratches and pits.

Other conclusions can be made from Table 6.3. The source of material and its quality (within the range tested), as measured using an extended Laue x-ray back reflection method [6.50, 51], have no appreciable effect on propagation loss. Doping the LiNbO$_3$ with 0.01 mole-% MgO also produced no effect.

Table 6.3. Acoustic surface wave propagation loss on YZ LiNbO$_3$ related to surface finish quality

Delay line	Vacuum attenuation interpolated to 1866 MHz [dB/cm]	Surface finish and crystalline quality[f]
A 5[a]	7.83	Completely featureless with no directional texture. Occasional random appearance of small dirt particles or pits. Well crystallized material free of faults
A 49[b]	8.15	Flaw-free surface, free of work damage. Material shows variation along lengths of specimen. Small angle boundaries present
A 47[c]	8.27	Generally good surface with some shallow circular depressions. Some random scratches up to 0.15 µm wide. Some evidence of work damage. Material contains small angle boundaries and some extremely poor areas
A 32[d]	8.28	—
A 36[d]	8.44	Very well prepared surface showing only a very, few random scratches (0.05 µm wide). Very uniform and well crystallized lithium niobate with no significant flaws or imperfections
A 48[e]	8.45	—
B 46[a]	8.71	Covered over most of the area with fine scratches (0.1 µm). Some evidence of work damage. Very uniform crystal free of imperfections
B 6[a]	9.48	Covered with a series of directional scratches about 0.1 µm in width. Random small pits and subsurface imperfections. Well crystallized material free of faults.
B 4[a]	10.72	Covered by shallow scratches 0.1 µm wide spaced 1–2 µm apart. Incompletely polished with a fine abrasive (insufficient to remove remnants of previous coarse operation). Well crystallized material free of strains, twins, or imperfections
B 4[a]	11.20	Different section of delay line described directly above
C 50[a]	11.71	Partially polished surface containing an array of undulating scratches 0.25 µm–0.5 µm wide. Most likely subsurface work damage. Very well crystallized material free of any structural defects.
C 50[a]	14.05	Different section (closer to ends of crystal) of delay line described directly above

Material Source: [a] Crystal Technology (Linobate).
   [b] Union Carbide (Acoustic Grade).
   [c] Crystal Technology (Acoustic Grade).
   [d] Crystal Technology (0.01 mole-% MgO Doped Linobate).
   [e] Harshaw
   [f] [6.50, 51].

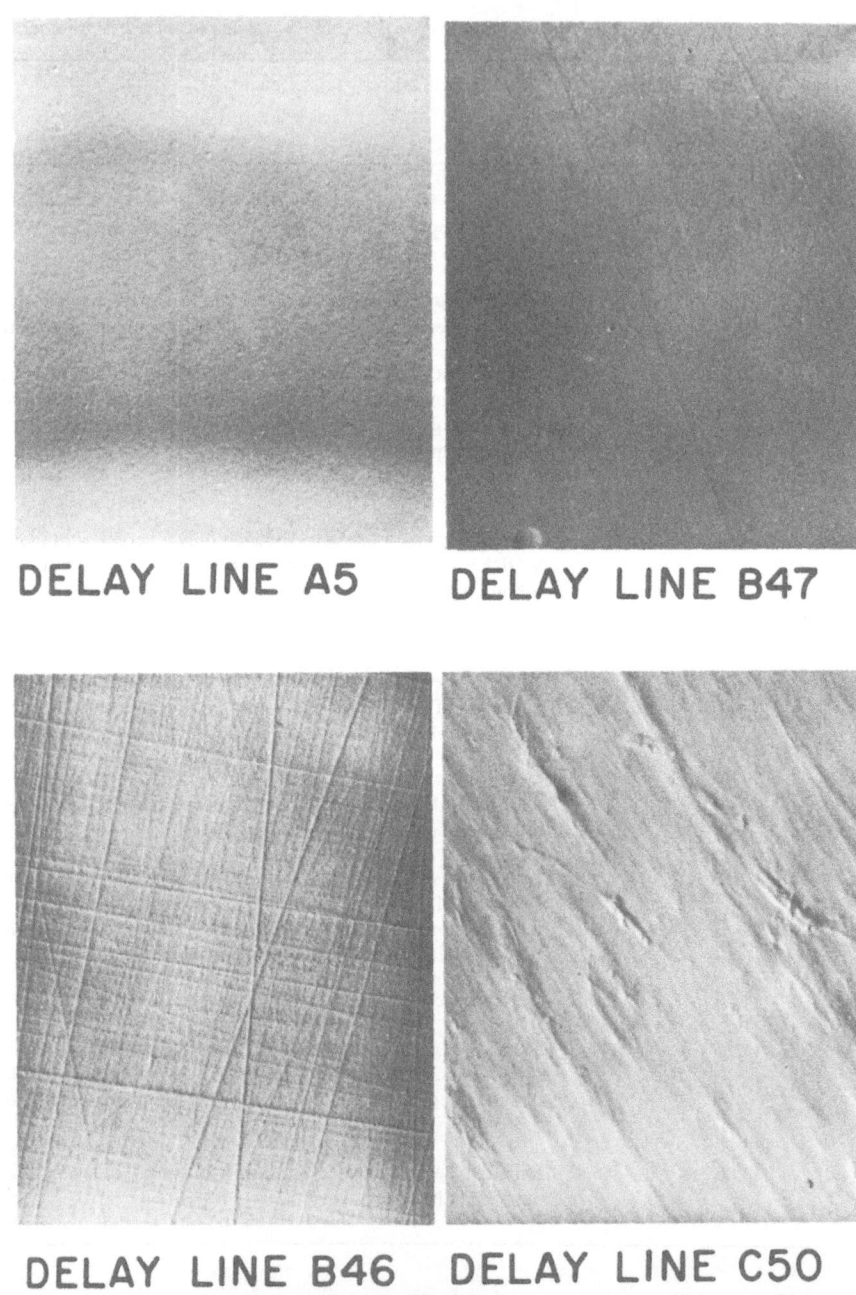

DELAY LINE A5   DELAY LINE B47

DELAY LINE B46   DELAY LINE C50

⊢――――⊣
1 μ m

Fig. 6.22. Photographs from electron microscope studies illustrating surface quality of typical areas of delay lines indicated (from [6.50,51])

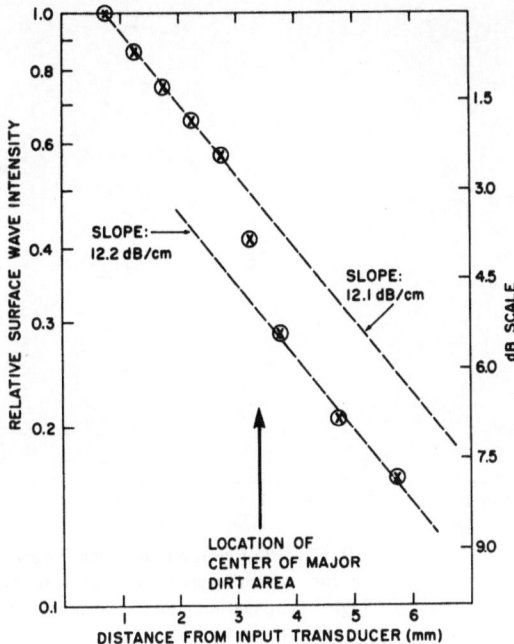

Fig. 6.23. Attenuation versus distance plot traversing area of dirt pictured in Fig. 6.24. Data were taken at 1897 MHz on delay line B4

This latter result is in agreement with earlier volume wave results of *Smith* et al. [6.53]. They showed that MgO doping could reduce the temperature-independent component of loss but not the temperature-dependent part. Since the temperature- dependent component has been shown in Section 6.2.2 to be the major contributor to surface wave attenuation on YZ LiNbO$_3$, the present data are consistent with the earlier work [6.53].

The effect of dirt on surface wave propagation loss is also of interest and is therefore illustrated in Fig. 6.23. The dirt causing the anomalously high attenuation evident in Fig. 6.23 is pictured [6.51] in Fig. 6.24. This photograph shows spots of some sort of stain or residue. These spots are very thin and are approximately 0.25 μm in diameter. Cracks, deep scratches, and deep pits also cause anomalously high attenuation. This is indicated in Fig. 6.25 in which a very deep pit was located within a localized area of serious scratches and cracks, which are in turn illustrated in Fig. 6.26.

In summary, for ultimate surface wave performance, surface finishes which are scratch free under 15000X can and should be used, and delay lines should be kept perfectly clean. From a more practical standpoint, however, scratches of the order of 0.1 μm or less will cause no substantial performance degradation for moderate time delays at frequencies below 1 GHz.

A similar study was undertaken [6.54] on two of the higher coupling orientations of GaAs: 211 cut, 111 propagating, and 100 cut, 100 propagating to evaluate differences between chemical and mechanical polishes on this material.

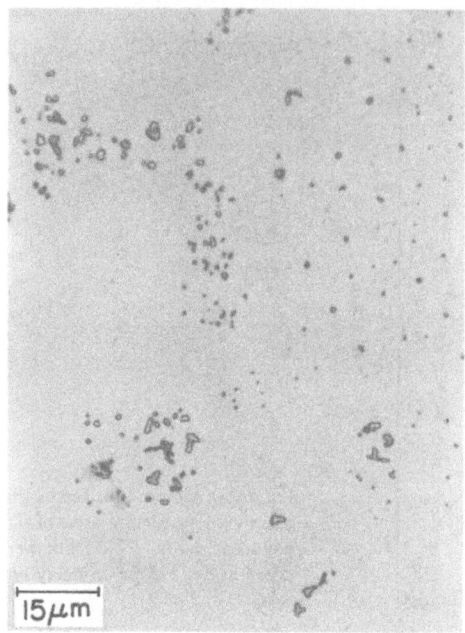

15μm

Fig. 6.24. Photograph illustrating the area of dirt which caused the anomalously high propagation loss indicated in Fig. 6.23 (from [6.51])

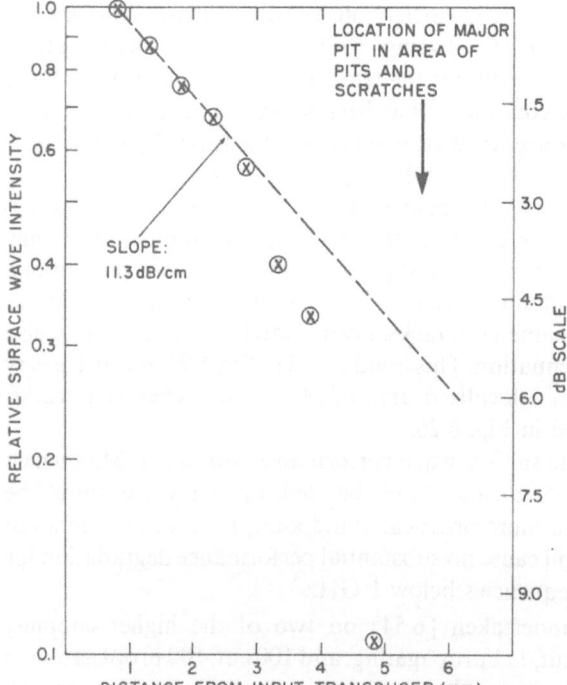

Fig. 6.25. Attenuation versus distance plot near and within area of cracks, scratches, and pits pictured in Fig. 6.26

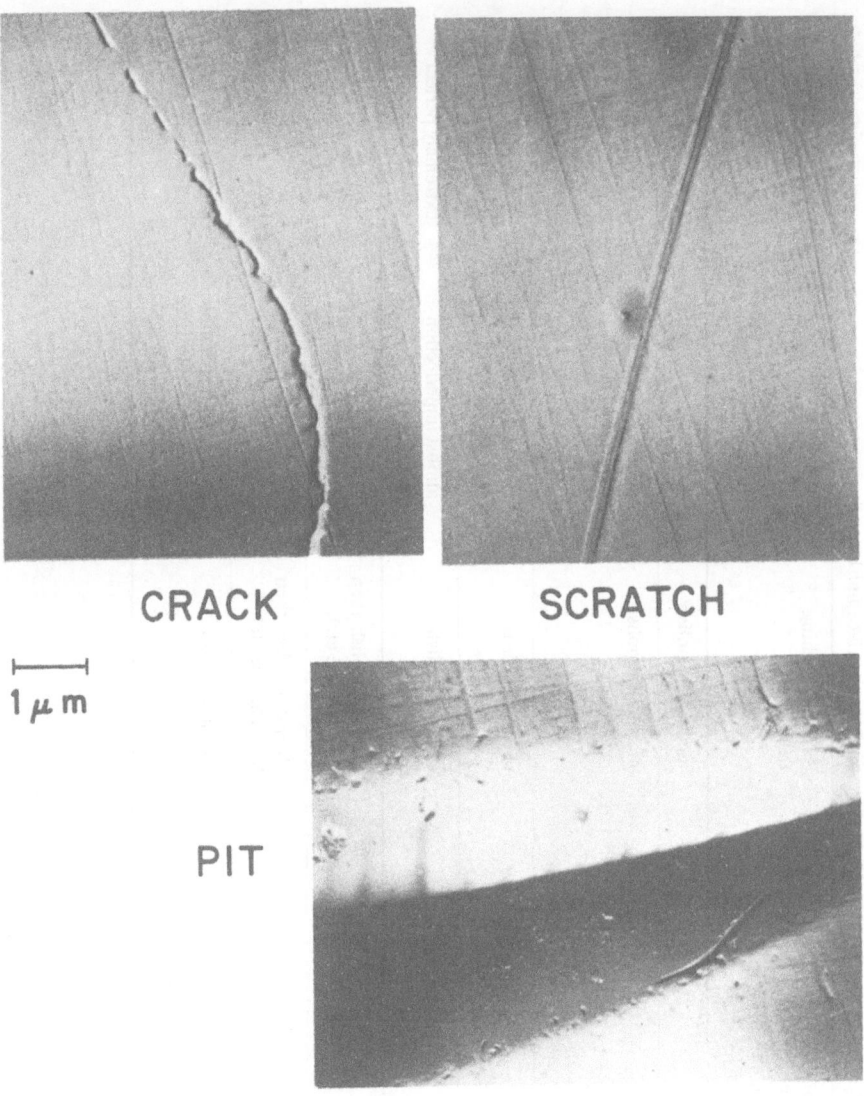

CRACK   SCRATCH

1 μm

PIT

Fig. 6.26. Photograph illustrating the cracks, scratches, and major pit which caused the anomalously high propagation loss indicated in Fig. 6.25

Results are summarized in Table 6.4. Data taken on virtually imperfection-free, chemically polished substrates yielded no lower loss values than those taken on scratched samples. It should be noted, of course, that although the 0.1 μm scratches are large when compared to the defects on a chemically polished surface, they are only approximately 4 % of an acoustic wavelength. Finally, note that 211 cut GaAs has both lower vacuum attenuation and lower air loading than the 110 cut orientation tested.

Table 6.4. Acoustic surface wave propagation loss on gallium arsenide related to surface finish quality

	Frequency [MHz]	Surface wave propagation loss in air [dB/cm]	Air loading [dB/cm]	Surface polishing technique	Surface finish quality	Subsurface quality
211 cut, 111 propagating GaAs	945	12.5	—	mechanical polish	—	—
	975	13.3	—	mechanical polish	covered with scratches the order of 0.1 μm deep across entire surface	no unusual features
	975	13.3	1.00	chemical polish	extremely fine back-ground scratches	some small-angle boundaries (~10 min)
	1005	13.7	—	chemical polish	almost imperfection free	extremely perfect crystal
110 cut, 100 propagating	980	—	1.38	chemical polish then light chrome etch	—	—
	1044.5	17.5	—	chemical polish then light chrome etch	—	—
	1059	16.7	—	chemical polish	background of shallow depressions indicative of polishing marks	perfect crystal with some strained areas

6.3 Diffraction and Beam Steering

6.3.1 Review of the Theories

Diffraction and beam steering have been discussed in Chapter 2 and again reviewed briefly in the context of this chapter in Section 6.1. Important quantities were also defined in Section 6.1.

The purpose of this section is to outline in a quantitative manner how material properties influence which diffraction theory should be used in a given situation, and how diffraction and beam steering themselves vary with material parameters. This variation strongly influences the choice of material in a given device application.

The parabolic theory originally developed by *Cohen* [6.55] is useful for calculating diffraction fields when the velocity anisotropy on or near pure-mode axes can be approximated by a parabola. By using a small angle approximation, *Cohen* [6.55] showed that for certain cases, the higher orders of the expression for the velocity could be neglected past the second order. That is,

$$\frac{v(\theta)}{v_0} \approx 1 + \frac{\gamma}{2}(\theta - \theta_0)^2 \tag{6.15}$$

where γ is an anisotropy parameter and θ_0 is the angular orientation of the pure-mode axis. Furthermore, for these conditions the acoustic power flow vector deviates from the pure-mode axis by the angle

$$\phi = \gamma(\theta - \theta_0) \equiv \gamma\theta' . \tag{6.16}$$

By comparing these approximations to an exact solution for electromagnetic diffraction in uniaxially anisotropic media, he showed that the diffraction integral reduces to Fresnel's integral with the following effective change in the distances from the aperture:

$$\hat{Z}' = \hat{Z}|1 + \gamma| . \tag{6.17}$$

Here, *Szabo* and *Slobodnik* [6.12] introduced the absolute magnitude signs to account for those materials having $\gamma < -1$. Note also that all "hatted" quantities refer to wavelength-scaled parameters. In other words, diffraction is either accelerated or retarded depending on the value and sign of γ.

As an example of the application of the parabolic theory, consider the velocity surface of Fig. 6.27 near the pure-mode axis at $\theta = 0$, where it has the appearance of a concave upward parabola. By using a least squares fit to relative velocity values computed to seven significant places within a range of $\pm 5°$ of $\theta' = 0$, the velocity surface can be shown [6.12] to be an excellent parabola with $\gamma = 0.366$. In addition, as expected within this range, the power flow angle curve is linear with a positive slope of $\partial\phi/\partial\theta = 0.366$, in accordance with (6.16). For this particular material and orientation, excellent agreement is to be expected between the parabolic theory and experiment as a consequence of the good parabolic fit.

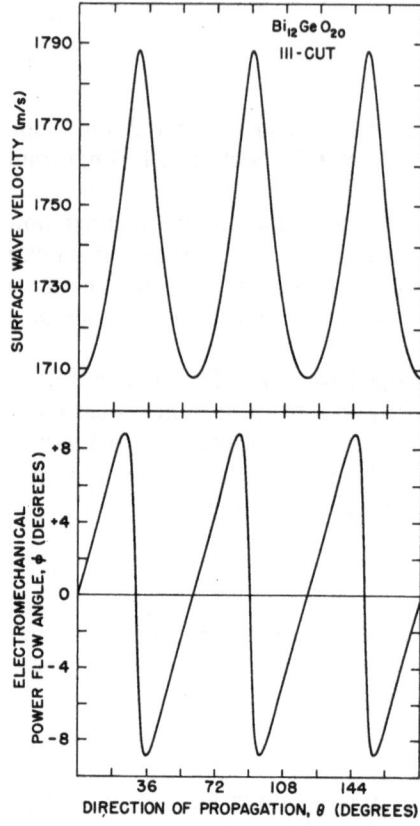

Fig. 6.27. Surface wave velocity and electro-mechanical power flow angle as a function of angle from 110 axis on 111 cut bismuth germanium oxide

To confirm this fact, a transducer pair was fabricated [6.12] on 111 cut 110 propagating ($\theta = 0$) $Bi_{12}GeO_{20}$ with transducer widths of $\hat{L} = 61$. Frequency of operation was 510 MHz and the laser probe technique described in Section 6.1 was used to obtain the experimental beam profiles. Figure 6.28 compares the measured profiles shown in the last column against several computer-drawn theoretical curves. Patterns in the first column were computed using an isotropic diffraction theory and, as expected, they do not agree with experiment. Profiles in the second column were calculated from the parabolic anisotropic theory and they show that the sequence of diffraction patterns has accelerated by a factor of $|1 + \gamma|$ over the isotropic patterns. Agreement here with experiment is excellent. The third column demonstrates profiles obtained from the more complex angular spectrum of waves theory discussed in Chapter 2. This theory gives equivalent results. Since this theory is more complex, it should not be used where the parabolic theory suffices. Quantitative guidelines for the choice between the two theories will be provided in Section 6.3.2.

The real power of the exact anisotropic theory can only be illustrated by its ability to predict even the fine structure of a diffraction pattern on a highly nonparabolic velocity surface, including profile asymmetry due to beam

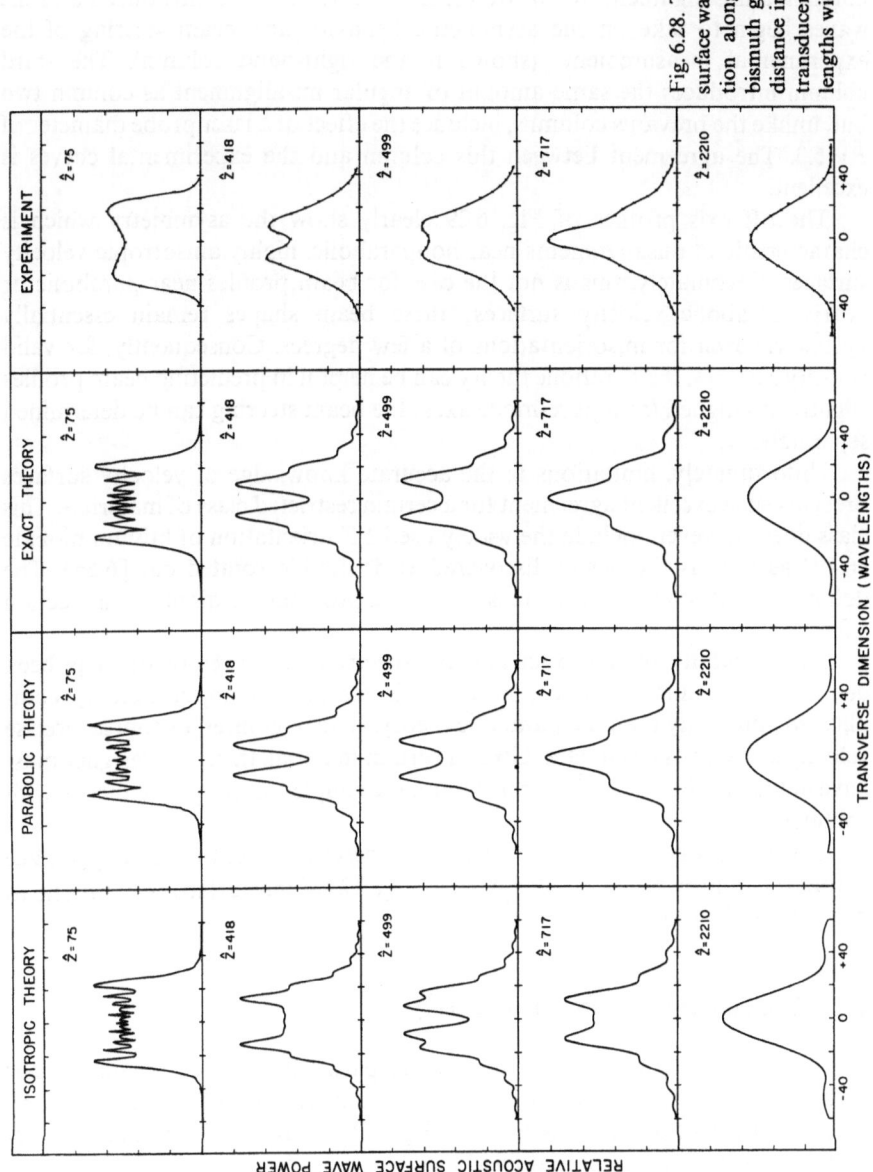

Fig. 6.28. Theoretical and experimental surface wave profiles illustrating diffraction along the 110 axis of 111 cut bismuth germanium oxide. $\hat{Z}$ indicates distance in wavelengths from the input transducer. Transducers were 61 wavelengths wide at frequency of 510 MHz

steering. An example of the full generality of this theory is shown in Fig. 6.29. The case studied [6.12] concerns surface waves launched nearly along the 111 axis of 211 cut gallium arsenide at a frequency of 280 MHz. Transducer widths were $\hat{L} = 51$. This orientation was chosen because the velocity is nonparabolic and changes very rapidly with direction. The first column of Fig. 6.29 shows profiles for waves propagating exactly along the pure-mode $1\bar{1}1$ axis, a direction corresponding to $\phi = 0$. Also note that the smoothing effect of the laser probe $\hat{P}$ wavelengths in diameter has not yet been included ($\hat{P} = 0$). For the second column, a misalignment of 0.6° from the $1\bar{1}1$ axis has been introduced and the waves begin to take on the asymmetric behavior and beam steering of the experimental measurements (shown in the right-hand column). The third column introduces the same amount of angular misalignment as column two but, unlike the previous columns, includes the effect of a laser probe diameter of $\hat{P} = 5.3$. The agreement between this column and the experimental curves is excellent.

The off axis profiles of Fig. 6.29 clearly show the asymmetry which is characteristic of misalignments near nonparabolic, highly anisotropic velocity surfaces. Fortunately, this is not the case for beam profiles near parabolic or nearly parabolic velocity surfaces; these beam shapes remain essentially symmetric even for misorientations of a few degrees. Consequently, for valid velocity surfaces, the parabolic theory can be helpful in predicting beam profiles slightly misaligned from pure-mode axes; the beam steering can be determined separately.

Unfortunately, limitations in the accurate knowledge of velocity surfaces prevents such excellent agreement for a certain restricted class of materials. This class does, however, include the widely used YZ orientation of lithium niobate [6.49] as well as the newly discovered 16.5° double rotated cut [6.56]. The definition of this class and the reasons for the problem are discussed in Section 6.3.3.

The versatility of the exact angular spectrum of waves theory has been demonstrated; this approach, however, is far more computationally complicated and costly than the parabolic theory. It also requires, of course, precise velocity surfaces as input data. Given a certain material, then, the designer must have guidelines from which he can choose the simplest appropriate theory with confidence.

Cohen's theory [6.55], which relies on a description of the velocity surface within 0.1 rad (or 6°) of $\theta' = 0$, works well [6.12] if the surface is parabolic as decribed in detail in Section 6.3.2.

6.3.2 Limitations to the Parabolic Theory

The closeness of a given velocity surface to a parabolic curve can be determined by fitting the surface to a parabola and noting any deviation. In fact, anisotropic velocity information such as that of Fig. 6.27 is available for nearly every popular surface wave material [6.57].

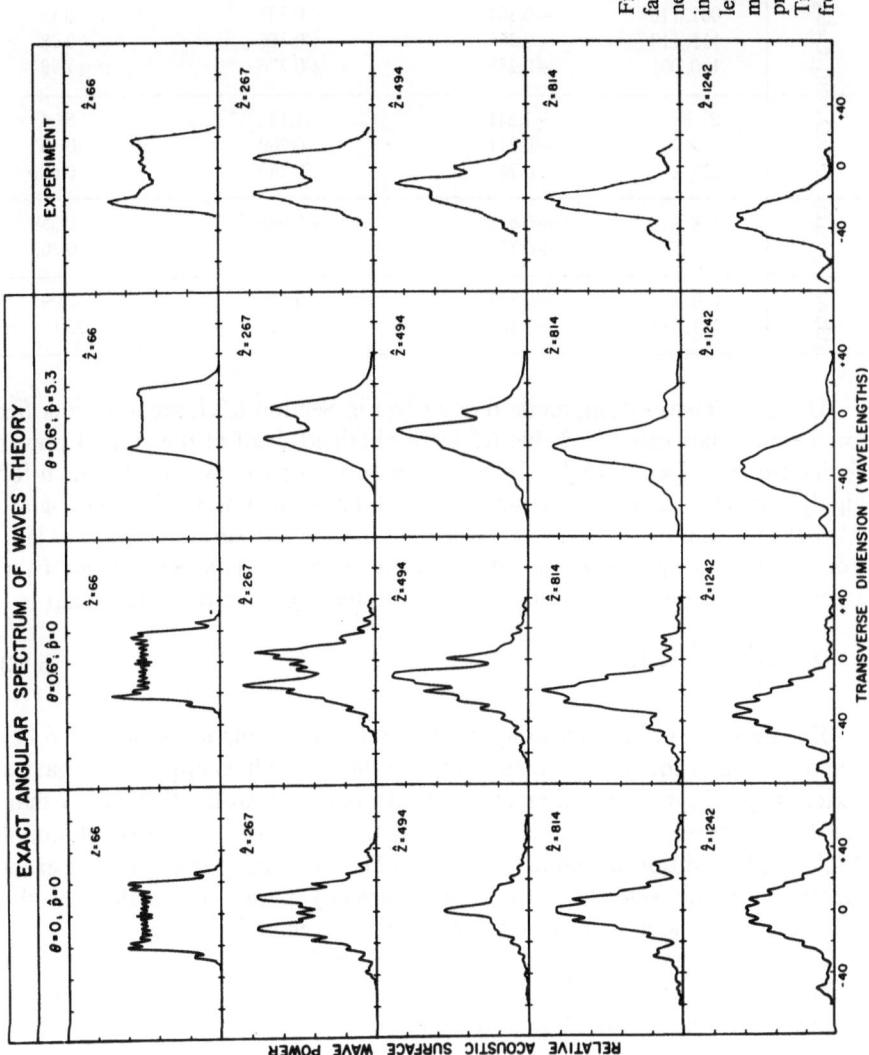

Fig. 6.29. Theoretical and experimental surface wave profiles illustrating diffraction near 111 axis of 211 cut gallium arsenide. $\hat{Z}$ indicates distance for propagation in wavelengths from input transducer. θ gives the misorientation from 111 axis, and $\hat{P}$ is laser probe diameter in acoustic wavelengths. Transducers were 51 wavelengths wide at frequency of 280 MHz

Table 6.5. Summary of data for use with parabolic theory. Parabolic theory can be used with full confidence for $0 \leqq |\delta_m| \lesssim 0.2$, it can be used with care for $0.2 \lesssim |\delta| \lesssim 2.0$ and cannot be used for $2.0 \lesssim |\delta_m| < \infty$

	Material orientation	Recommended γ for use with parabolic theory $(\partial\phi/\partial\theta)$	γ derived from least squares fit to parabola	Deviation factor $|\delta_m|$
LiNbO$_3$	Y, Z	-1.083	-0.906	7.87
	$16\frac{1}{2}$ double	-1.087	-1.00	3.99
	$41\frac{1}{2}°, X$	-0.445	-0.454	0.57
	Z, X	$+0.192$	$+0.193$	0.05
	$Y, 21.8°$ cut	$+0.393$	$+0.390$	0.5
Bi$_{12}$GeO$_{29}$	001, 110	-0.304	-0.304	0.14
	111, 110	$+0.366$	$+0.366$	0.08
	110, 001	$+0.236$	$+0.276$	1.93
LiTaO$_3$	Z, Y	-1.241	-1.13	5.04
	Y, Z	-0.211	-0.209	0.14
	$22°, X$	0.765	0.753	0.600
Quartz	YX	$+0.653$	$+0.645$	0.359
	ST, X	$+0.378$	$+0.373$	0.205
GaAs	110, X	-0.537	-0.536	0.097
	211, 111	-2.58	-2.11	20.1

Using the curve-fitting method described in Section 6.2.1, second-order fits for the materials listed in Table 6.5 were obtained. Half of the second power coefficient of θ' [see (6.15)] is listed as the anisotropy parameter γ, which for highly parabolic surfaces is the same as the slope of the power flow angle $\partial\phi/\partial\theta$. Also, the maximum deviation of the fit from the velocity surface is listed as $|\delta_M|$. For comparative purposes, this deviation is expressed as a percentage of the actual velocity and for convenience is multiplied by a factor of 10^5. That is

$$|\delta_M| = \left| \frac{v_{fit} - v}{v} \right| \times 10^5 . \tag{6.18}$$

For velocity surfaces *deviating* from a perfect parabola it was found [6.12] that a value of γ equal to the actual slope of the power flow angle $\partial\phi/\partial\theta$ gave a better range of agreement when used with the parabolic diffraction theory than did the parameter derived from the fit. This slope is listed in the first column of Table 6.5. In addition, a complete study [6.12] of diffraction loss using the exact theory on many velocity surfaces not perfectly parabolic resulted in the following conclusion. Anisotropy may be conveninetly grouped into two categories: *parabolic* $(0 < |\delta_M| \lesssim 2.0)$ and *nonparabolic* $(2.0 \lesssim |\delta_M| < \infty)$. Higher order terms discarded in Cohen's [6.55] approximation [see (6.15)] become significant for the nonparabolic surfaces.

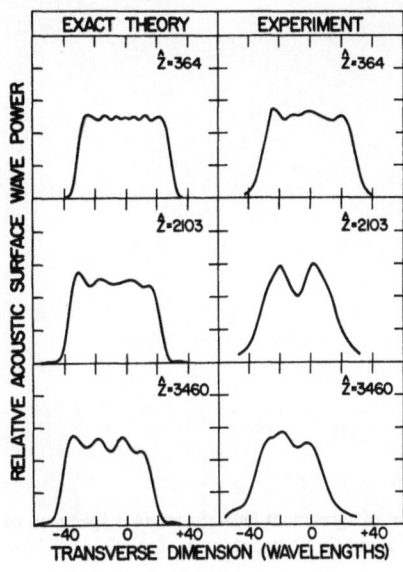

Fig. 6.30. Illustration of disagreement between angular spectrum of waves theory using the calculated velocity surface and YZ LiNbO$_3$ experiment. $\hat{L} = 64.5$, $f = 900$ MHz

For those surfaces which are highly *parabolic* such as 111 cut, 110 propagation Bi$_{12}$GeO$_{20}$; Z cut X propagation LiNbO$_3$ and Y cut Z propagation LiTaO$_3$, the γ calculated from the second-order parabolic fir agrees nearly exactly with $\partial\phi/\partial\theta$ as expected from (6.16). Differences between the exact and approximate theory appear as minor changes in beam profiles as the limit $|\delta_M| \approx 2.0$ is approached. Below this limit, differences in diffraction loss calculations for identical cases using the two theories are typically less than 0.05 dB.

6.3.3 Limitations in Using the Angular Spectrum of Waves Theory

The *nonparabolic* group has profiles which show marked departures from those obtainable from the approximate theory. Obvious examples are the experimental cases of GaAs, *YZ* LiNbO$_3$, and 16.5° double rotated LiNbO$_3$ [6.56]. To predict the correct pattern requires the exact angular spectrum of waves theory. If, however, one is interested in a gross idea of the losses – especially those which are significant and occur in the transition to and within the far field – the recommended value of γ, that is, $\partial\phi/\partial\theta$, can be helpful. Physically, the use of $\partial\phi/\partial\theta$ can be justified by the fact that at distant field points in the far field, less and less of the velocity surface contributes until at infinity the source appears as a point.

The basic limitation in using the angular spectrum of waves approach lies not in the theory itself but in the requirement for accurate knowledge of the velocity surface of *interest*. For nonparabolic matreials having $\partial\phi/\partial\theta \approx -1$ the

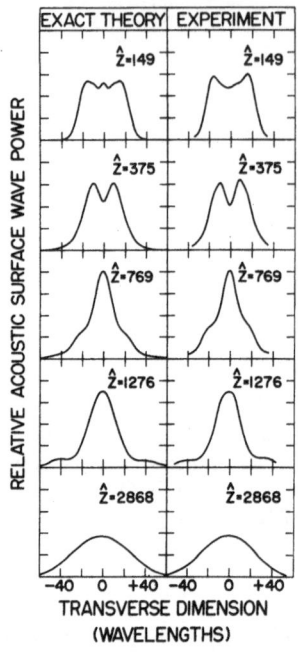

RELATIVE ACOUSTIC SURFACE WAVE POWER

TRANSVERSE DIMENSION
(WAVELENGTHS)

Fig. 6.31. Illustration of agreement between angular spectrum of waves theory and Y cut $21.8°$ propagating $LiNbO_3$. $\hat{L} = 56.4$, $f = 966\,MHz$

diffraction patterns are excedingly sensitive to any uncertainty in the shape of the velocity surface. A complete study [6.12] of YZ $LiNbO_3$ and $16.5°$ double rotated cut $LiNbO_3$ has indicated that the velocity surfaces (or equivalently the material constants) have not yet been determined accurately enough to predict the correct diffraction patterns. This is illustrated in Fig. 6.30. The theoretical curves using the normal Y cut surface do not duplicate the experimental data. Of particular note is the inability to reproduce the "double-peak" patterns. This double-peak is a recurring experimental phenomenon over widely varying conditions and could never be duplicated using the theory [6.58]. Using the $LiNbO_3$ constants measured by *Smith* and *Welsh* [6.59] rather than those measured by *Warner* et al. [6.60] and even assuming temperature variations were equally unsuccessful [6.12], crystalline misorientations could also not account for the discrepancy.

On the other hand, this same sort of comparison also on Y cut $LiNbO_3$ but now $21.8°$ from the Z axis (where $\partial\phi/\partial\theta = +0.393$), yielded excellent agreement with experiment, as illustrated in Fig. 6.31. It is also worth recalling that excellent theoretical agreement was obtained for the experimental use of GaAs (Fig. 6.29), an orientation having $\partial\phi/\partial\theta = -2.58$, more than twice as steep as the troublesome YZ $LiNbO_3$ orientation but away from the -1 point. The main fact to emphasize here is that a given uncertainty in velocity surface has a far more drastic effect on predicted profiles when $\partial\phi/\partial\theta$ approaches -1 [6.58]. It is hoped that recently developed techniques for direct measurement of velocity surfaces [6.61, 62] will soon remove this restriction.

6.3.4 Material Dependence of Diffraction and Beam Steering

Having experimentally determined the limits of validity of the parabolic and exact anisotropic diffraction theories, it is now possible to use them to generate diffraction loss design curves. This was done [6.12] by integrating the complex wave amplitude and phase over the width of the receiving transducer, taking the magnitude of this quantity, squaring, and referencing the result to zero loss exactly at the input transducer (which is assumed identical to the output transducer).

It has been established that, for velocity surfaces having $|\delta_M| \lesssim 2.0$, parabolic diffraction theory yields highly accurate results. Thus for all materials meeting this criterion, diffraction patterns are exactly equivalent in form but merely scaled in distance by the factor $|1 + \gamma|$. Thus, a universal diffraction loss curve [6.12] can be calculated in terms of this scale factor and the acoustic aperture $\hat{L}$ in wavelengths.

This curve is illustrated in Fig. 6.32, which is a plot of diffraction loss versus the parameter $(\hat{Z}/\hat{L}^2)|1 + \gamma|$. Similar curves for nonuniform excitation functions can also be generated [6.63]. The curve of Fig. 6.32 allows the determination of loss for any combination of transducer width and separation for all parabolic anisotropic velocity surfaces. In the Fresnel region the loss never exceeds 1.6 dB, which is the loss at the far-field length $\hat{Z}_F$ (where the final peak in the beam profile has started its descent to a far-field pattern). The distance and transducer width at which a given loss will occur can always be given in far-field lengths. For example, the 3 dB loss point is

$$\hat{Z}_{3dB} = 1.769\,\hat{Z}_F = \frac{1.769\,\hat{L}^2}{|1 + \gamma|}. \tag{6.19a}$$

In the far field the loss mechanism is spreading of the beam with a slope of 10 dB/decade. The far-field loss can be approximated by [6.12]

$$\text{Loss[dB]} = -10\log\frac{\hat{Z}}{\hat{Z}_F}. \tag{6.19b}$$

It is interesting to note that for nonparabolic velocity surfaces the far-field loss slope deviates from the 10 dB/decade slope as the part of the velocity surface that contributes to the field point changes with separation distance [6.58].

When beam steering is included with diffraction the situation cannot be dealt with as concisely as in Fig. 6.32. Combined beam steering and diffraction loss is not simply the sum of its separate constituents, and thus design curves must be more specific than the universal information possible when each loss mechanism is considered individually.

Diffraction is a fixed physical phenomenon for a given material, while beam steering (see Fig. 6.6) can be controlled by precise x-ray alignment at the expense of increased device cost. Both, however, influence the choice of acoustic surface wave substrate. An example [6.64] of how the combined loss of beam steering

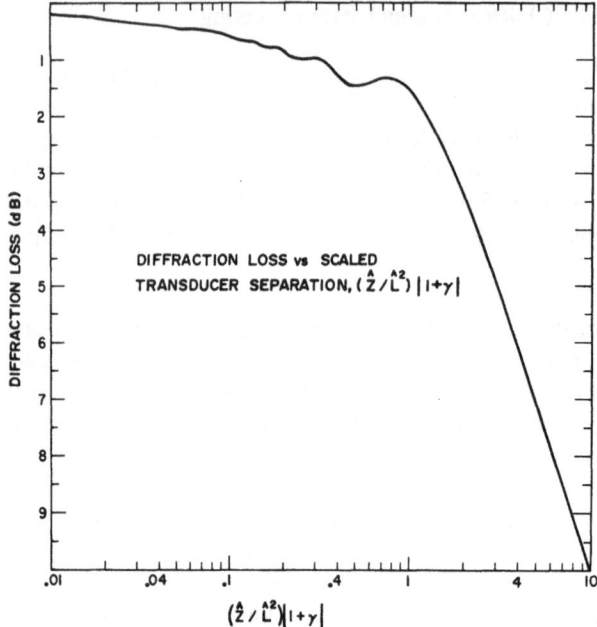

Fig. 6.32. Universal diffraction loss curve for all parabolic materials as a function of $(\hat{Z}/\hat{L}^2)|1+\gamma|$. To convert to actual wavelengths on the horizontal scale simply insert $\hat{L}$, width of your transducer in wavelengths, and γ (from Table 6.5) appropriate to your material. Beam steering here is assumed negligible

and diffraction varies among materials is illustrated in Fig. 6.33 where the loss is given as a function of $\gamma = \partial\phi/\partial\theta$, the slope of the power flow angle. As above, loss has been calculated by integrating the complex acoustic amplitude over the aperture of the receiving transducer using the parabolic diffraction theory for identical unapodized input and output transducers.

For Fig. 6.33 the acoustic aperture is $\hat{L} = 80$ wavelengths, the distance between input and output transducers is $\hat{Z} = 5000$ wavelengths, and the misalignment from the desired pure-mode axis, or the beam steering (BS) angle, is BS $\measuredangle = 0.1°$. To use these data for practical situations, it is only necessary to insert the slope of the power flow angle appropriate to the type and cut under consideration. It is also useful to note that

$$\hat{Z} = tf \tag{6.20}$$

where t is the time delay and f the frequency of the device of interest.

Several important features can be noted with reference to Fig. 6.33. Diffraction loss goes to 0 for those materials having $\gamma = -1.0$ and, as expected, the combined loss curve agrees exactly with the beam steering loss curve. Those materials having $\gamma = 0$ correspond to locally isotropic cases and beam steering goes to 0. Here, diffraction accounts for the total loss. Diffraction loss

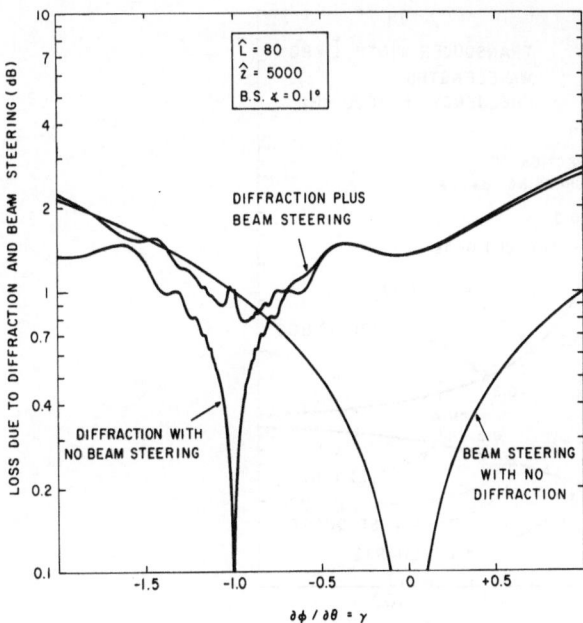

Fig. 6.33. Loss due to diffraction and beam steering as a function of slope or power flow angle for parabolic materials. $\hat{L}$ represents width of transducers in wavelengths, $\hat{Z}$ the distance between transducers in wavelengths, and BS $\not\chi$ the beam steering angle

alone is symmetric about $\gamma = -1.0$ and beam steering loss about $\gamma = 0$, while the combined curve is clearly nonsymmetric.

The results illustrated in Fig. 6.33 are of major importance in choosing a material for a particular application. For example, where diffraction is potentially a very serious problem, as in highly apodized filters, a material having $\gamma \approx -1.0$ would be most desirable [6.16, 17]. This subject will be treated in detail in Section 6.3.5.

A more graphic (but less accurate) manner of presenting the inherent trade-off between beam steering and diffraction is to take them as occurring independently of each other. Again for those materials approximated by perfect parabolic velocity surfaces ($\partial v_S/\partial\theta = 0$, $\partial^2 v_S/\partial\theta^2 \neq 0$ and all higher order $\partial^n v_S/\partial\theta^n = 0$), it is possible, with additional assumptions, to generate analytic functions (and therefore, continuous design curves) for the estimation of the beam steering diffraction trade-offs. For example, from (6.19a), the length of time delay available before 3 dB of diffraction loss is incurred can be defined by

$$C[\mu s] = \frac{1.769\,\hat{L}^2}{|1 + \partial\phi/\partial\theta|f} \tag{6.21}$$

where, as usual, $\hat{L}$ is the input-output transducer width (or finger overlap) in wavelengths, f is the frequency in MHz, and beam steering is assumed negligible. It is interesting to comment here on the $\hat{L}^2$ and inverse frequency dependence of

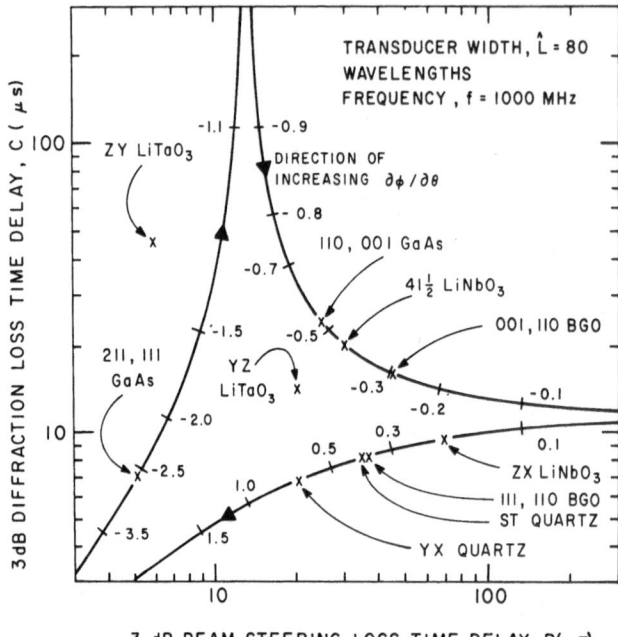

Fig. 6.34. Design curves estimating trade-off between diffraction loss and beam steering loss for ideal parabolic materials. Curves yield time delays available before 3 dB of loss is suffered from given cause. Running parameter is slope of power flow angle $\partial\phi/\partial\theta$. Discrete points show exact diffraction calculations for many materials of practical interest and include μ-type misorientations where necessary. Materials on upper right-hand curve yield optimum performance

the diffraction loss time delay. Conversion to a physical (rather than wavelength) transducer width shows linear frequency dependence. As far as beam steering is concerned, simple geometric arguments and an assumption of $0.1°$ misalignment from the desired pure-mode axis (where $\phi = 0$) result in the following definition for the 3 dB beam steering loss time delay

$$B[\mu s] = \frac{(1 - 1/\sqrt{2})\hat{L}}{f \tan 0.1|\partial\phi/\partial\theta|} \tag{6.22}$$

where again f is in MHz and now diffraction is assumed negligible. Again it is of interest to notice the functional dependence of this expression on $\hat{L}$ and f. For a physical transducer width, B is independent of frequency. Note that the assumptions inherent in (6.21) and (6.22) prevent their direct application other than as a simple, rapid illustration of the material dependence of losses.

Figure 6.34 illustrates an estimate of the beam steering, diffraction trade-off obtained by simultaneously plotting (6.21) and (6.22), assuming an 80 wavelength wide transducer at 1000 MHz. The running parameter is, of course, the slope of the power flow angle $\partial\phi/\partial\theta$. Superimposed on the continuous curves are

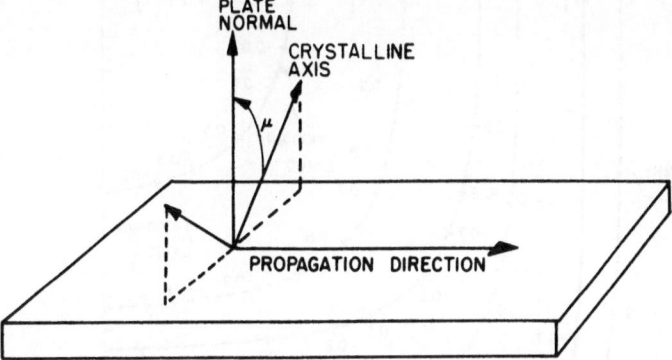

Fig. 6.35. Diagram illustrating the manner in which μ is defined for the purposes of describing the effect of plate-normal variations on beam steering

calculations of discrete points corresponding to practical materials of interest. Where differences occur they can generally be traced to $\partial\phi/\partial\mu \neq 0$ affecting beam steering calculations. Here the quantity μ refers to variation of the plate normal about the direction of propagation. This is illustrated in Fig. 6.35. (To simulate realistic conditions, a $0.2°$ misorientation of the plate normal is assumed which, of course, does not affect results when $\partial\phi/\partial\mu = 0$.) This phenomenon can, of course, be taken into account approximately by an analytic expression similar to (6.22) but, since it adds an additional parameter into consideration, it does not lend itself to the generation of demonstration curves. Also, in a few cases the angular spectrum of waves diffraction calculations differ substantially from the ideal parabolic cases of the continuous curves. No points for either YZ or 16.5 double rotated $LiNbO_3$ are provided since as stated above current lack of very precise knowledge of the velocity surfaces of these low diffraction orientations prevents accurate, exact calculations from being made.

Figure 6.34 illustrates several important points. 1) For optimum operation it is necessary to utilize materials on the upper right-hand portion of the curve. This demonstrates why 001, 110 $Bi_{12}GeO_{20}$ is superior to the 111, 110 $Bi_{12}GeO_{20}$ orientation and also shows the usefulness of the $41.5°$ $LiNbO_3$ cut [6.65]. 2) For materials having $\partial\phi/\partial\theta \approx -1$, very long time delays can be attained before diffraction loss is suffered. 3) For $\partial\phi/\partial\theta \approx 0$, very low beam steering loss need be suffered as this value of $\partial\phi/\partial\theta$ implies a locally isotropic material. These conclusions agree, of course, with the evaluation of results from Fig. 6.33.

Design curves similar to those of Fig. 6.34 are provided in Fig. 6.36 for a variety of frequencies and transducer widths. Only the optimum section of the curve is given. The upper right-hand curves in Fig. 6.36 clearly illustrate why beam steering and diffraction are negligible sources of loss at low frequencies (below 200 MHz) unless extremely long time delays are being considered.

Returning now to *combined* loss curves, the importance of the beam steering angle is illustrated in Fig. 6.37, in which the combined loss curve of Fig. 6.33 is

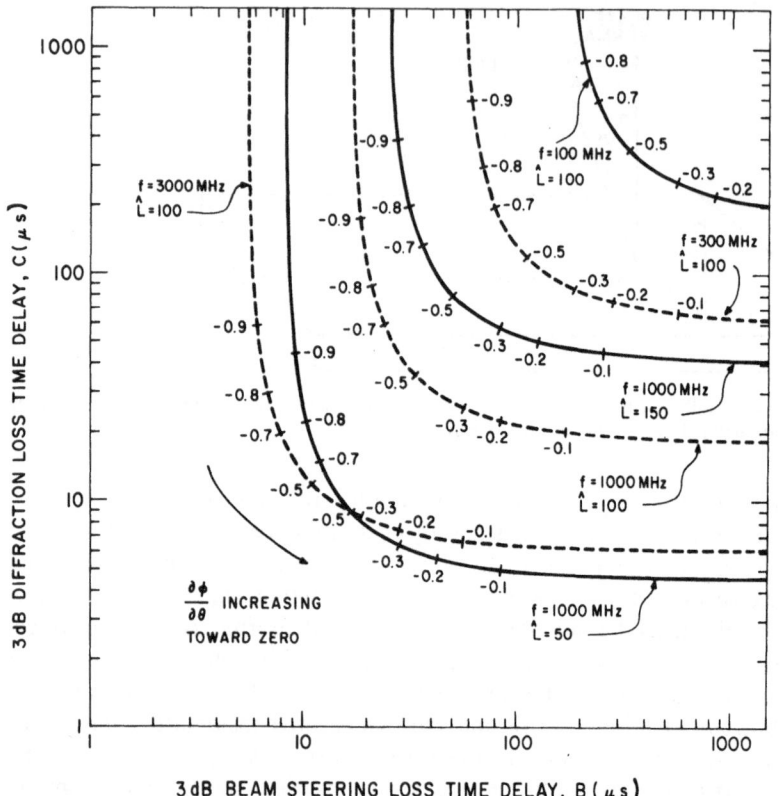

Fig. 6.36. Optimum sections of design curves estimating trade-off between diffraction and beam steering for ideal parabolic materials for several frequencies f and transducer widths $\hat{L}$ in wavelengths. Running parameter is $\partial\phi/\partial\theta$, which is slope of power flow angle

repeated along with two other curves for different beam steering conditions. From Fig. 6.37 it is evident again that whenever beam steering angles are expected to be moderate or large, as, for example, in mass-produced devices where cost limits x-ray alignment precision, a material having $\gamma \approx 0$ should be chosen. Figure 6.37 also shows by means of the drastic changes in the shapes of the curves that universal beam steering diffraction loss curves are not possible. This same point is illustrated in Fig. 6.38, which also yields some practical loss data for ST quartz [6.4] ($\gamma = +0.378$) and 001, 110 $Bi_{12}GeO_{20}$ ($\gamma = -0.304$). The shape of the $\gamma = -0.304$, BS $\measuredangle = 1.0$ curve is clearly different from that of the $\gamma = -0.304$, BS $\measuredangle = 0.0$ curve, and thus they could not be combined or superimposed.

Figure 6.39 illustrates combined beam steering and diffraction loss versus the time-delay-frequency parameter $\hat{Z}$. These are the final design curves directly intended to aid in the choice of material. It is of interest to point out that the loss is very high for the $\hat{Z} = 75000$ curve near $\gamma = -1.0$. For this large distance, beam

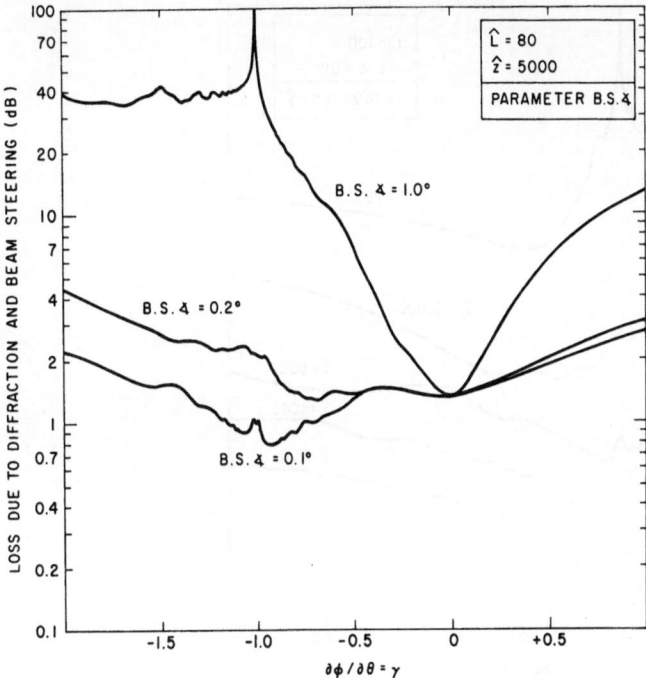

Fig. 6.37. Loss due to diffraction and beam steering as a function of slope of power flow angle with beam steering angle as parameter. Beam steering angle is defined as misalignment of center line between transducers from desired pure-mode axis

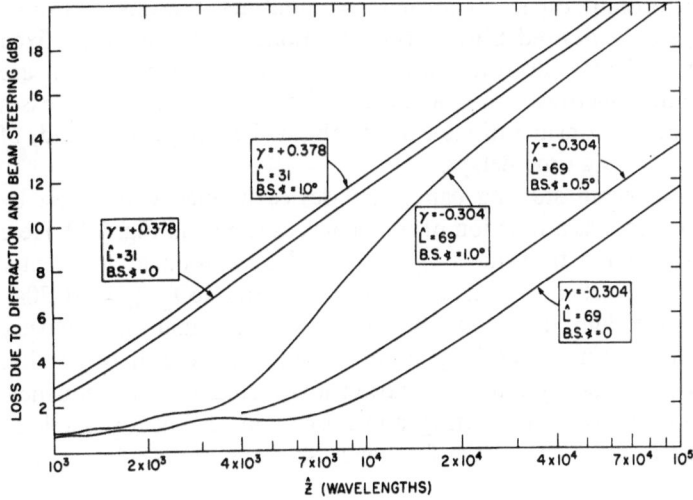

Fig. 6.38. Loss due to diffraction and beam steering as a function of distance in wavelengths between transducers. Value of $\gamma = +0.378$ for slope of power flow angle corresponds to ST quartz, while $\gamma = -0.304$ corresponds to 001, 110 $Bi_{12}GeO_{20}$

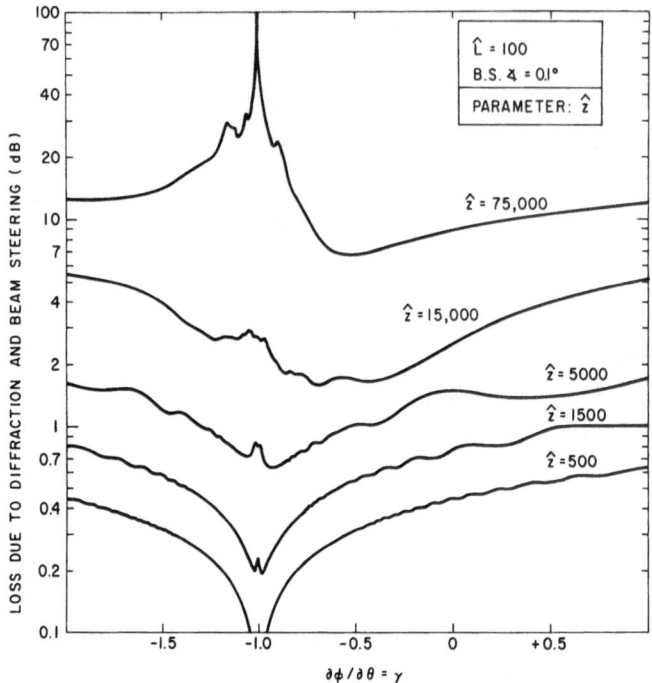

Fig. 6.39. Loss due to diffraction and beam steering as a function of slope of power flow angle with distance in wavelengths between transducers, $\hat{Z}$ as parameter

steering is very important, especially for narrow undiffracted beams, and some beam spreading is to be desired. Since $\hat{Z}$ is proportional to frequency (for fixed time delay), Fig. 6.39 also illustrates why beam steering and diffraction are considered UHF and microwave frequency design problems. Significant losses and material trade-off considerations exist at the higher frequencies and, of course, also for very long time delays.

Diffraction and beam steering can affect device frequency response by increasing insertion loss as a function of frequency, as shown in Fig. 6.40. Here $\hat{L}_0$ and $\hat{Z}_0$ represent the transducer aperture and separation at the center frequency f_0 for ST quartz ($\gamma = +0.378$) and 001, 110 $Bi_{12}GeO_{20}$ ($\gamma = -0.304$). The change in loss is slight and would be considered important only for very wide bandwidths or extreme cases (large $\hat{Z}_0$ or small $\hat{L}_0$). In general, diffraction loss decreases with frequency but propagation loss increases with frequency. These effects are partially compensating, and for certain cases it is possible to design the device so as to cancel the frequency skewing effects of these two mechanisms.

For the ST quartz case depicted in Fig. 6.40, the variation of all loss mechanisms (including transducer effects) will be explored in some detail in Section 6.4.

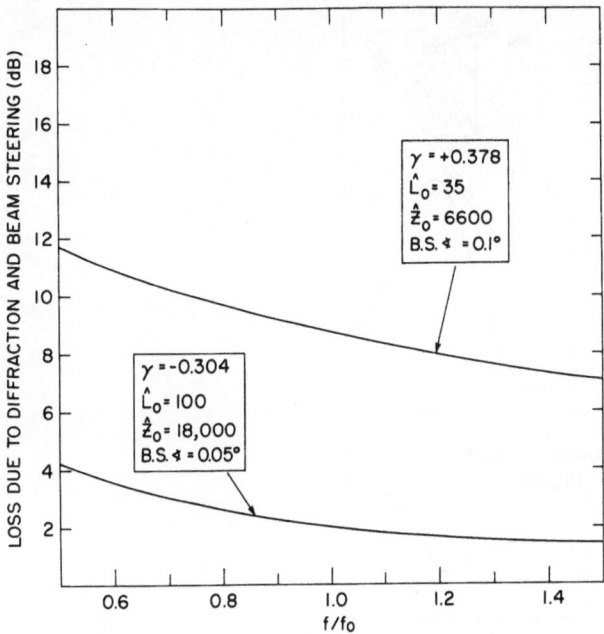

Fig. 6.40. Loss due to diffraction and beam steering corresponding to fixed transducer pair having an acoustic aperture at center frequency of $\hat{L}_0$, a separation at center frequency of $\hat{Z}_0$, and operated over indicated bandwidth. That is, $\hat{L} = \hat{L}_0(f/f_0)$ and $\hat{Z} = \hat{Z}_0(f/f_0)$

6.3.5 Minimal Diffraction Cuts

Much of the recent growth in SAW devices can be traced to the success with which first-order design theories can be used to obtain highly satisfactory device performance [6.66–68]. However, at the present time, to realize any substantial improvement in device performance it is essential to be able to quantitatively understand second-order effects [6.69] such as bulk-wave generation, acoustic reflections, and diffraction.

In general, compensation for second-order effects has never been an easy design problem. Therefore, it is highly desirable to minimize their effects to such an extent that they prove negligible. Two excellent examples of this approach are the recent discoveries of the use of dummy electrodes [6.70, 71] to eliminate the phase front distortion of waves propagating through apodized interdigital transducers and the use of split electrodes [6.72, 73] to minimize acoustic reflections within transducers. These two corrective measures are described in greater detail in Section 3.2.2.

Another second-order effect which can prove to be a severe limitation in the design of highly apodized acoustic surface wave filters and long-time-delay devices is diffraction or beam spreading. The progress previously reported in this section in the quantitative understanding of this phenomenon has allowed

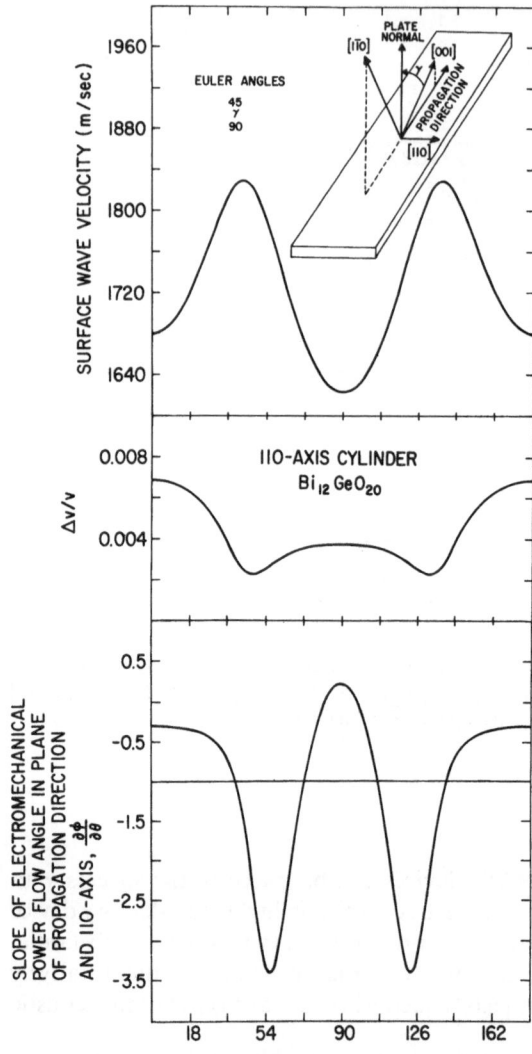

.Fig. 6.41. Surface wave velocity $\Delta v/v$ and *slope* of the power flow angle (with respect to θ variations) for simultaneous rotation of the direction of propagation and the plate normal in a plane perpendicular to the 110 axis of $Bi_{12}GeO_{20}$ as indicated. The power flow angle ϕ itself is identically zero for all values of γ

efforts to proceed towards its minimization and its elimination from consideration when designing devices [6.16, 17].

In particular, the purpose of this subsection is twofold: first, to describe techniques for theoretically determining orientations having minimal acoustic surface wave diffraction and, second, to describe an orientation [6.17] for surface wave propagation on bismuth germanium oxide [6.16, 17] $(Bi_{12}GeO_{20})$ on which diffraction is retarded by at least a factor of 100 over isotropic materials and by at least a factor of 70 over the popular, [001], [110] orientation. It is expected that for this orientation and for others in the same class, which are

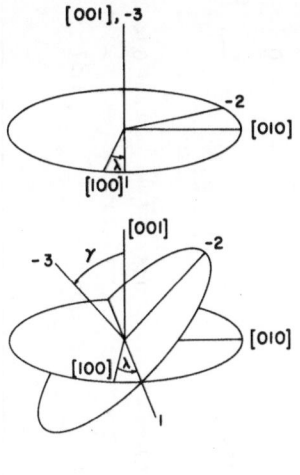

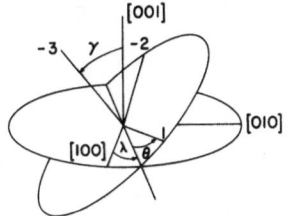

Fig. 6.42. Coordinate system used to define acoustic surface wave propagation. The phase velocity vector lies along the $+1$ axis, while the plate normal lies along the -3 axis. The crystalline axes are given by 100, 010, and 001, while the Euler angles are λ, γ, and θ (after [6.75])

called minimal diffraction cuts (MDC), diffraction loss can be virtually ignored for a large new class of devices.

To have a reasonable chance of success in finding an MDC orientation, several conditions must be satisfied. The first two prerequisites can be deduced from Fig. 6.33. They are: a) that the velocity surface be parabolic (Fig. 6.33 is valid only for parabolic orientations) within the constraints on $|\delta_M|$ given in Section 6.3.2 near the pure-mode axis ($\phi=0$) chosen, and b) that $\partial\phi/\partial\theta = -1$. Three other conditions should also be satisfied: c) a continuum of possible cuts all having $\phi=0$ must be available to be easily searched by computer for $\partial\phi/\partial\theta = -1$; d) material constants must be extremely accurately known for theoretical predictions to translate into successful practice, and e) parabolic surfaces should be known to exist in the vicinity of the search area for the material chosen.

Since $Bi_{12}GeO_{20}$ was known [6.12, 57, 74] to satisfy conditions c–e), it was decided [6.17] to first search for MDC orientations on this material. Several computer runs were performed on sets of orientations meeting conditions c), the most successful of which is shown in Fig. 6.41. The data plotted in Fig. 6.41 were obtained by varying the second Euler angle [6.75] from 0° to 180° while holding the first at 45° and the third at 90°. This is shown in the inset to Fig. 6.41 and is called a 110 axis cylinder rotation [6.57, 76]. (A brief explanation of the Euler angle notation is given in Fig. 6.42; full details can be found in [6.57].)

As is seen in Fig. 6.41, the slope of the power flow angle does in fact go through the -1 point at Euler angles of 45, 40.04, 90° and 45, 72.53, 90°. In

Table 6.6 Complete summary of acoustic surface wave properties of 45°, 40.04°, 90°; 45°, 72.53°, 90°; and 001, 110 $Bi_{12}GeO_{20}$. Based mainly on its low value of $|\delta_{Ml}|$. MDC orientation No. 1 was chosen over orientation No. 2 as having an absolute minimum of diffraction spreading

Source of material constants	MDC orientation No. 1		MDC orientation No. 2		Standard 001 cut, 110 propagating	
	Slobodnik and Sethares [6.74]	Kraut et al. [6.77]	Slobodnik and Sethares [6.74]	Kraut et al. [6.77]	Slobodnik and Sethares [6.74]	Kraut et al. [6.77]
Euler angles [deg]						
λ	45	45	45	45	0	0
γ	40.04	39.84	72.53	72.84	0	0
θ	90	90	90	90	45	45
Free-surface velocity, $v_{s\infty}$	1827	1831	1667	1668	1679	1681
Coupling parameter estimate, $\Delta v/v$	0.0031	0.0031	0.0035	0.0035	0.0068	0.0068
Slopes of the power flow angle $\frac{\partial\phi}{\partial\theta}$	$-$ 1.000	$-$ 1.000	$-$ 1.000	$-$ 1.000	$-$ 0.302	$-$ 0.304
$\frac{\partial\phi}{\partial\mu}$	$+$ 0.538	$+$ 0.539	$+$ 0.337	$+$ 0.332	0	0
$\frac{\partial\phi}{\partial\gamma}$	0	0	0	0	0	0
MDC sensitivity to misorientation, $\partial^2\phi/\partial\theta\,\partial\gamma$	$-$ 0.103	$-$ 0.093	$+$ 0.147	$+$ 0.153	—	—
Parabolic deviation factor, $|\delta_{Ml}|$	1.53	1.44	13.8	13.9	0.15	0.14
Slow shear bulk-wave velocity, $v_{\text{slow shear}}$	2062	2077	1732	1732	1756	1757

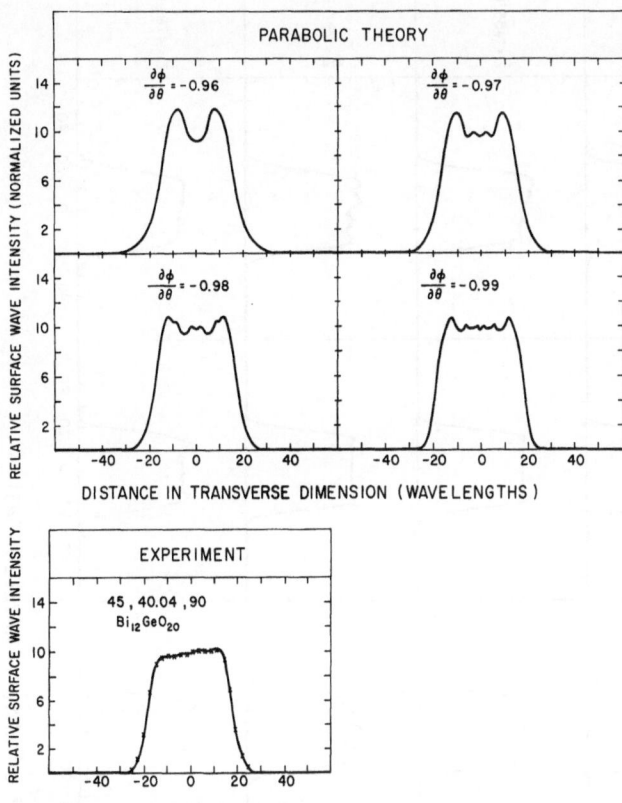

Fig. 6.43. Experimental and theoretical acoustic surface wave profiles at a distance of $\hat{Z}=4783.8$ wavelengths from the input. Frequency is $f=380\,\text{MHz}$ with an acoustic aperture of $\hat{L}=40.56$ wavelengths. A laser probe size of $\hat{P}=8.3$ wavelengths was used for the theoretical calculation. Experimental agreement with the $\partial\phi/\partial\theta = -0.99$ theoretical curve indicates that diffraction for the 45°, 40.04°, 90° orientation of $Bi_{12}GeO_{20}$ is retarded by at least two orders of magnitude over isotropic materials

addition to this most important information, the velocity v and the coupling parameter [6.2] $\Delta v/v$ are also given in Fig. 6.41. A complete summary of the properties of the two potential MDC orientations is given in Table 6.6 along with equivalent information for the popular 001, 110 orientations. Calculations were made using material constants measured by both *Slobodnik* and *Sethares* [6.74] and by *Kraut* et al. [6.77]. Results are in very substantial agreement, confirming that condition d) is satisfied for $Bi_{12}GeO_{20}$.

Several of the results listed in Table 6.6 deserve further comment. The surface wave coupling for both MDC orientations is certainly adequate as it is only approximately a factor of 2 lower than that for the 001, 110 orientation. The fact that $\partial\phi/\partial\mu \neq 0$ for the MDC orientations implies a more stringent requirement on crystallographic x-ray alignment of the plate normal in order to minimize

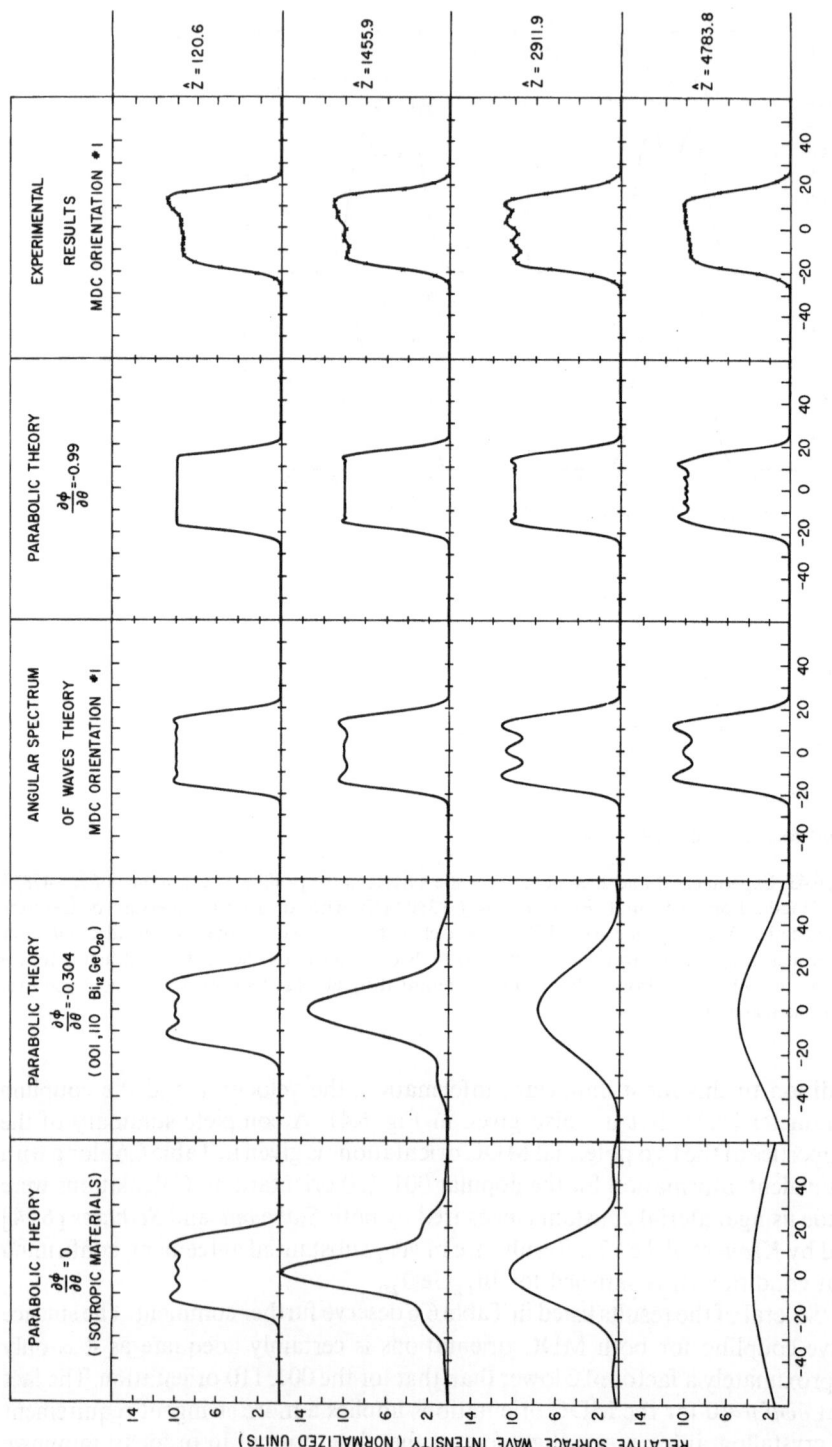

beam steering, since $\partial\phi/\partial\mu$ is defined as the change in power flow angle with respect to plate normal rotations about a fixed direction of propagation, as was illustrated in Fig. 6.35. However, this is certainly not a severe limitation. Of the two MDC orientations, MDC No. 1 is less sensitive to a γ-type angular misorientation; that is, $\partial^2\phi/\partial\theta\partial\gamma$ is lower than that for MDC No. 2. The slow shear bulk-wave velocity is also included in Table 6.6 for comparison.

The most important result listed in Table 6.6 is, however, the value of $|\delta_M|$ for MDC orientation No. 1. Since $|\delta_M| = 1.53 < 2.0$, it can be concluded that the predicted value of $\partial\phi/\partial\theta = -1$ actually implies ultralow diffraction. Because of this fact, MDC orientation No. 1 was chosen for experimental investigation.

To confirm the prediction of minimal diffraction for MDC No. 1, experimental acoustic surface wave beam profile measurements were made [6.17] on a delay line oriented according to the inset in Fig. 6.41 and operating at a frequency of 380 MHz. The laser probe technique described in Section 6.1 was used with the laser spot size focused small with respect to the overall beam width. The acoustic aperture or transducer width was $L = 195\,\mu m$ (or $\hat{L} = 40.56$ wavelengths). An experimental profile taken at a distance of $\hat{Z} = 4783.8$ wavelengths from the input is compared in Fig. 6.43 to theoretical profiles having $\partial\phi/\partial\theta = -0.96$, -0.97, -0.98, and -0.99.

Direct superposition of the experimental data on the four theoretical profiles indicated [6.17] that the one with $\partial\phi/\partial\theta = -0.99$ gave the best match in beam width and features. This comparison indicates that the experimental diffraction is retarded by a factor of at least 100, compared to the isotropic case.

A complete series of experimental profiles taken at different distances, $\hat{Z}$, from the input is compared in Fig. 6.44 with various calculations. The significant reduction in diffraction over the isotropic and 001, 110 $Bi_{12}GeO_{20}$ cases is clearly illustrated. Good agreement is also indicated between experiment and the exact angular spectrum of waves theory using the velocity surface of Fig. 6.45. Finally, as expected, full comparison with the parabolic case of $\partial\phi/\partial\theta = -0.99$ again confirms that this is indeed a minimal diffraction cut.

6.4 Optimum Transducer Design in the Presence of Material Limitations

After an optimum material has been chosen for a given delay line application, the number of interdigital periods N and the optimum acoustic aperture can be determined. In practical design situations dealing with losses and real elements, the choices of these parameters are interdependent. Thus, for example, optimum

---

◄ Fig. 6.44. Experimental and theoretical acoustic surface wave profiles at various distances, $\hat{Z}$, in wavelengths from the input. Frequency is $f = 380\,MHz$ with an acoustic aperture of $\hat{L} = 40.56$ wavelengths. A laser probe size of $\hat{P} = 8.3$ wavelengths was used for the theoretical calculations. The diffraction properties of the 45°, 40.04°, 90° orientation of $Bi_{12}GeO_{20}$ are clearly superior to both the isotropic case and to 001, 110, $Bi_{12}GeO_{20}$

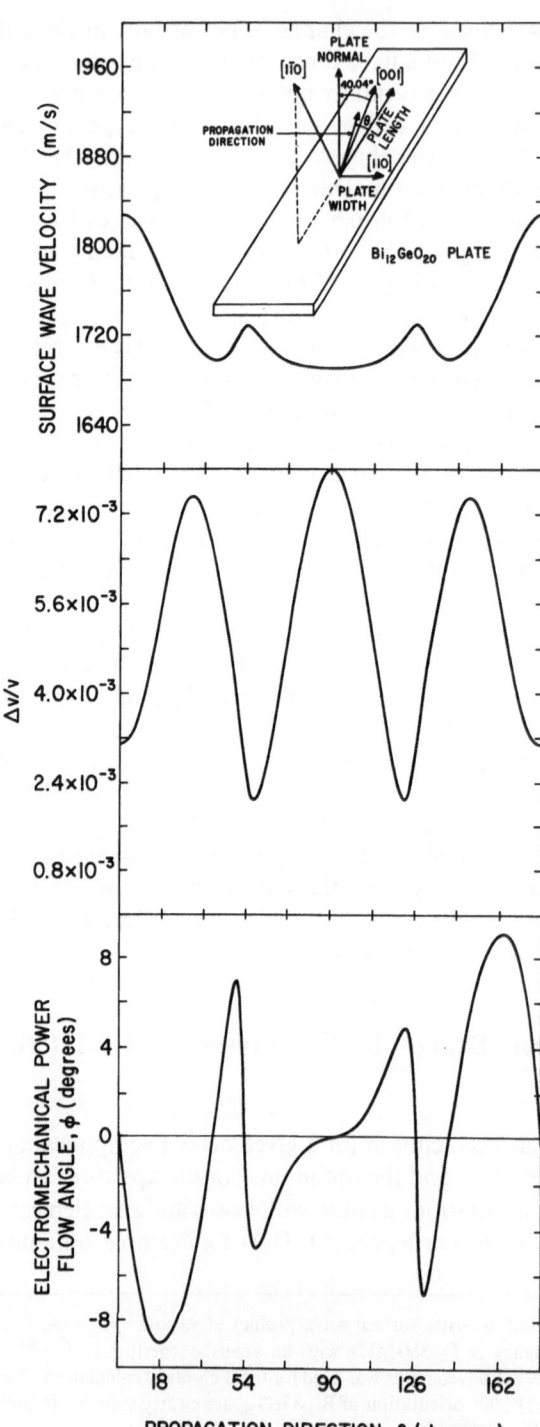

Fig. 6.45. Surface wave velocity $\Delta v/v$ and power flow angle on the 45°, 40.04°, 90° plane of $Bi_{12}GeO_{20}$ as functions of the angle from the plate axis as shown. Note that at $\theta = 0°$ the slope of the power flow angle curve is -1.0. This is a minimal diffraction cut

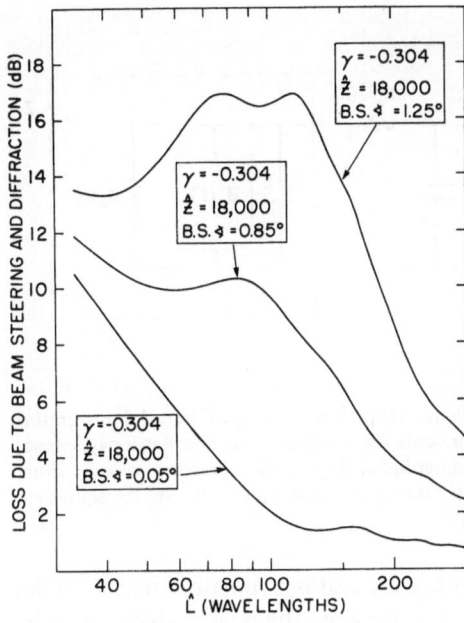

Fig. 6.46. Loss due to diffraction and beam steering as a function of acoustic aperture $\hat{L}$ in wavelengths. Curves correspond to 001, 110 $Bi_{12}GeO_{20}$

apertures should be determined for several values of N and the absolute optimum finally chosen. A reasonable starting value for N is [6.66, 67]

$$N_{OPT}^2 = \frac{\pi}{4k^2} \tag{6.23}$$

where k^2 is the well-known electromechanical coupling coefficient [6.3]. Once the value of N is fixed, the best value of acoustic aperture depends on transducer and tuning element losses, parasitic elements, and beam steering and diffraction losses.

To reduce beam steering and diffraction losses it is necessary to use the widest possible acoustic apertures, as illustrated in Fig. 6.46. Unfortunately, electrical matching considerations limit the extent to which increased finger overlap can be used to reduce overall device insertion loss. To demonstrate this effect and to develop optimum delay line design procedures, it is necessary to investigate transducer insertion loss as a function of the various design parameters, particularly acoustic aperture.

To accomplish this task we shall first review some of the transducer design techniques discussed in Chapter 3.

We adopt the generalized [6.13–15] or extended Stanford [6.66, 67] equivalent circuit model, which includes lossy fingers, lossy tuning inductor, and the effects of shunt capacitance. This model is illustrated in Fig. 6.47.

If only transducer effects are considered, overall delay line insertion loss in dB is given by

$$IL\,[dB] = -10\log_{10}(TE)^2. \tag{6.24}$$

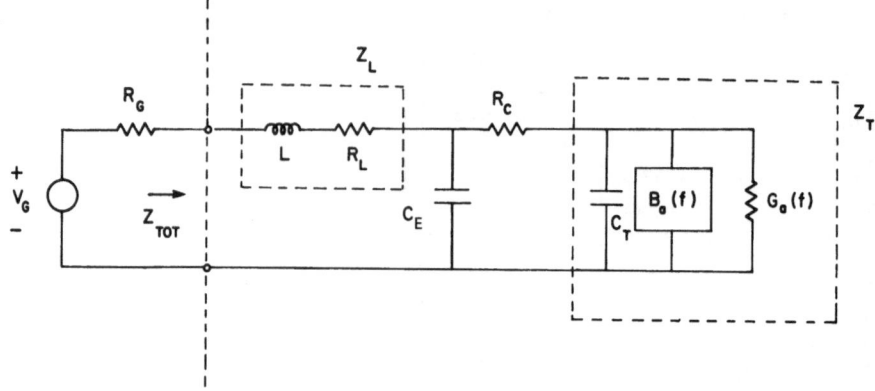

Fig. 6.47. Generalized equivalent circuit of a periodic, unapodized interdigital transducer operating in a matched transmission line system. Z_T represents the usual acoustic and finger-capacitance elements, R_C represents the ohmic loss in the interdigital fingers, C_E represents parasitic shunt capacitance, and Z_L is the impedance of the lossy tuning inductor. V_G and R_G are the equivalent circuit elements of the generator

Here TE is the individual transducer efficiency and is a function of each of the circuit elements of the model. It is the ratio between the power which could be delivered to a load (transducer) from a matched signal generator and the actual power leaving one acoustic port. With reference to Fig. 6.48 this can be expressed as

$$TE = \tfrac{1}{2} [\tfrac{1}{2} \operatorname{Re}\{(Z_T)|I_T|^2\}] \left(\frac{8R_G}{V_G^2} \right), \tag{6.25}$$

where the factor of $\tfrac{1}{2}$ accounts for bidirectionality and

$$Z_{TOT} = Z_L + \frac{1}{j\omega C_E + (R_C + Z_T)^{-1}}. \tag{6.26}$$

Equations (6.24)–(6.26) can be used to generate a wide variety of delay line design curves [6.78]; by including expressions such as (6.11)–(6.13), extension to lossy propagation paths is straightforward [6.78]. Two such design curves are presented in Figs. 6.49 and 6.50. Beam steering and diffraction effects are not included.

In Fig. 6.49, the effect of shunt capacitance is investigated for a 60.44 µs delay line at 300 MHz on 001, 110 $Bi_{12}GeO_{20}$. As can be seen, increasing the shunt capacitance C_E increases the insertion loss and reduces the bandwidth. A smaller tuning inductor is also required to obtain absolute minimum insertion loss. Here the number of interdigital periods is $N = 5$, the acoustic aperture (transducer width) is $\hat{L} = 100$ wavelengths, the sheet resistivity of the transducer fingers is $\varrho/t = 0.345\,\Omega$/square, and the tuning inductor is assumed to have a Q of 35.

Figure 6.50 represents a collection of curves defining the optimum insertion loss, bandwidth trade-off under the same conditions listed for Fig. 6.49 with $C_E = 1.0\,\mathrm{pF}$ and N varying as shown. For the BW at Min IL curve, the tuning

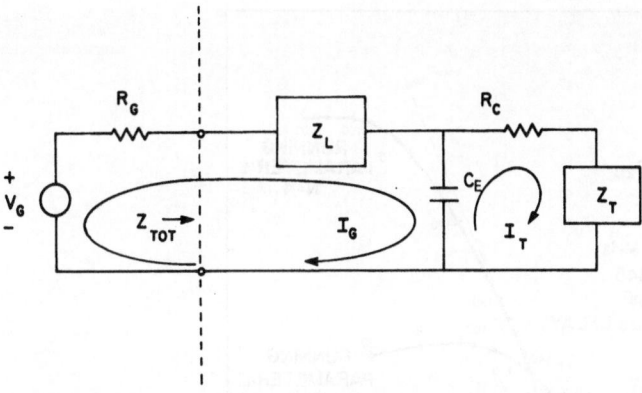

Fig. 6.48. Equivalent circuit of Fig. 6.47 with loop currents necessary to define transducer efficiency

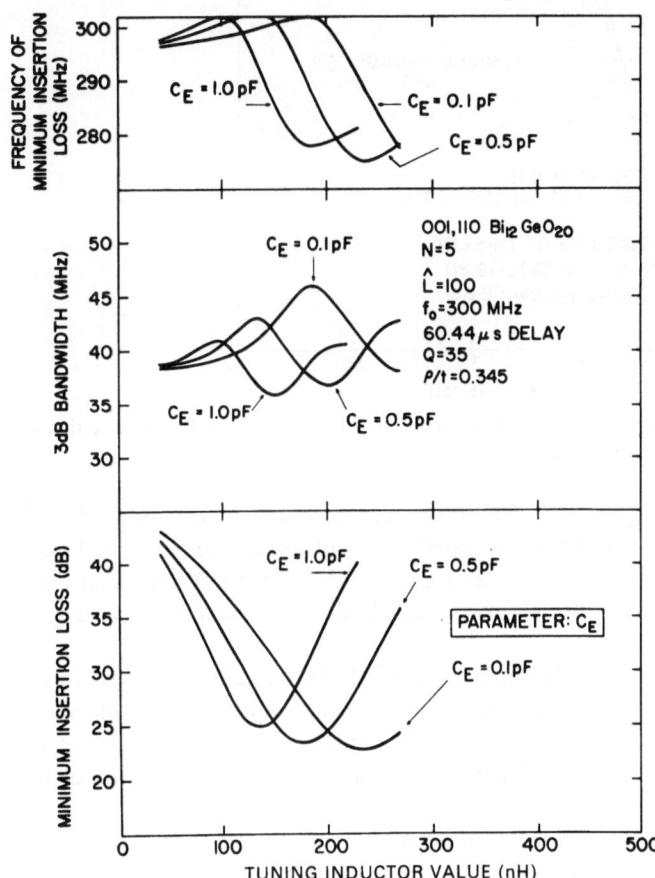

Fig. 6.49. Periodic transducer acoustic surface wave delay line design curves illustrating the effects of tuning inductor value with shunt capacitance C_E as the parameter

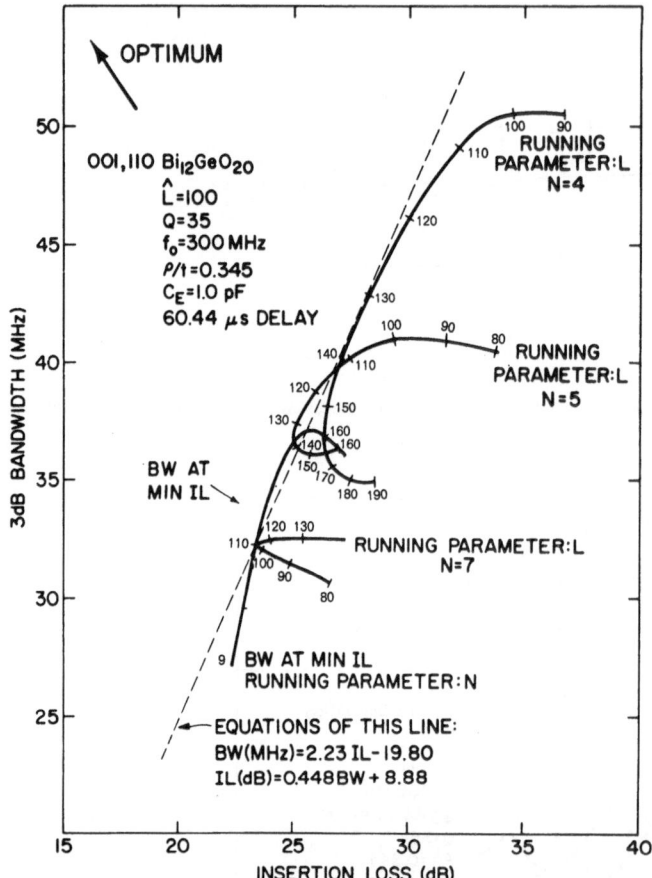

Fig. 6.50. Design curves illustrating the optimum trade-off between bandwidth and insertion loss for the specified delay line parameters

inductor is chosen to yield the absolute minimum insertion loss. Beyond $N = 5$ this particular curve no longer represents the optimum trade-off. That is, by using a smaller tuning inductor L and a larger N value than that resulting in absolute minimum insertion loss, one can obtain more bandwidth at a given insertion loss.

Another, perhaps more familiar, type of design curve is illustrated in Fig. 6.51. Insertion loss vs transducer width $\hat{L}$ is given for ST quartz and a specific set of realistic circuit element values [6.78]. The value of the tuning inductor was varied for each value of $\hat{L}$ to obtain the lowest value of insertion loss of that particular aperture.

Curve 3 represents transducer and circuit element effects alone, i.e., does not include information similar to (6.11)–(6.13) or diffraction or beam steering effects. When propagation loss at 660 MHz is included (Curve 2), the overall loss increases substantially but the optimum value of acoustic aperture yielding

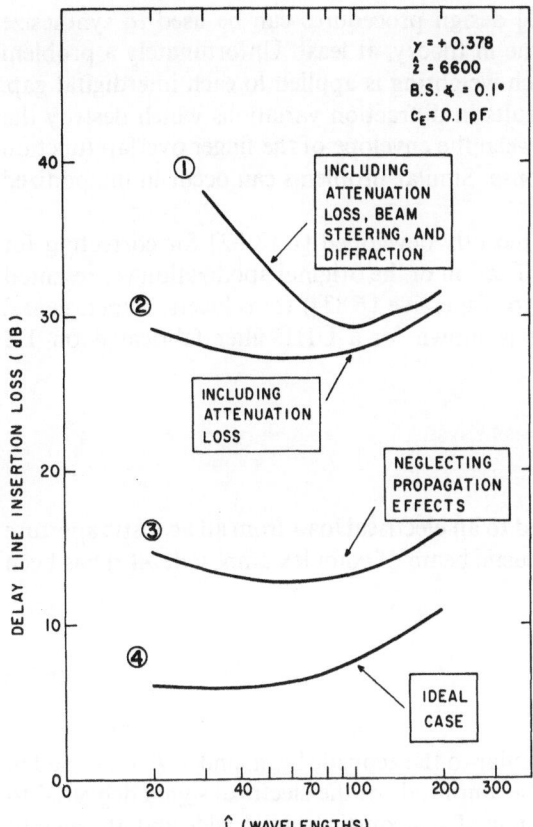

$y = + 0.378$
$\hat{z} = 6600$
B.S. $\measuredangle = 0.1°$
$C_E = 0.1$ pF

① INCLUDING
ATTENUATION
LOSS, BEAM
STEERING, AND
DIFFRACTION

② INCLUDING
ATTENUATION
LOSS

NEGLECTING
PROPAGATION
EFFECTS

③

④ IDEAL
CASE

DELAY LINE INSERTION LOSS (dB)

$\hat{L}$ (WAVELENGTHS)

Fig. 6.51. Delay line insertion loss vs acoustic aperture curves used to choose the optimum (minimum insertion loss) acoustic aperture. Top curve includes all significant sources of loss

minimum insertion loss remains the same. The final result of all our efforts is the top curve of Fig. 6.51 which represents the totally optimum design information. Beam steering and diffraction loss have been included. Overall minimum insertion loss is obtained using an acoustic aperture of $\hat{L} = 100$ wavelengths. Finally, Curve 4 for the case corresponding to the ideal design procedure [6.66, 67] is also presented for comparison. Here $R_L = 0, R_C = 0, C_E = 0$, BS $\measuredangle = 0$, and attenuation and diffraction losses are neglected. The substantial difference is easily seen.

6.5 Diffraction Compensation in Periodic SAW Filters

6.5.1 Introduction

Surface acoustic wave bandpass filters offer the advantages of small size and weight together with low cost and high reproducibility. Since an interdigital transducer can be made to be an excellent transversal filter, Fourier transform pair and digital filter [6.79, 80] design procedures can be used to synthesize frequency responses. This is true in theory, at least. Unfortunately a problem arises from the manner in which weighting is applied to each interdigital gap. Finger overlap apodization results in diffraction variations which destroy the Fourier transform relation between the envelope of the finger overlap function and the device frequency response. Similar problems can occur in unapodized transducers as well [6.81].

A direct *synthesis* method recently developed [6.18, 82] for correcting for these diffraction effects by modification of the original apodization is presented here for periodic (as opposed to dispersive [6.83]) transducers. Experimental confirmation of this technique is shown for a UHF filter fabricated on YZ LiTaO$_3$.

6.5.2 Basic Theory

The signal amplitude transferred to an electrical load from an acoustic aperture of width $\hat{L}$ irradiated by an acoustic beam of complex amplitude $A(x)$ has been given by *Waldron* [6.84] as

$$S = \frac{T(\hat{L})}{\sqrt{\hat{L}}} \int_{-\hat{L}/2}^{\hat{L}/2} A(x)dx. \tag{6.27}$$

Here x is the direction perpendicular to the acoustic beam and $T(\hat{L})$ is defined in the following manner [6.84]. The amplitude of the electrical signal delivered to the load is $T(\hat{L})$ times the amplitude of an acoustic beam $\hat{L}$ wide and of constant amplitude and phase centered on the transducer at normal incidence.

Under the conditions for which an interdigital transducer can be directly represented as a transversal filter, it has been shown [6.85] that $T(\hat{L})$ is directly proportional to $\sqrt{\hat{L}}$. In other words

$$T(\hat{L}) = C\sqrt{\hat{L}} \tag{6.28}$$

where C depends on other physical and geometrical parameters of the delay line [6.85] but is independent of $\hat{L}$. This is an important result as it means that the C associated with a given individual finger pair in an apodized transducer is independent of the finger overlap $\hat{L}_N$ and thus is the *same* for all gaps (for periodic transducers).

Equation (6.27) can thus be rewritten as

$$S = C \int_{-\hat{L}/2}^{L/2} A(x)dx \tag{6.29}$$

which forms the basis of the following development.

6.5.3 Correction for Diffraction

Consider an acoustic surface wave delay line having one apodized transducer and one uniform launching aperture as illustrated in Fig. 6.52. Under the condition of *no diffraction*, the voltage across the load due to the Nth finger pair having overlap $\hat{L}_N$, with respect to the voltage across the load due to the widest finger pair having overlap $\hat{L}_0$, is, from (6.29)

$$\frac{\int_{-\hat{L}_N/2}^{\hat{L}_N/2} A(0)dx}{\int_{-\hat{L}_0/2}^{\hat{L}_0/2} A(0)dx} = \frac{\hat{L}_N}{\hat{L}_0} \equiv R_{\text{NOD}} \text{ (ideal case — no diffraction)}. \tag{6.30}$$

Note that each finger pair is treated separately and superposition is used to generate the total result.

To synthesize the desired frequency response *in the presence of diffraction* it is necessary to achieve this *same* ratio. That is

$$R_{\text{D}} \equiv \frac{\int_{-\hat{L}_N/2}^{\hat{L}_N/2} A(x, \hat{Z}_N)dx}{\int_{-\hat{L}_0/2}^{\hat{L}_0/2} A(x, \hat{Z}_0)dx} \text{ (diffraction present)} \tag{6.31}$$

must be set equal to R_{NOD}. Or, combining (6.30) and (6.31),

$$\frac{\int_{-\hat{L}_N'/2}^{\hat{L}_N'/2} A(x, \hat{Z}_N)dx}{\int_{-\hat{L}_0/2}^{\hat{L}_0/2} A(x, \hat{Z}_0)dx} = \frac{\hat{L}_N}{\hat{L}_0}. \tag{6.32}$$

In these equations, $\hat{L}_N'$ is the *unknown* aperture in the presence of diffraction of the Nth finger pair. It is located a distance $\hat{Z}_N$ from the launching aperture, as indicated in Fig. 6.52. For the present analysis we have arbitrarily set the widest overlap ($\hat{L}_0$) in the presence of diffraction to be equal to the widest overlap if no diffraction were present and in addition have taken it to be located a distance $\hat{Z}_0$ from the launching transducer.

In practice, then, a set of $\hat{L}_N$ is generated using ideal Fourier transform or digital design techniques [6.79, 80]. The actual delay line is, however, fabricated using values of $\hat{L}_N'$ determined according to a modified verstion of (6.32), as described below [6.18]. This form is more convenient for synthesis calculations and is obtained by first taking the magnitude squared of both sides of (6.32), dividing both numerator and denominator of the right-hand side by $\hat{L}_0^2|A(0)|^2$, rearranging terms and multiplying numerator and denominator by $\hat{L}_0\hat{L}_N'$, and finally by taking logarithms and recognizing that one term represents diffraction

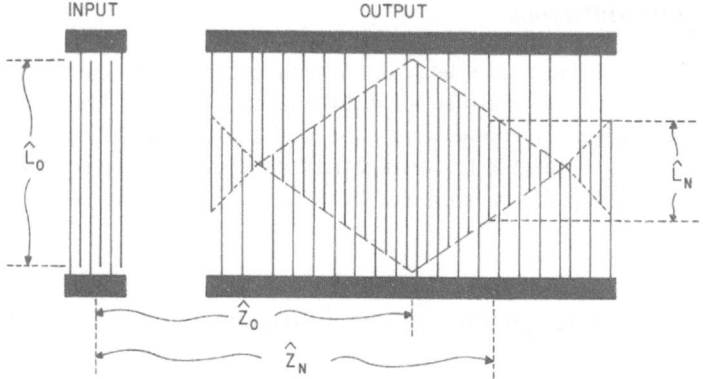

Fig. 6.52. Illustration of interdigital transducers with definition of terms used in diffraction correction derivations

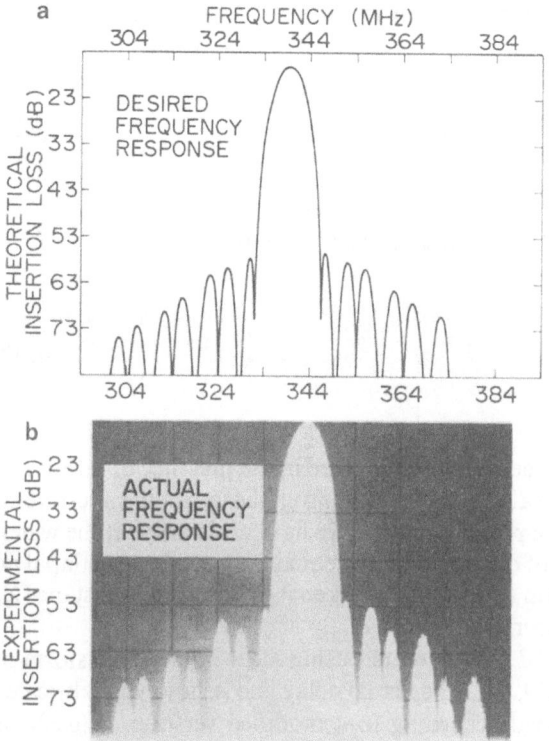

Fig. 6.53. (a) Computer generated frequency response including second-order transducer effects in the absence of diffraction. That is, using L_N. It is desired to synthesize this ideal response in the presence of diffraction. (b) Experimental frequency response of actual LiTaO₃ surface wave filter. Scales: 10 dB/division on the vertical axis with the top line representing 13 dB and 10 MHz/division on the horizontal axis with the center cross-hatched line equaling 344 MHz

loss between two equal transducers [6.12] of width $\hat{L}_0$ and separated by a distance $\hat{Z}_0$.

$$-10\log_{10}\frac{\left|\int_{-\hat{L}_N/2}^{\hat{L}_N/2}A(x,\hat{Z}_N)dx\right|^2}{\hat{L}_0\hat{L}_N'|A(0)|^2}-10\log_{10}\left(\frac{\hat{L}_N'\hat{L}_0}{\hat{L}_N^2}\right)$$

$$=\begin{Bmatrix}\text{Diffraction loss in dB for two}\\\text{equal }\hat{L}_0\text{ transducers}\end{Bmatrix}.\tag{6.33a}$$

Computer iterative solution [6.18] of (6.33a) provides the corrected overlap values, $\hat{L}_N'$.

It should be emphasized that by correcting for diffraction using (6.33a), *only* the magnitude of the desired waveform is correctly achieved. If phase correction is necessary [6.86], a return to (6.32) would in general be required and each $\hat{Z}_N$ would become an unknown. However, use of an alternate (approximate) technique developed by *Szabo* [6.87] is more convenient. Here, amplitude correction is first achieved as described above, then phase correction is obtained by analyzing the new filter with $\hat{L}_N'$ placed at $\hat{Z}_N$ to determine the relative phase $\zeta(N)$ at each finger pair. Then by setting

$$\hat{Z}_N'=\hat{Z}_N+\frac{\zeta(N)}{2}\tag{6.33b}$$

a good approximation to phase corerction results [6.87]. However, for the purpose of illustration in the present case let us return to (6.33a) and consider only magnitude correction.

6.5.4 Experimental Investigation

The design to be tested consists of an unapodized transducer having 69 fingers and an apodized transducer having 137. The particular set of $\hat{L}_N$ chosen for this example has the desired (no diffraction) frequency response shown in Fig. 6.53a. The apodization used, although similar in appearance to cosine-squared-on-a-pedestal characteristics [6.88], followed no analytic function and was specified individually for each finger. The computer program [6.89] used to generate this plot accounts for electrical loading and back voltages as well as the use of double electrodes. Apodization is treated using the segmented approach [6.89] which, although costly in computer time, produces highly accurate results.

Correction for diffraction, that is, determination of $\hat{L}_N'$, requires specification of the following parameters:

Material: YZ LiTaO$_3$

$\hat{L}_0=23.33$

$\hat{Z}_0=577.69$

$\Lambda_s=v_s/f_0=\dfrac{3230\,\text{m s}^{-1}}{340\,\text{MHz}}=9.5\,\mu\text{m}.$

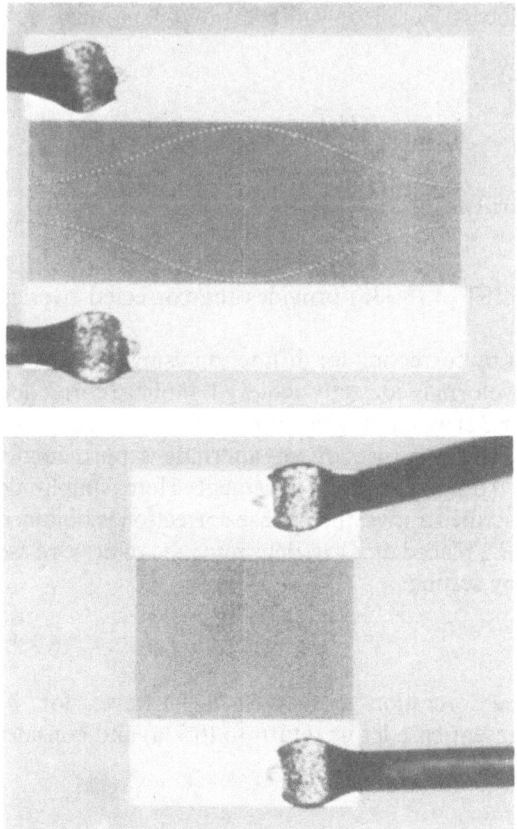

Fig. 6.54. Photographs of actual transducers fabricated on YZ LiTaO$_3$ using electron beam techniques [6.90]. The finger overlap values for the double electrode pairs were specified by the calculated values of L'_N

(Here Λ_S is the acoustic wavelength at center frequency f_0 and v_S is the surface wave velocity.)

Using these parameters in (6.33a), a corrected set of $\hat{L}'_N$ was determined [6.18] and an actual delay line fabricated [6.18] on YZ LiTaO$_3$. Photographs of both transducers are presented in Fig. 6.54. Since split electrodes were used to minimize electrical and mass loading, electron beam photolithographic techniques [6.90] were used to obtain the 1.188 µm line widths and gap spacings.

The experimental insertion loss versus frequency response of the actual delay line is illustrated in Fig. 6.53b. Except for a small difference in insertion loss of 3 dB and a slight shift in overall center frequency of 1.4 % as well as a minor shift between the center frequencies of the two experimental transducers, due probably to variations in electron beam fabrication technique, agreement with the desired response is excellent.

This agreement can be confirmed quantitatively in a manner independent of the slight differences observed above. The first two nulls on either side of the central resonance are due to the particular apodization present in the long transducer [6.88]. Since these nulls can be measured very accurately (3 parts in 10^4), a very sensitive measure of apodization is available. For the experimental case

$$2\frac{f_U - f_L}{f_U + f_L} = 0.04274 \text{ (experimental)} \tag{6.34}$$

where f_U represents the frequency of the upper null and f_L the lower. This figure is in excellent agreement with that for the theoretical case.

$$2\frac{f_U - f_L}{f_U + f_L} = 0.04288 \quad \text{(theory)}. \tag{6.35}$$

6.6 Coupling Efficiency and Temperature Coefficients

6.6.1 Introduction

Other parameters besides those causing propagation loss must, of course, be considered when arriving at any given overall design. For example, it does no good to have excellent beam steering and diffraction properties if the electromagnetic input cannot be efficiently converted to acoustic energy thus allowing wide bandwidth and low insertion loss devices to be realized. The best method for estimating this coupling efficiency is to compute the well-known $\Delta v/v$ parameter developed by *Campbell* and *Jones* [6.2].

The coupling constant of (6.23) can be computed [6.3] from $\Delta v/v$ as follows:

$$k^2 = 2[1 + (\varepsilon_{PR}^T)^{-1}]\frac{\Delta v}{v}\left(1 - \frac{\Delta v}{v}\right)^{-1}. \tag{6.36}$$

Here the parameter ε_{PR}^T is the relative equivalent dielectric constant given by [6.3]

$$\varepsilon_P^T = (\varepsilon_{11}\varepsilon_{33} - \varepsilon_{13}^2)^{1/2} = \varepsilon_0 \varepsilon_{PR}^T, \tag{6.37}$$

and ε_{11}, ε_{33}, and ε_{13} are actual dielectric constants at constant stress with the 1 direction being the direction of propagation of the surface wave. ε_0 is the permittivity of free space.

The revised edition of the *Microwave Acoustics Handbook* [6.57] is a convenient reference for a large number of $\Delta v/v$ values. Computations of surface wave velocity and electromechanical power flow angle, and estimates of surface wave coupling to interdigital transducers, are given for various orientations of the following surface wave substrate materials: $Ba_2NaNb_5O_{15}$, $Bi_{12}GeO_{20}$,

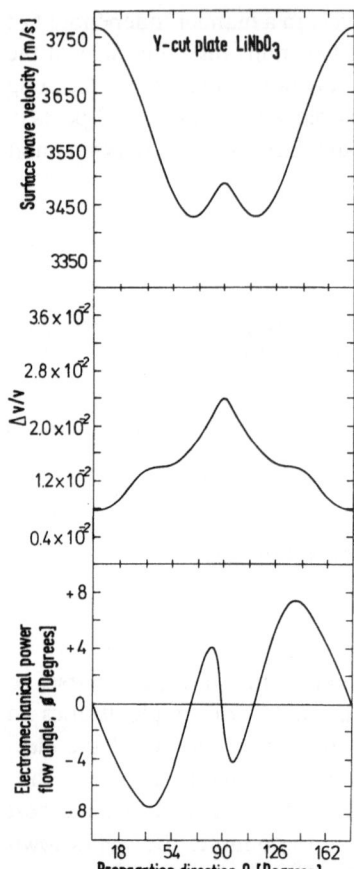

Fig. 6.55. Typical surface wave velocity $\Delta v/v$ and electromechanical power flow angle curves from the *Microwave Acoustics Handbook* [6.57]

CdS, diamond, $Eu_3Fe_5O_{15}$, gadolinium gallium garnet, GaAs, germanium, InSb, InAs, PbS, $LiNbO_3$, MgO, quartz, rutile, sapphire, silicon, spinel, TeO_2, YAG, YGaG, YIG, and ZnO. Typical data curves are illustrated in Figs. 6.55 and 6.56. Particular cuts of interest are then chosen for more detailed numerical calculations of mechanical and electrical parameters governing acoustic wave propagation in crystalline media. Similar data are given for common metals. A list of material constants and a bibliography of 520 surface wave papers are also included [6.57].

Another important parameter in many applications is temperature sensitivity [6.4, 5]; for example, *Carr* et al. [6.91] have shown that the principal limitation on the application of surface wave encoders and decoders to multiple-access, secure communications systems is the degradation of the peak-to-side-lobe ratio of the autocorrelation function due to temperature differences. Additionally, the temperature stability of the center frequency of surface wave bandpass filters is a direct function of the temperature coefficients of the material and orientation being used.

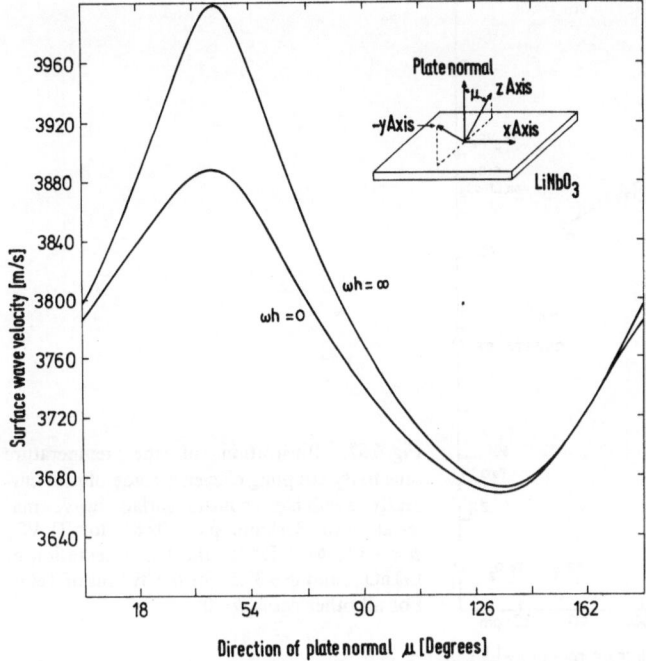

Fig. 6.56. Surface wave velocities along the X axis of $LiNbO_3$ as functions of the direction of the plate normal. $\omega h = \infty$ corresponds to a free surface while $\omega h = 0$ corresponds to a surface coated with an infinitesimally thin perfect conductor. The direction of time-average power flow coincides with the phase-velocity vector for all directions of plate normal and is identically zero for both $\omega h = \infty$ and $\omega h = 0$. Note the unusually high $\Delta v/v$ value at $\mu = 41.5°$ [6.65].

6.6.2 Temperature Coefficient Computations

The temperature coefficients of surface wave velocity and delay have previously been tabulated [6.92] for several piezoelectric materials of interest. The main results of this study and others [6.93–95] are summarized in Fig. 6.57 along with $\Delta v/v$ information. Further details can be found in [6.92–95]. An outline of procedures which can be used for these computations is as follows:

1) Computation of material constants at 15, 25, and 35 °C using

$$X(T) \approx X(T_0)\left[1 + \frac{1}{X(T_0)} \frac{\partial X}{\partial T}(T - T_0) + \frac{1}{2X(T_0)} \frac{\partial^2 X}{\partial T^2}(T - T_0)^2\right]. \tag{6.38}$$

Here X is the desired material constant, T_0 is room temperature, and $[1/X(T_0)](\partial X/\partial T)$ and $[1/2X(T_0)](\partial^2 X/\partial T^2)$ are the first- and second-order normalized temperature coefficients, respectively. Where direct measurements of the temperature coefficients of the density are not available, these coefficients

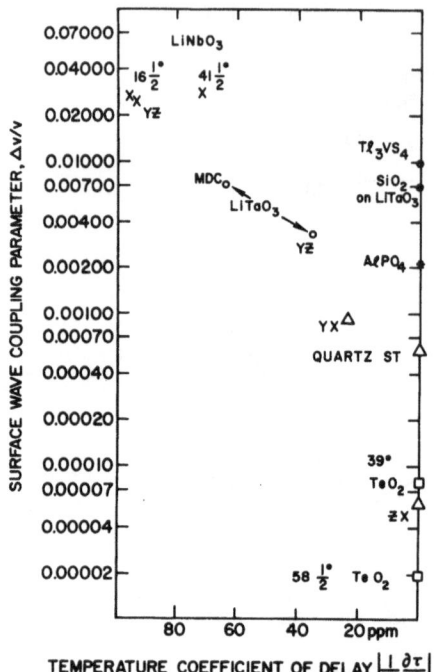

Fig. 6.57. Illustration of the temperature sensitivity-coupling efficiency trade-offs for currently available acoustic surface wave materials. For Berlinite $\phi = -0.68°$; for Tl_3VS_4 $\phi \approx -17$; $\phi = 1.55°$ for the 112° orientation of $LiTaO_3$; and $\phi = 37.5°$ for the 39° cut of TeO_2. For all other cases $\phi = 0°$

can be calculated [6.92] from the coefficients of thermal expansion α_{ii} according to

$$\frac{1}{\varrho(T_0)} \frac{\partial \varrho}{\partial T} = -(\alpha_{11} + \alpha_{22} + \alpha_{33}). \tag{6.39}$$

2) This step consists of computing the surface wave velocities at 15, 25, and 35 °C and using the results of Step 1) in conjunction with a computer program similar to that of *Campbell* and *Jones* [6.2].

3) Use of the following approximation to compute the temperature coefficient of velocity:

$$\frac{1}{v_S} \frac{\partial v_S}{\partial T}\bigg|_{25\,°C} \approx \frac{1}{v_S(25\,°C)} \frac{v_S(35\,°C) - v_S(15\,°C)}{20\,°C}. \tag{6.40}$$

For many delay line and other signal processing applications, the actual parameter of interest is not the temperature coefficient of velocity but is in fact the change in delay time with temperature. The first-order temperature coefficient of delay is given by

$$\frac{1}{\tau} \frac{\partial \tau}{\partial T} = \left(\frac{l}{v_S}\right)^{-1} \frac{\partial}{\partial T} \left(\frac{l}{v_S}\right) = \frac{1}{l} \frac{\partial l}{\partial T} - \frac{1}{v_S} \frac{\partial v_S}{\partial T} = \alpha - \frac{1}{v_S} \frac{\partial v_S}{\partial T}, \tag{6.41}$$

where $\tau = l/v_s$ is the delay time, l is the distance between two material points, and α is the coefficient of thermal expansion.

4) Computation of the temperature coefficient of delay can be accomplished directly by using the previously determined value of $(1/v_s)(\partial v_s/\partial T)$ in conjunction with the appropriate coefficient of thermal expansion and (6.41).

6.6.3 Temperature Coefficient, Coupling Trade-Offs, and Conclusions

Ideally one desires zero temperature coefficient of delay and high coupling. At present this is not possible as illustrated in Fig. 6.57, thus requiring design trade-offs. ST cut quartz [6.4] has the advantage of zero temperature coefficient, low cost, and the ready availability of large substrates. Its coupling is however quite low. *Berlinite* [6.93] may prove to be a useful substitute as high quality crystals become available. The 41.5° orientation [6.65] of $LiNbO_3$ appears to have the best properties of that popular material. Tellurium dioxide (TeO_2) has two orientations with zero temperature coefficients of delay [6.92]. Unfortunately, one orientation has a substantial amount of beam steering ($\phi = 37.5$) and the other highly accelerated diffraction (approximately 40 times worse than an isotropic material). Tl_3VS_4 [6.94] also suffers from substantial beam steering ($\phi \approx -17$). The composite structure of SiO_2 on $LiTaO_3$, although promising [6.95], does not possess the simplicity and high frequency advantages of single-crystal materials. Finally, $LiTaO_3$ seems to offer a reasonable current compromise between coupling, temperature sensitivity, and high quality substrate availability. However, rapid advances are expected in this fast moving research area [6.96].

6.7 Nonlinear Effects

6.7.1 Introduction

The nonlinear elastic effects associated with microwave-acoustic *volume* wave propagation have been extensively studied by several workers [6.97–100]. However, rather high input power is required to obtain the necessary acoustic power densities for the observation of appreciable nonlinearities.

In this section, an experimental investigation of nonlinear effects in microwave-acoustic *surface* waves [6.33, 101] will be described. In the case of surface waves, energy is confined to within a few wavelengths of the crystalline surface, and thus high power densities are possible with only moderate total input power. Although this fact will definitely limit the power handling capability of many proposed surface wave devices, it also offers the promise of several practical new devices such as harmonic generators, mixers, and parametric amplifiers.

The fundamental surface wave or any of its harmonics can be monitored by placing the laser probe photomultiplier tube at an angle corresponding to the proper wavelength. Since the higher diffraction orders of the fundamental coincide with the first diffraction orders of the harmonics, it will be necessary to rely on additional information to positively identify the harmonics. Using these procedures, nonlinear effects have been investigated [6.33] as a function of distance by scanning the laser along the direction of propagation of the surface wave, and as a function of microwave input power. For a fundamental-frequency surface wave at 905 MHz, second-, third-, and fourth-harmonic generation, buildup, and subsequent decay with distance have been detected [6.33]. The effects of nonlinear phenomena on the fundamental have also been observed. For example, under certain conditions harmonic power is returned to the fundamental causing a region of negative attenuation [6.33] or "gain".

6.7.2 Nonlinear Effects in the Fundamental Frequency Wave

Experimental results illustrating nonlinear effects as a function of both distance and power are shown in Fig. 6.58. These two curves were obtained by scanning the laser along the acoustic path, while monitoring the light deflected from the fundamental-frequency surface wave. The conditions under which these two curves were generated differ *only* in the magnitude of the acoustic power density at the input transducer. The high power case corresponds to 83 mW mm^{-1} while the low power curve corresponds to only 0.68 mW mm^{-1}.

The low power case illustrates the familiar linear dB versus distance curve characteristic of the exponential decay of energy. The high power curve is obviously nonlinear, having regions of greater apparent attenuation, points of zero slope, and a region of negative attenuation or "gain". Since the fundamental decay is no longer linear with distance, one concludes that the power entering the harmonics is comparable to that in the fundamental.

6.7.3 Harmonic Generation

The origin of the regions of higher and lower attenuation in the fundamental can be demonstrated with reference to Fig. 6.59. In addition to reproducing the high power fundamental curve of Fig. 6.58, the behaviors of the second-, third-, and fourth-harmonics versus distance are also shown. The regions of *increased* attenuation of the fundamental correspond to the *growth* of the harmonics. Power is being drawn from the fundamental into the harmonics. The region of "gain" corresponds to the rapid *decay* of the harmonics; power is being returned from the harmonics back into the fundamental.

The growth and decay of the harmonics demonstrate that the light detected at the appropriate angles is not due to higher diffraction orders from the fundamental [$m = -2, -3, \ldots$ in (6.1)] but to the different Λ_s's associated with the higher frequency harmonics. Higher diffraction orders of the fundamental

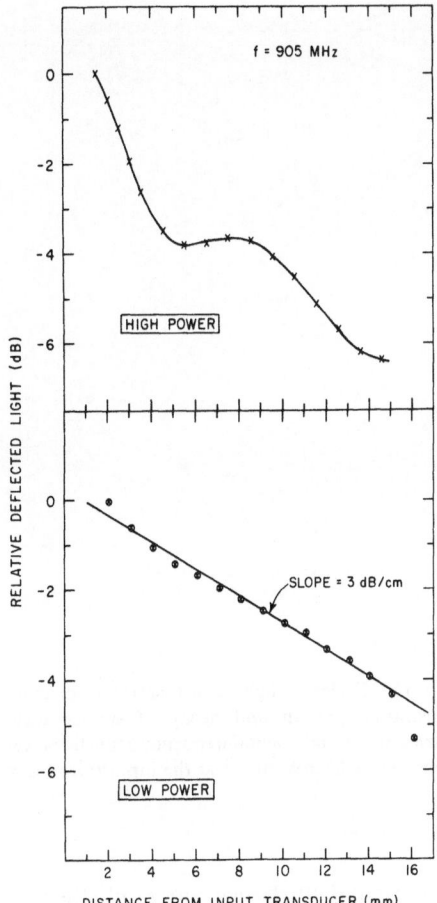

Fig. 6.58. Light deflected from a 905 MHz fundamental frequency acoustic surface wave as a function of distance from the input transducer. The high power case corresponds to an acoustic power density of $83\,\mathrm{mW\,mm^{-1}}$ at the input transducer, while the low power curve corresponds to $0.68\,\mathrm{mW\,mm^{-1}}$

would, of course, have the same spatial behavior as the fundamental itself. In addition, the growth of the harmonics also demonstrates that nonlinearities are occurring along the propagation path in the crystalline medium, and not merely at the generating transducer.

Approximate quantitative information on the relative amplitudes of the harmonics can be obtained by using the *Bechmann-Spizzichino* [6.26] theory developed for purely sinusoidal surfaces, to relate the deflected light intensity measurements of Fig. 6.59 to the amplitudes of the surface wave disturbances. By making several other approximations [6.26], all of which are fairly well satisfied under present experimental conditions, it can be shown that the ratio of the intensity of light deflected into side lobes to that specularly reflected from a *smooth* surface of the same material is given by [6.26]

$$I(\theta_0, \theta_m) = \sec^2 \theta_0 \left[\frac{1 + \cos(\theta_0 + \theta_m)}{\cos\theta_0 + \cos\theta_m}\right]^2 J_m^2 \left[\frac{2\pi h}{\lambda}(\cos\theta_0 + \cos\theta_m)\right] \qquad (6.42)$$

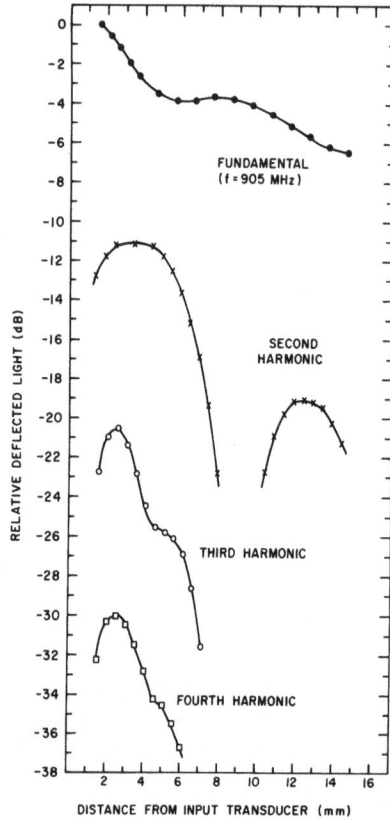

Fig. 6.59. Deflected light as a function of distance illustrating growth and decay of surface wave harmonics. Fundamental frequency acoustic power density was $83\,\mathrm{mW\,mm^{-1}}$ at the input transducer

where h is the amplitude of the sinusoidal surface disturbance. Since only the first diffraction order ($m = -1$) of extremely small amplitude surface waves will be considered, the Bessel function may be approximated to within a high degree of accuracy by one-half its argument. This means

$$I(\theta_0, \theta_{-1}) = \sec^2\theta_0 \,[1 + \cos(\theta_0 + \theta_{-1})]^2 \left(\frac{\pi h}{\lambda}\right)^2. \qquad (6.43)$$

The ratio of (6.43), evaluated at the appropriate angles and wavelengths for the fundamental, to the appropriate expression for the second harmonic indicates that the *peak* second harmonic amplitude is approximately 32% of the fundamental at the same distance from the transducer. Recall again that use of the Bechmann-Spizzichino theory neglects the non sinusoidal character of the actual surface wave disturbances. Similar calculations indicate that the third-harmonic *peak* surface amplitude is approximately 31% of the peak of the second harmonic, while the fourth harmonic is approximately 31% of the amplitude of the third harmonic. By further allowing these sinusoidal amplitudes to be directly related to the mechanical displacement component u_3 of a pure-mode

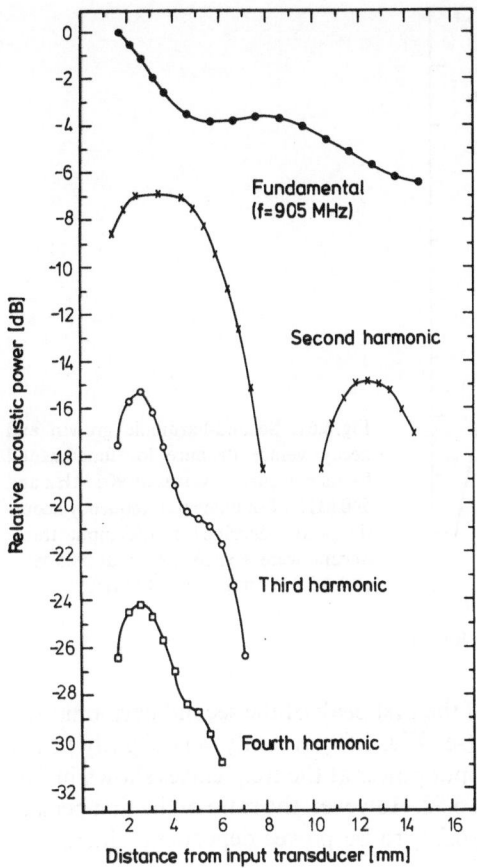

Fig. 6.60. The relative deflected light measurements of Fig. 6.59 converted to relative acoustic power through the use of (6.43), and the computer programs of *Campbell* and *Jones* [6.2]

surface wave [6.2], some estimate of the acoustic power density ratios can be made. The results are shown in Fig. 6.60. Recently a theory developed by *Adler* and *Nassar* [6.102] based on the dispersive properties of a damage layer on the substrate surface has yielded excellent agreement with the curves of Fig. 6.60.

One final comment on the use of the Bechmann-Spizzichino theory. This theory assumes that the deflected light is caused by an actual surface disturbance and not by any modulation of the index of refraction which might be caused by the surface wave. The former effect is indeed predominant for the conditions of the present experiments, as demonstrated by the results of *Salzmann* and *Weismann* [6.103].

Data have also been taken [6.33] to illustrate the effect of frequency on second-harmonic growth. The results given in Fig. 6.61 for two samples of LiNbO$_3$ at 905 and 540 MHz indicate that the peak of the second harmonic occurs farther away from the transducer at lower frequencies. Thus trend agrees with the results of *Lopen* [6.104] at 9 MHz, since at this low frequency he was able to observe only second-harmonic growth but no peak or decay. It has also

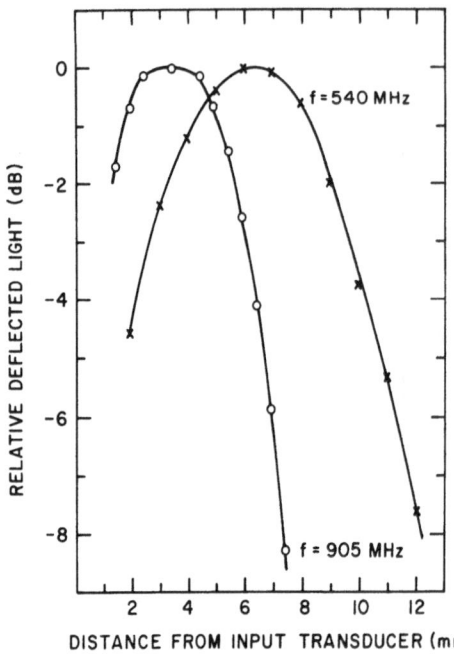

Fig. 6.61. Second-harmonic growth and decay versus distance for fundamental frequency surface waves at 905 MHz and 540 MHz. Fundamental frequency acoustic power densities at the input transducers were $83\,mW\,mm^{-1}$ at 905 MHz and $157\,mW\,mm^{-1}$ at 540 MHz

been found [6.20] that the *position* of the first peak of the second harmonic and the *position* of its second maximum (see Fig. 6.59) are only very slightly power dependent for a 10 dB reduction in input power at the frequencies shown in Fig. 6.61, as illustrated at 905 MHz in Fig. 6.62. However, the ratio of the first peak to the second maximum does appear [6.20] to be power dependent—increasing with increasing power.

6.7.4 Detailed Power Dependence

Additional information on nonlinear effects in surface wave delay lines can also be obtained by monitoring deflected light at one point along the acoustic path while changing the microwave input power [6.33]. In particular, the *continuous* transition from the linear to the nonlinear case illustrated in Fig. 6.58 can be observed. An example of one such study [6.33] is given in Fig. 6.63. For low input power levels there is a linear, unity slope (within experimental error) relation between microwave input power and deflected light. All the energy is essentially confined to the fundamental surface wave. However, since the harmonics increase according to power law relations (see Fig. 6.65), they build up more rapidly than the fundamental with increasing input power. When the harmonics began to draw an appreciable amount of the input power, saturation of the fundamental occurs as is clearly illustrated in Fig. 6.63.

The data given in Fig. 6.63 can be used to determine the useful linear region of operation for $LiNbO_3$ surface wave devices. Below an acoustic power density of

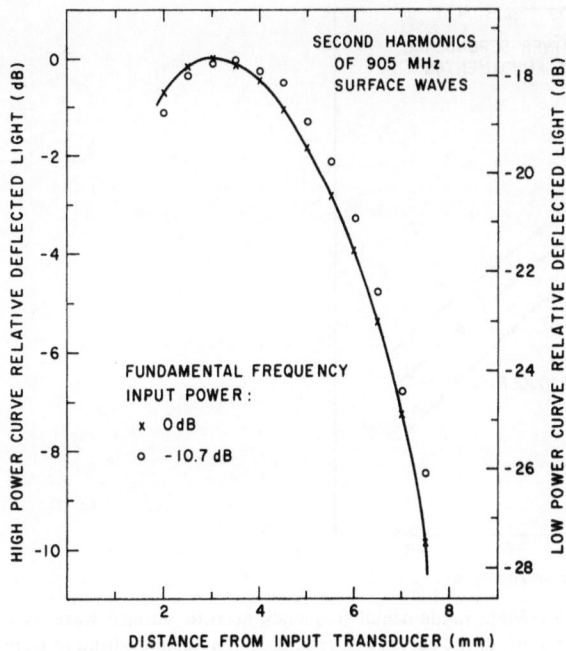

Fig. 6.62. The effect of fundamental frequency input power level on the growth and decay of the second harmonic

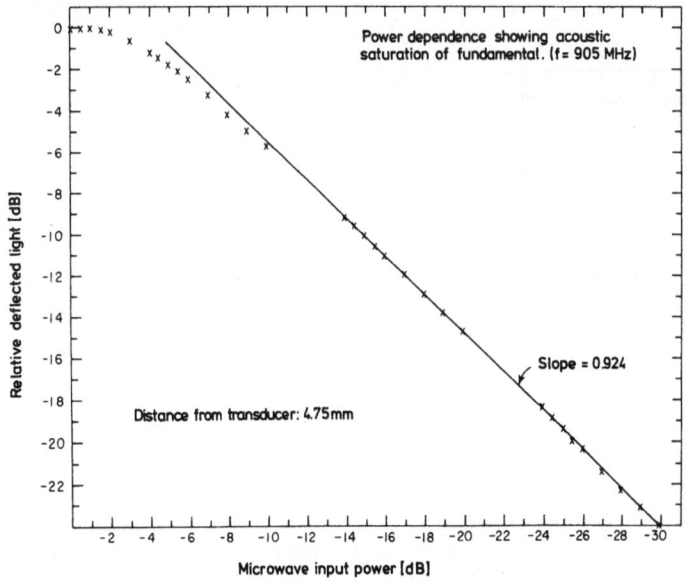

Fig. 6.63. Light deflected from a 905 MHz fundamental frequency acoustic surface wave as a function of microwave input power. Acoustic power density at the 0 dB level of input power was 123 mW mm^{-1} at the transducer. The deviation from unity slope at low power levels is attributed to experimental error [6.33]

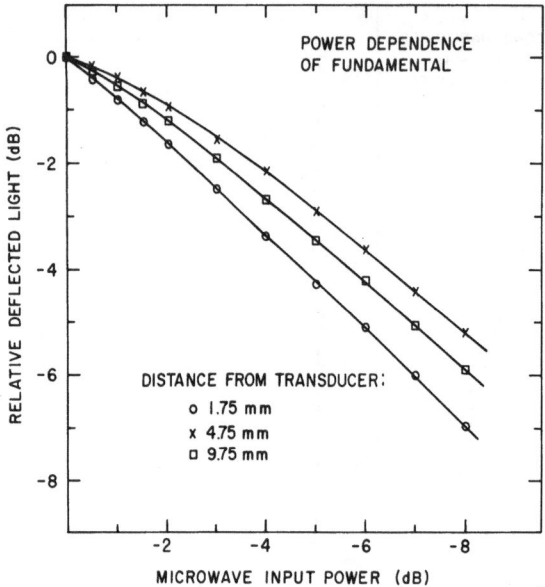

Fig. 6.64. Light deflected from a 905 MHz fundamental frequency acoustic surface wave as a function of microwave input power. Each of the curves was generated at a different distance from the input transducer. Acoustic power density at the 0 dB level of input power was 78 mW mm⁻¹ at the transducer

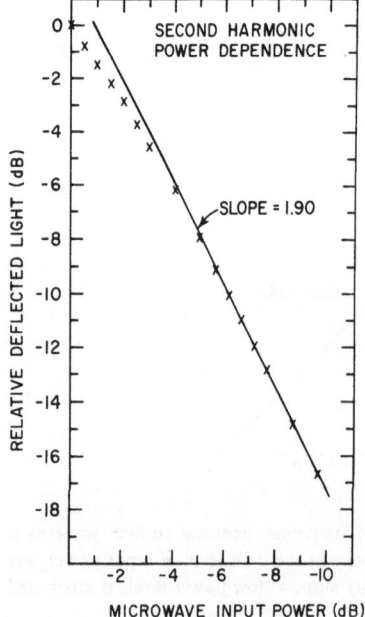

Fig. 6.65. Light deflected from the second harmonic of a 905 MHz acoustic surface wave as a function of the fundamental frequency microwave input power. Fundamental frequency acoustic power density at the 0 dB level of input power was 83 mW mm⁻¹ at the transducer

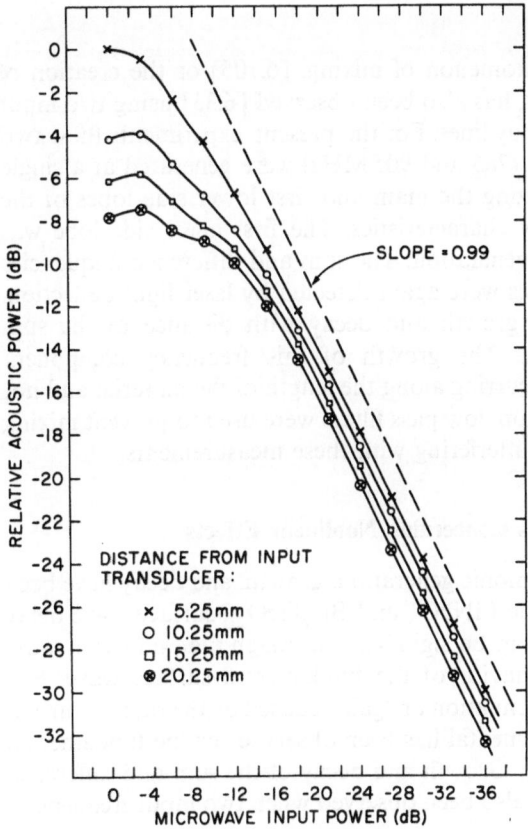

Fig. 6.66. One-way acoustic power on 001, 110 $Bi_{12}GeO_{20}$ as a function of microwave input power at various distances from the input transducer. Frequency was 320 MHz, and 0 dB of microwave power corresponds to 570 mW mm^{-1} of acoustic power at the input transducer

approximately 10 mW mm^{-1}, operation may be considered to be in the linear region, while above this power level nonlinearities begin to become significant.

Since the transition shown in Fig. 6.58 is from a linear (with distance) low power curve to a nonlinear (with distance) high power curve, the power dependence at different points along the acoustic path must be different. These effects have indeed been observed [6.33] and are illustrated in Fig. 6.64.

The power dependence of the second harmonic can also be investigated by choosing the proper angle for light detection. This is illustrated in Fig. 6.65 where the distance from the transducer was chosen to correspond to the peak of the second harmonic. For low levels of fundamental frequency input power, behavior is linear, this time with the expected square-law dependence (within experimental error). Acoustic saturation again becomes evident at higher power levels.

Data [6.8] for 001 cut, 110 propagating $Bi_{12}GeO_{20}$ are shown in Fig. 6.66. Linear operation ceases at approximately 9 mW mm^{-1} at 320 MHz since 0 dB of input power corresponds to 570 mW mm^{-1} of peak fundamental frequency acoustic power at the transducer.

6.7.5 Mixing

The additional nonlinear phenomenon of mixing, [6.105] or the creation of sum and difference frequencies, has also been observed [6.33] using two input frequencies on surface wave delay lines. For the present experiments these two input frequency surface waves (785 and 905 MHz) were generated at a single interdigital transducer by utilizing the main and first lower side lobes of the insertion loss versus frequency characteristics. The first *lower* side lobe was chosen to avoid volume wave generation. The sum and difference frequencies resulting from the nonlinearities were again detected by laser light deflection.

Figure 6.67 illustrates the growth and decay with distance of the sum frequency wave at 1690 MHz. The growth of this frequency component demonstrates that mixing is occurring along the length of the material and not just at the transducer. In addition, low pass filters were used to prevent mixing in the microwave circuit from interfering with these measurements.

6.7.6 Summary and Conclusions Concerning Nonlinear Effects

Second-, third-, and fourth-harmonic generation, growth, and decay have been observed on microwave acoustic $LiNbO_3$ and $Bi_{12}GeO_{20}$ surface wave delay lines. These nonlinear effects can be significant in magnitude and can cause measurable changes in the behavior of the fundamental surface wave. For example, an area of negative attenuation or "gain" caused by the rapid return of harmonic power into the fundamental has been observed for the fundamental component of the surface wave. Growth and decay of the sum and difference surface wave components have also been observed when two input frequencies are used (mixing).

The relative ease with which nonlinear effects can be excited in surface wave delay lines has two important consequences for practical surface wave devices. First, it seems clear that many new devices can be realized, including convolvers, correlators, acoustic harmonic generators, mixers, parametric amplifiers, and limiters. The principles underlying some of these devices are presented in Section 4.6. On the other hand, the observed nonlinearities will definitely limit the power handling capability of other proposed devices, such as delay lines and signal processors. For example, at 905 MHz on Y cut, Z propagating $LiNbO_3$, nonlinearities become appreciable above $10 \, mW \, mm^{-1}$ while at 320 MHz on 001, 110 $Bi_{12}GeO_{20}$ the corresponding figure is $9 \, mW \, mm^{-1}$.

6.8 Summary and Properties of Materials

It is the purpose of this section to summarize in a qualitative manner some of the points discussed in detail in this chapter as well as to present a convenient quantitative summary of some of the properties of popular materials which

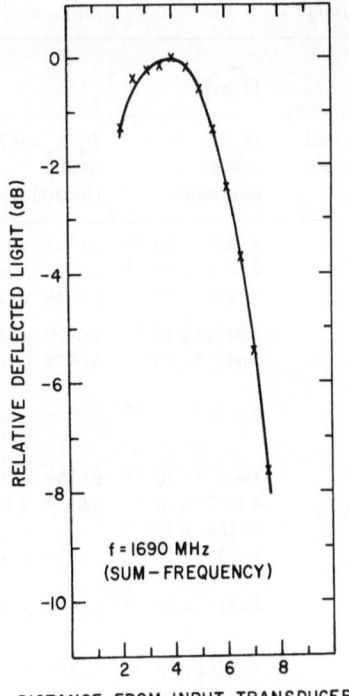

f = 1690 MHz
(SUM – FREQUENCY)

RELATIVE DEFLECTED LIGHT (dB)

DISTANCE FROM INPUT TRANSDUCER (mm)

Fig. 6.67. Growth and decay versus distance of the sum frequency surface wave component generated with input frequencies of 905 MHz and 785 MHz

affect device design. Table 6.7 provides this latter information. An asterisk (∗) indicates experimental data.

Quantitative propagation loss expressions as a function of frequency, such as (6.11)–(6.13), can be obtained from Table 6.7 by taking

$$\text{Loss}[\text{dB}\,\mu\text{s}^{-1}] = (\text{VAC})F^2 + (\text{AIR})F \tag{6.44}$$

where F is in GHz.

Diffraction and beam steering properties of materials can be determined from the slope of the power flow angle $\partial\phi/\partial\theta$ and $|\delta_M|$. Recall that when $|\delta_M| < 2.0$, the parabolic diffraction theory is highly accurate and predicts minimal diffraction for $\partial\phi/\partial\theta = -1$. Minimal beam steering and isotropic diffraction are predicted for $\partial\phi/\partial\theta = 0$; thus we note the beam steering, diffraction trade-off.

The remainder of Table 6.7 is devoted to providing parameters which are useful in interdigital transducer models. Both computed and experimental [6.4] values for the coupling constant k^2 are provided. The theoretical value is computed by using $\Delta v/v$ and ε_{PR}^T in (6.36). ε_{PR}^T itself is computed from (6.37). Capacitance per unit length C_{FF} for a single interdigital *period* is also provided for finger width to gap spacing ratios of $D_X/L_X = 0.5$ and 0.75. This quantity is computed [6.106] from (6.45)

$$C_{FF} = 2(\varepsilon_{PR}^T + 1)[6.5(D_X/L_X)^2 + 1.08(D_X/L_X) + 2.37] \times 10^{-12}. \tag{6.45}$$

Table 6.7. Acoustic surface wave and interdigital transducer design data. Set text for definition of all

Material	Orientation	v_S [m/s]	$\Delta v/v_\infty$	k^2		C_{FF} [F/m]	
				Calculated from (6.36)	Measured by *Schulz* and *Matsinger*, 1972	$D_x/L_x=0.50$ single electrodes	$D_x/L_x=0.75$ double electrodes
LiNbO$_3$	Y, Z	3488.	0.0241	0.0504	0.045	4.6438×10^{-10}	7.0003×10^{-10}
	16-1/2 DR	3503.	0.0268	0.0562	0.048	4.6438×10^{-10}	7.0003×10^{-10}
	41-1/2, X	4000.	0.0277	0.0578	0.057	6.1857×10^{-10}	9.3246×10^{-10}
Bi$_{12}$GeO$_{20}$	001, 110	1681.	0.0068	0.0140	0.015	4.04522×10^{-10}	6.0979×10^{-10}
	111, 110 45	1708.	0.0082	0.0169	0.017	4.04522×10^{-10}	6.0979×10^{-10}
	40.04 90	1827.	0.0031	0.0064	—	4.04522×10^{-10}	6.0979×10^{-10}
LiTaO$_3$	Z, Y	3329.	0.0059	0.0121	0.0093	4.43523×10^{-10}	6.6859×10^{-10}
	Y, Z	3230.	0.0033	0.0068	0.0074	4.43523×10^{-10}	6.6859×10^{-10}
	Y, X	3148.	0.00037	0.00075	—	4.7164×10^{-10}	—
	Z, X 0	3205.	0.00114	0.00233	—	4.4352×10^{-10}	6.6859×10^{-10}
	166.65* 90	3378.	0.0075	0.0154	—	4.4352×10^{-10}	6.6859×10^{-10}
Quartz	Y, X	3159.	0.0009	0.0022	0.0023	5.00664×10^{-11}	7.54722×10^{-11}
	ST, X	3158.	0.00058	0.0014	0.0016	5.03385×10^{-11}	7.58824×10^{-11}

$\hat{L}_{OPT}$ represents the acoustic aperture (in wavelengths) providing the minimum insertion loss available under the specified tuning conditions from a simple, periodic interdigital transducer (IDT) of N_{OPT} periods [6.78]. Beam steering and diffraction effects are not included. Q applies to the tuning inductor and ϱ/t to the IDT finger sheet resistivity.

Some of the topics discussed in this chapter which are not summarized in Table 6.7 include the highly useful nature of the laser probe as a basic SAW measurement tool, particularly at high frequencies; the effect of surface quality on SAW attenuation; and the ease with which nonlinear effects can be generated and observed in SAW devices.

It is hoped that this chapter has provided some insight concerning SAW materials and their influence on performance.

Acknowledgment. Much of the information contained in this chapter is the results of original research performed by members of the Microwave Acoustics Branch, Electromagnetic Sciences Division, Deputy for Electronic Technology, Rome Air Development Center. In compiling this review, the author has drawn heavily from the work of his RADC/ET colleagues and is very deeply indebted to them for the opportunity of utilizing results of their research. Particularly thanks go to *T. L. Szabo*, *A. J. Budreau*, and Dr. *P. H. Carr* without whom this chapter could never have been written.

quantities

ε^T_{PR}	N_{OPT} calculated from (6.23)	$\hat{L}_{OPT}$ (with tuning inductor)		$\hat{L}_{OPT}$ (no tuning inductor)	$\frac{\partial\phi}{\partial\theta}=\gamma$	$|\delta_M|$	VAC [dB/μs]	AIR [dB/μs]	$\frac{1}{v}\frac{\partial v}{\partial T}$	$\frac{1}{\tau}\frac{\partial\tau}{\partial T}$
		$Q=\infty$ $\varrho/t=0$	$Q=100$ $\varrho/t=0.345$	$\varrho/t=0$						
50.2	4.0	107	76	479	-1.083	7.87	0.88	0.19	-87	94
50.2	4.0	113	79	476	-1.087	3.99	0.94	0.21	-88	96
67.2	3.5	88	63	357	-0.445	0.57	0.75	0.30	-57	72
43.6	7.0	88	69	664	-0.304	0.14	1.45	0.19	—	—
43.6	7.0	98	76	651	$+0.366$	0.08	1.45	0.19	—	—
43.6	11.00	35	29	390	-1.000	1.44	—	—	—	—
47.9	9.0	25	22	238	-1.241	5.04	0.77	0.23	-52	69
47.9	10.5	21	18	211	-0.211	0.14	0.94	0.20	-31	35
51	32.5	—	—	—	$+0.159$	—	—	—	-33	49
47.9	18.5	—	—	—	$+0.450$	—	—	—	-50	66
47.9	7.0	—	—	—	-0.95^*	2.60	—	—	-50	64
4.52	18.5	59	46	1087	$+0.653$	0.359	2.15	0.45	38	-24
4.55	22.0	41	31	910	$+0.378$	0.205	2.62	0.47	14	0 .

References

6.1 A.J.Slobodnik,Jr.: (Proc. IEEE **64**, 581 (1976); Proc. of the Symp. on Optical and Acoustical Micro-Electronics, Vol. 23 (Polytechnic Press, New York, 1974) p. 205
6.2 J.J.Campbell, W.R.Jones: IEEE Trans. SU-**15**, 209 (1968)
6.3 M.B.Schulz, J.H.Matsinger: Appl. Phys. Lett. **20**, 367 (1972)
6.4 M.B.Schulz, J.H.Matsinger, M.G.Holland: J. Appl. Phys. **41**, 2755 (1971)
6.5 M.B.Schulz, M.G.Holland: IEEE Trans. SU-**19**, 381 (1972)
6.6 A.J.Slobodnik,Jr., P.H.Carr, A.J.Budreau: J. Appl. Phys. **41**, 4380 (1970)
6.7 A.J.Budreau, P.H.Carr: Appl. Phys. Lett. **18**, 239 (1971)
6.8 A.J.Slobodnik,Jr., A.J.Budreau: J. Appl. Phys. **43**, 3278 (1972)
6.9 A.J.Slobodnik,Jr., E.D.Conway: International Microwave Symposium Digest, 314 (1970)
6.10 M.S.Kharusi, G.W.Farnell: J. Acoust. Soc. Am. **48**, 665 (1970)
6.11 M.S.Kharusi, G.W.Farnell: IEEE Trans. SU-**18**, 35 (1971)
6.12 T.L.Szabo, A.J.Slobodnik,Jr.: IEEE Trans. SU-**20**, 240 (1973)
6.13 D.B.Armstrong: *Research to Develop Microwave Acoustic Surface Wave Delay Lines*, AFCRL-72-0378, (Litton Industries, San Carlos, California, 1972)
6.14 H.Gerard, M.Wauk, R.Weglein: *Large Time-Bandwidth Product Microwave Delay Line*, ECOM-03852, (Hughes Aircraft Company, Fullerton, California, 1970)
6.15 R.D.Weglein, E.D.Wolf: International Microwave Symposium Digest, 120 (1973)
6.16 A.J.Slobodnik,Jr., T.L.Szabo: Electron. Lett. **9**, 149 (1973)
6.17 A.J.Slobodnik,Jr., T.L.Szabo: J. Appl. Phys. **44**, 2937 (1973)
6.18 T.L.Szabo, A. J. Slobodnik,Jr.: IEEE Trans. SU-**21**, 114 (1974)
6.19 A.J.Slobodnik,Jr.: Proc. IEEE **58**, 488 (1970)

6.20 A.J.Slobodnik,Jr.: *A Laser Probe for Microwave Acoustic Surface Wave Investigations*, AFCRL-70-0404, (Air Force Cambridge Research Laboratories, Hanscom AFB, Massachusetts, 1970)

6.21 B.A.Richardson, G.S.Kino: Appl. Phys. Lett. **16**, 82 (1970)

6.22 J.Krokstad: *Scattering of Light by Ultrasonic Surface Waves in Crystals*, (Norwegian Institute of Technology, Trondheim, 1967)

6.23 A.Korpel, L.J.Laub, H.C.Sievering: Appl. Phys. Lett. **10**, 295 (1967)

6.24 D.C.Auth, W.G.Mayer: J. Appl. Phys. **38**, 5138 (1967)

6.25 E.Salzmann, T.Plieninger, K.Dransfeld: Appl. Phys. Lett. **13**, 14 (1968)

6.26 P.Bechmann, A.Spizzichino: *The Scattering of Electromagnetic Waves from Rough Surfaces*, (MacMillan, New York, 1963)

6.27 J.D.DeLorenzo, E.S.Cassedy: IEEE Trans. AP-**14**, 611 (1966)

6.28 A.N.Broers, E.G.H.Lean, M.Hatzakis: Appl. Phys. Lett. **15**, 98 (1969)

6.29 V.V.Zhdanova, V.P.Klyuev, V.V.Lemanov, I.A.Smirnov, V.V.Tikhonov: Sov. Phys. Solid State **10**, 1360 (1968)

6.30 V.V.Lemanov, G.A.Smolenskii, A.B.Sherman: Sov. Phys. Solid State **11**, 524 (1969)

6.31 A.A.Maradudin, D.L.Mills: Phys. Rev. **173**, 881 (1968)

6.32 E.G.H.Lean, C.C.Tseng, C.G.Powell: Appl. Phys. Lett. **16**, 32 (1970)

6.33 A.J.Slobodnik,Jr.: J. Acoust. Soc. Amer. **48**, 203 (1970)

6.34 C.P.Wen, R.F.Mayo: Appl. Phys. Lett. **9**, 135 (1966)

6.35 A.B.Smith, M.Kestigian, R.W.Kedzie, M.I.Grace: J. Appl. Phys. **38**, 4928 (1967)

6.36 W.Rehwald: J. Appl. Phys. **44**, 3017 (1973)

6.37 E.G.Spencer, P.V.Lenzo, A.A.Ballman: Proc. IEEE **55**, 2074 (1967)

6.38 I.A.Viktorov, E.K.Grishehanko, T.M.Kaekina: Sov. Phys. Acoust. **9**, 131 (1963)

6.39 R.M.Arzt, E.Salzmann, K.Dransfeld: Appl. Phys. Lett. **10**, 165 (1967)

6.40 G.Bradfield: Electron. Lett. **4**, 95 (1968)

6.41 A.A.Oliner: IEEE Trans. MTT-**17**, 812 (1969)

6.42 A.J.Slobodnik,Jr.: J. Appl. Phys. **43**, 2565 (1972)

6.43 R.M.Arzt: Private communication

6.44 J.J.Campbell, W.R.Jones: IEEE Trans. SU-**17**, 71 (1970)

6.45 K.Dransfeld, E.Salzmann: *Physical Acoustics*, Vol. 7 (Academic, New York, 1970) p. 219

6.46 P.M.Morse, K.U.Ingard: *Theoretical Acoustics* (McGraw-Hill, New York, 1968) p. 227

6.47 A.J.Slobodnik,Jr.: Proc. Ultrasonics Symp. (1973) p. 369

6.48 A.J.Slobodnik,Jr.: Electron. Lett. **10**, 233 (1974)

6.49 J.H.Collins, H.M.Gerard, H.J.Shaw: Appl. Phys. Lett. **13**, 312 (1968)

6.50 E.P.Warekois, S.A.Kulin: *Examination of Surface Wave Plate Specimens* (Manlabs, Inc., Cambridge, Massachusetts, 1972)

6.51 E.P.Warekois, S.A.Kulin: *Characterization of Lithium Niobate Samples* (Manlabs, Inc., Cambridge, Massachusetts, 1973)

6.52 E.D.Wolf: Private communication (1972)

6.53 A.B.Smith, J.C.Worley, M.Kestigian et al.: *Microwave Acoustical Loss Mechanism in Crystals*, AFML-TR-69-200, (Sperry Rand Research Center, Sudbury, Massachusetts, 1969)

6.54 A.J.Slobodnik,Jr.: Electron. Lett. **8**, 307 (1972)

6.55 M.G.Cohen: J. Appl. Phys. **38**, 3821 (1967)

6.56 A.J.Slobodnik,Jr., T.L.Szabo: Electron. Lett. **7**, 257 (1971)

6.57 A.J.Slobodnik,Jr., E.D.Conway, R.T.Delmonico: *Microwave Acoustics Handbook, Volume 1A, Surface Wave Velocities*, AFCRL-TR-73-0597, (Air Force Cambridge Research Laboratories, Hanscom AFB, Massachusetts, 1973)

6.58 T.L.Szabo, A.J.Slobodnik,Jr.: *Acoustic Surface Wave Diffraction and Beam Steering*, AFCRL-TR-73-0302, (Air Force Cambridge Research Laboratories, Hanscom AFB, Massachusetts, 1973)

6.59 R.T.Smith, F.S.Welsh: J. Appl. Phys. **42**, 2219 (1971)

6.60 A.W.Warner, M.Onoe, G.A.Coquin: J. Acoust. Soc. Am. **42**, 1223 (1967)

6.61 F.Piro, M.Sinou: Proc. Ultrasonics Symp. (1975) p. 492

6.62 H.K.Wickramasinghe, E.A.Ash: Proc. Ultrasonics Symp. (1975) p. 496
6.63 T.L.Szabo: Proc. Ultrasonics Symp. (1975) p. 116
6.64 A.J.Slobodnik,Jr., T.L.Szabo: IEEE Trans. MTT-**22**, 458 (1974)
6.65 A.J.Slobodnik,Jr., E.D.Conway: Electron. Lett. **6**, 171 (1970)
6.66 W.R.Smith, H.M.Gerard, J.H.Collins, T.M.Reeder, H.J.Shaw: IEEE Trans. MTT-**17**, 856 (1969)
6.67 W.R.Smith, H.M.Gerard, J.H.Collins, T.M.Reeder, H.J.Shaw: IEEE Trans. MTT-**17**, 865 (1969)
6.68 R.H.Tancrell, M.G.Holland: Proc. IEEE **59**, 393 (1971)
6.69 W.S.Jones, C.S.Hartmann, T.D.Sturdivant: IEEE Trans. SU-**19**, 368 (1972)
6.70 R.H.Tancrell, R.C.Williamson: Appl. Phys. Lett. **19**, 456 (1971)
6.71 H.M.Gerard, G.W.Judd, M.E.Pedinoff: IEEE Trans. MTT-**20**, 188 (1972)
6.72 T.W.Bristol, W.R.Jones, P.B.Snow, W.R.Smith: Proc. Ultrasonics Symp. (1972) p. 343
6.73 P.H.Carr: Proc. IEEE **60**, 1103 (1972)
6.74 A.J.Slobodnik,Jr., J.C.Sethares: J. Appl. Phys. **43**, 247 (1972)
6.75 H.Goldstein: *Classical Mechanics* (Addison-Wesley, New York, 1970)
6.76 L.P.Solie: J. Appl. Phys. **44**, 619 (1973)
6.77 E.A.Kraut, B.R.Tittmann, L.J.Graham, T.C.Lim: Appl. Phys. Lett. **17**, 271 (1970)
6.78 A.J.Slobodnik,Jr.: *UHF and Microwave Frequency Acoustic Surface Wave Delay Lines: Design*, AFCRL-TR-73-0538, (Air Force Cambridge Research Laboratories, Hanscom AFB, Massachusetts, 1973)
6.79 C.S.Hartmann, D.T.Bell, R.C.Rosenfeld: IEEE Trans. MTT-**21**, 162 (1973)
6.80 R.H.Tancrell: IEEE Trans. SU-**21**, 12 (1974)
6.81 R.S.Wagers: IEEE Trans. SU-**23**, 249 (1976)
6.82 A.J.Slobodnik,Jr., T.L.Szabo: International Microwave Symposium Digest (1974) p. 247
6.83 J.D.Maines, G.L.Moule, N.R.Ogg: Electron. Lett. **8**, 431 (1972)
6.84 R.A.Waldron: IEEE Trans. SU-**19**, 448 (1972)
6.85 W.R.Smith, H.M.Gerard, W.R.Jones: IEEE Trans. MTT-**20**, 458 (1972)
6.86 P.J.Hagon, K.M.Lakin: Proc. Ultrasonics Symp. (1974) p. 341
6.87 T.L.Szabo: Private communication (1975)
6.88 C.E.Cook, M.Bernfeld: *Radar Signals* (Academic Press, New York, 1967)
6.89 R.H.Tancrell, F.Sandy: *Analysis of Interdigital Transducers for Acoustic Surface Wave Devices*, AFCRL-TR-73-0030, (Raytheon Research Division, Waltham, Massachusetts, 1973)
6.90 F.S.Ozdemir, E.D.Wolf, C.R.Buckey: IEEE Trans. ED-**19**, 624 (1972)
6.91 P.H.Carr, P.A.DeVito, T.L.Szabo: IEEE Trans. SU-**19**, 357 (1972)
6.92 A.J.Slobodnik,Jr.: *The Temperature Coefficients of Acoustic Surface Wave Velocity and Delay on Lithium Niobate, Lithium Tantalate, Quartz and Tellurium Dioxide*, AFCRL-72-0082, (Air Force Cambridge Research Laboratories, Hanscom AFB, Massachusetts, 1971)
6.93 P.H.Carr, R.M.O'Connell: Proc. 30th Annual Symp. on FREQ Control (1976)
6.94 R.W.Weinert, T.J.Isaacs: Proc. 29th Annual Symp. on FREQ Control (1975) p. 139
6.95 T.E.Parker, M.B.Schulz: Proc. Ultrasonics Symp.. (1975) p. 261
6.96 G.R.Barsch, R.E.Newnham: *Piezoelectric Materials with Positive Elastic Constant Temperature Coefficients*, AFCRL-TR-75-0163, (Pennsylvania State University, University Park, Pennsylvania 16802, 1975)
6.97 P.H.Carr: Phys. Rev. **169**, 718 (1968)
6.98 D.H.McMahon: J. Acoust. Soc. Am. **44**, 1007 (1968)
6.99 A.L.VanBuren, M.A.Breazeale: J. Acoust. Soc. Am. **44**, 1014 (1968)
6.100 B.A.Richardson, R.B.Thompson, C.D.W.Wilkinson: J. Acoust. Soc. Am. **44**, 1608 (1968)
6.101 E.G.Lean, C.C.Tseng, C.G.Powell: Appl. Phys. Lett. **16**, 32 (1970)
6.102 E.L.Adler, A.A.Nassar: Proc. Ultrasonics Symp. 268 (1973)
6.103 E.Salzmann, D.Weismann: J. Appl. Phys. **40**, 3408 (1969)
6.104 P.O.Lopen: J. Appl. Phys. **39**, 5400 (1968)
6.105 E.G.H.Lean, C.G.Powell, L.Kuhn: Appl. Phys. Lett. **15**, 10 (1969)
6.106 G.W.Farnell, I.A.Cermak, P.Silvester, S.K.Wong: IEEE Trans. SU-**17**, 188 (1970)

7. Fabrication Techniques for Surface Wave Devices[1]

H. I. Smith

With 8 Figures

Most surface acoustic wave devices are made by the planar fabrication technique illustrated in Fig. 7.1, or some variation thereof. The exposure of the pattern in the polymer film is the most crucial step in ensuring desired device geometry, dimensional control, and freedom from pattern distortion. The various methods of exposing patterns in polymer films include: optical projection, conventional contact printing, conformable-photomask contact printing, holographic recording, scanning electron beam lithography, projection electron lithography, and x-ray lithography.

Etching of relief structures in substrate surfaces or in films deposited on substrates is done by aqueous chemical etching, plasma etching, or ion bombardment. Doping of certain regions of surface wave substrates is a relatively new technique, but potentially very important. The formation of relief patterns on a substrate surface by deposition over a polymer film pattern, the so-called liftoff technique, is widely used and especially well suited to surface wave devices. Metallic relief patterns can also be formed by electroplating if the substrate is conducting or coated with a thin metal film. Electroplating has the advantage that no metal is deposited on top of the polymer.

The use of most of the above techniques in the fabrication of surface acoustic wave devices has been reviewed in detail in a journal article [7.1] and a book chapter [7.2]. Instead of attempting another review here, the major conclusions of the earlier reviews will be briefly summarized, and updated where appropriate. In addition, some new remarks are made in connection with conformable photomask lithography and doping techniques.

7.1 Techniques for Exposing Patterns in Polymer Films

Several methods for forming relief patterns in the polymer films over substrate surfaces are summarized in this section. Methods for transferring this pattern onto the substrate itself are considered in the next section.

In the case of exposure by short wavelength, visible and ultraviolet radiation, photoresists of the type AZ1350 (Shipley Company, Newton, Mass.) have proven to be the most useful for fabricating surface acoustic wave devices.

---

[1] This work is sponsored by the Department of the Army.

These photoresists are positive acting, that is, exposed regions are removed during development. They respond in a nonlinear fashion to exposure and development [7.3, 4], and this, in effect, leads to very high contrast. These properties, plus the ability to retain precise dimensions and record very high resolutions (spatial frequencies < 1000 Å), are the reasons such photoresists are preferred.

In the case of exposure by ionizing radiation, polymethyl methacrylate (PMMA) has demonstrated the highest contrast as well as the highest resolution (spatial frequencies < 1000 Å) and dimensional control. However, PMMA has relatively low sensitivity. A number of new high-speed radiation-sensitive polymers have been developed [7.5–7]. It remains to be seen, however, whether they will be favored over PMMA for the fabrication of surface acoustic wave devices.

The discussion in this section considers separately the several photo-lithographic methods, electron-beam lithography, and x-ray lithography. Of these, stress is placed on conformable-photomask contact printing and x-ray lithography because of their simplicity and economy, and because they are particularly well adapted to replicating the high resolution, high precision, repetitive grating type patterns that are characteristic of surface acoustic wave devices.

7.1.1 Optical Projection Printing

In optical projection printing, an optical system is used to image a pattern onto a photoresist-coated substrate, usually with some demagnification.

Although optical projection systems are capable of very high resolution (linewidths ∼0.4 μm) [7.8, 9] and have been used to expose patterns directly on device substrates [7.10, 11], their greatest utility is in generating photomasks having linewidths ≳1 μm. Accuracy and precision over large areas is best achieved by means of laser interferometer controlled stages.

7.1.2 Conventional Contact Printing

The most widely used photolithographic technique is conventional contact printing. In this technique, a pattern on a glass plate is held in contact with, or in close proximity to, a photoresist film on a substrate, and the pattern is transferred to the photoresist by passing collimated ultraviolet light through the glass plate. The glass plate with its pattern is called the photomask.

In the conventional contact printing technique, the air gap between a photomask and a photoresist-coated substrate is not specifically controlled. The diffraction which occurs in the air gap results in rounded photoresist profiles and loss of dimensional control. This effectively limits the minimum linewidths to ≳2 μm. Thus, conventional contact printing is not the preferred technique for high frequency devices, or where metal liftoff or ion beam etching is to follow the photolithography, since both of these require vertical sidewalls in the photoresist.

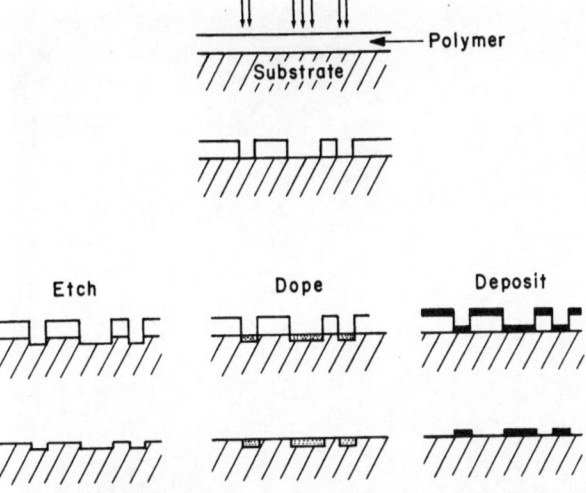

Fig. 7.1. Schematic illustration of the principles of the planar fabrication technique. A substrate surface coated with a radiation-sensitive polymer film is exposed to radiation in some desired pattern. Following exposure, a development step removes either the exposed or unexposed polymer, thereby leaving the pattern in relief on the substrate surface. The substrate is then patterned either by etching a relief structure in it, by chemically doping the patterned areas, or by depositing a material into the interstices of the polymer relief pattern

7.1.3 Conformable-Photomask Contact Printing

The principle of conformable-photomask lithography is to eliminate the gap between a mask and a substrate by making the former conform to the latter, and deform around any dust particles or surface asperities [7.12–14]. This, in turn, leads to the elimination of the deleterious effects of optical diffraction which normally occurs in the gap. A conformable photomask consists of a pattern in thin film chromium, Fe_2O_3, or other ultraviolet attenuator on 0211 type glass, 0.2 mm thick. With conformable photomasks in intimate contact, extremely high resolution patterns can be exposed in AZ1350 photoresists, as illustrated in Fig. 7.2. A number of other advantages are also realized through the use of conformable photomasks, as summarized in Table 7.1.

The third advantage refers to the fact that good dimensional control is possible even if narrow linewidths are not required. In surface wave resonators, for example, the ratio of linewidth to period must be repeatable to about 5% to achieve consistent performance. For linewidths down to about 2 µm this sort of control is possible using conformable-photomask lithography.

Of particular significance for surface acoustic wave devices is the large area capability, and the apparent absence of significant distortions (<1 part in 10^5). The burdens of providing dimensional control and freedom from distortion thus devolve to the pattern generation technique used to make the mask.

In the early work with conformable photomasks, intimate contact between mask and substrate was achieved by the application of mechanical pressure

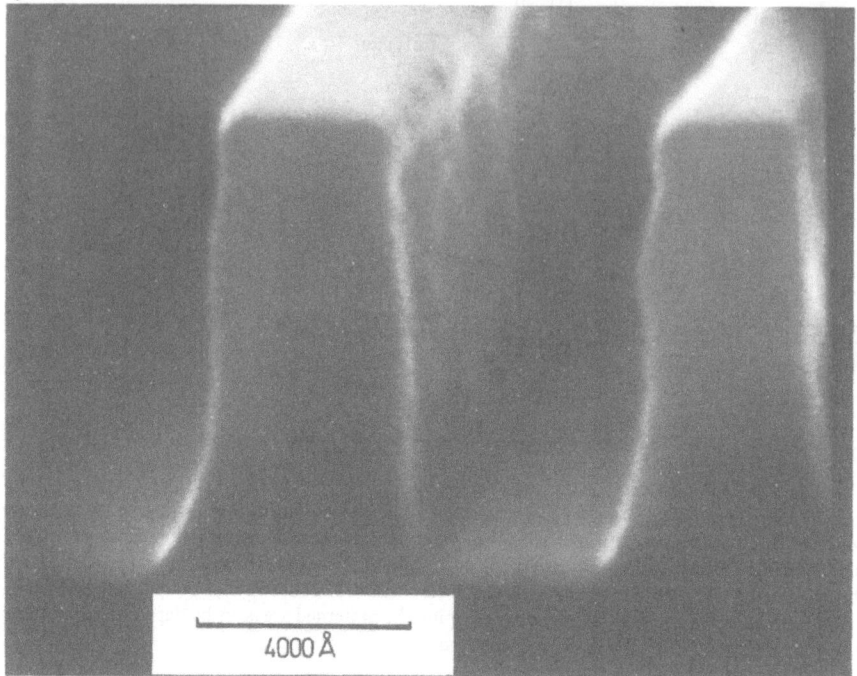

Fig. 7.2. Scanning electron micrograph of the cross section of a grating pattern exposed in AZ1350H photoresist using a high-pressure mercury arc lamp and the conformable photomask-intimate contact technique. The photoresist is 9800 Å thick. Because of the angles used in microscopy, vertical dimensions are foreshortened relative to lateral dimensions. The scale refers to lateral dimensions only. The photomask consisted of five slits in an 800 Å thick chromium film on 0211 glass. It was made by scanning electron beam lithography and chemical etching of the chromium, which caused irregularities along the edges of the slits. These small irregular features are replicated in the photoresist. The smaller space between two of the slits was in the original mask, and was caused by an error in the scanning electron beam pattern generator (micrograph by *R. Eager*)

[7.12]. This method is suitable for areas up to about 5 mm diameter but is clearly unsuitable for modern surface acoustic wave devices. Figures 7.3 and 7.4 illustrate a type of vacuum frame developed for large area devices. This type of frame permits intimate contact with a photomask to be achieved over an entire front surface of a substrate. Although Figs. 7.3 and 7.4 depict a rectangular substrate, the basic design has also been adapted to circular substrates. Substrate areas ranging in size from 1 × 1 cm to 2 × 15 cm have been accommodated with ease. Substrate thicknesses have ranged from 0.25 to 3.5 mm.

The thickness of the top plate in Figs. 7.3 and 7.4 is chosen so that when a substrate is held down on the base plate by means of vacuum 1, its top surface is 0.05–0.1 mm below the top surface of the top plate. This prevents contact between a photomask and a substrate, while the former is slid over the top plate for alignment with the substrate. Once a photomask has been aligned, it is

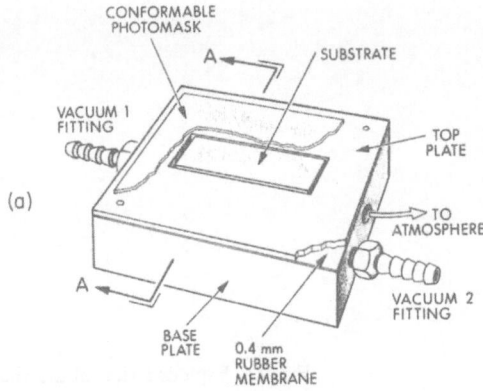

(a)

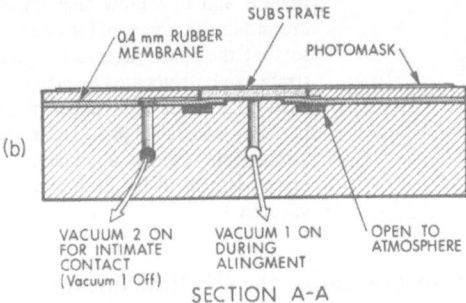

(b)

SECTION A-A

Fig. 7.3. (a) Vacuum frame which permits a conformable photomask to be aligned with respect to a substrate and then pulled into intimate contact. (b) Cross section of the vacuum frame. To obtain intimate contact, vacuum 2 is pulled. When vacuum 1 is released, the substrate is free to move up to the conformable photomask. The flexure of the photomask is thus removed and intimate contact can be achieved over the entire substrate surface

Table 7.1. Conformable photomask—Intimate contact printing

Advantages

1. Minimum linewidths $\lesssim 0.4\,\mu m$
2. Pattern area $\sim 100\,cm^2$ or greater
3. Linewidth controllable to $\sim 0.1\mu m$
4. Vertical or slightly undercut profiles – ideal for metal liftoff process and ion bombardment etching
5. Dust defects minimized
6. Long life of masks
7. Low distortion (<1 part in 10^5)
8. Wide exposure latitude. Threshold can be exceeded by a factor of 3 or more
9. High index of refraction and nonlinear behavior of photoresist suppress diffraction broadening
10. Processing independent of pattern resolution
11. Low cost

Disadvantages

1. Intimate contact required. Substrates must be smooth
2. High level of operator skill required, technique not suitable for mass production at present
3. Multilevel masking precision $\sim 1\,\mu m$

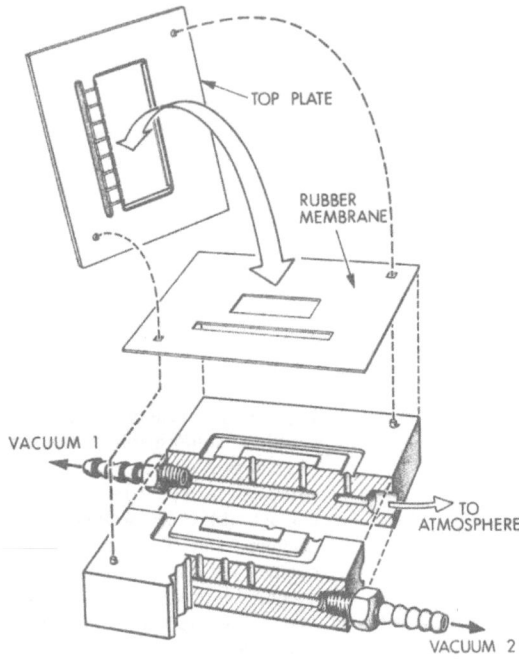

Fig. 7.4. Exploded view of the vacuum frame. The top plate has a series of grooves milled in its bottom surface and a shallow step milled around the perimeter of the opening so that the space around the substrate can be evacuated by means of vacuum 2. Under the rubber membrane an opening is provided to the atmosphere. During alignment the crystal is held to the base plate by vacuum 1

pulled into contact with the substrate by evacuating the substrate opening. As illustrated in Fig. 7.3, the top plate has grooves milled in its bottom surface and a shallow step milled around the perimeter of the substrate opening to permit evacuation of this space through the vacuum 2 fitting. The rubber membrane under the substrate seals its bottom surface. Vacuum 1, which holds the substrate to the base plate, is then released, and the substrate rises upward until the photomask is flat and in intimate contact over the entire substrate surface. Actually, with rectangular substrates there is frequently a buildup of photoresist at the edges and corners which can prevent intimate contact in these areas. However, with positive type photoresists, it is possible to remove the photoresist buildup by exposure and development prior to use of the vacuum frame.

Reference [7.14] describes a mask alignment apparatus suitable for use with conformable photomasks and the above vacuum frame.

7.1.4 Holographic Recording

Holographic techniques are ideal for exposing gratings in photoresist films. The grating is formed by interfering two laser beams. In a typical configuration, a beam from a short-wavelength visible or ultraviolet laser is passed through a spatial filter and beam expander, and split into two plane wave beams of roughly equal intensity. A pair of plane mirrors is then arranged so as to bring the two beams together at a specified angle of incidence on a photoresist-coated substrate. The grating period is given by $d = \lambda/2 \sin\alpha$, where λ is the wavelength, and α is half the angle between the two beams.

Holography is capable of producing accurate gratings with distortions less than 1 part in 10^{10}. Periodicities of 1100 Å have been produced using relatively simple fixtures and a HeCd laser ($\lambda = 3250$ Å) [7.15]. Gratings of 835 Å period have been made using an upconverted Nd:YAG laser [7.16]. For surface wave resonators operating at frequencies above about 1 GHz, holography is the preferred method for producing high Q gratings.

It is difficult to align holographically produced gratings with respect to other structures, such as transducers or waveguides, if the exposure is done directly on a device substrate. However, if the holographic technique is used instead to make a mask, the alignment problem is more tractable. Gratings of 7000 Å period have been replicated by conformable-photomask contact printing [7.13], and gratings of 3600 Å period have been replicated by x-ray lithography [7.17–19].

7.1.5 Scanning Electron Beam Lithography

In scanning electron beam lithography, a finely focused electron beam is scanned over a substrate coated with a radiation-sensitive polymer film in accordance with a desired pattern [7.20–23]. Figure 7.5 illustrates a simple system made from a modified scanning electron microscope. A system specifically tailored for making accurate, low-distortion, multielement arrays for surface acoustic wave devices is described in [7.22, 23]. Scanning electron beam lithography can be used either to expose patterns directly on device substrates, or to make masks whose patterns can then be transferred onto device substrates by photolithography, electron projection, or x-ray lithography. The mask-making alternative is preferred for fabricating surface acoustic wave devices.

Table 7.2 summarizes the advantages and disadvantages of scanning electron beam lithography. Again, the reader is referred to [7.1, 2] for more details.

Although an electron beam can be focused to less than 100 Å diameter, and linewidths less than 1000 Å can be achieved in certain cases where the detrimental influence of electron backscattering can be minimized, electron beams do not in themselves provide accuracy, precision, or freedom from distortion. Within a single scan field, these aspects are derived from local benchmarks which in turn are produced by optical means. A large-area pattern, which of necessity is made up of a montage of many scan fields, derives its accuracy, precision, and integrity from laser interferometers, since these are employed either directly in the electron beam pattern generator [7.24, 25], or indirectly in an optical projection system to produce a matrix of local benchmarks. In many cases it is undesirable to distribute benchmarks on a device substrate. This is seldom a problem in mask making.

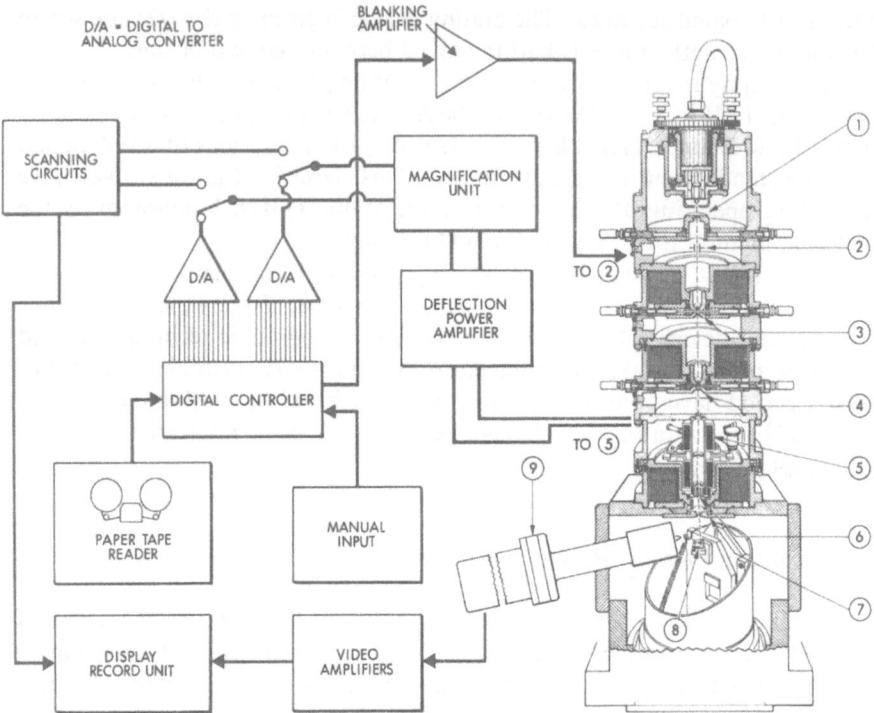

Fig. 7.5. A scanning electron microscope modified by the addition of an external digital controller so that scanning electron beam lithography can also be carried out. (*1*) electron source, (*2*) electrostatic beam blanking plates, (*3*) first lens, (*4*) second lens, (*5*) double deflection coils, (*6*) electrostatic astigmatism correctors, (*7*) final lens, (*8*) sample mounting platform, (*9*) photomultiplier. The deflection coils scan the beam over the sample in a pattern determined by the voltage functions from the digital controller and *D/A* converters. The physical size of the pattern (the field size) is controlled by the magnification unit which sets the voltage levels. The rate of scanning is controlled by a clock which drives the digital controller. Pattern information can be put in either manually or by paper tape

Table 7.2. Scanning electron beam lithography

Advantages

1. Resolution (~ 1000 Å)
2. Registration via electron microscopy
3. Large depth of focus
4. Programmability

Disadvantages

1. *Limited field of view* ($\sim 1 \times 1$ mm)
2. Electron scattering
3. Slow speed
4. Electrical charging
5. High cost and complexity

Pattern distortion goes up rapidly with the size of a scan field, and this argues in favor of working with relatively small fields ($< 1 \times 1$ mm). In mask making, small field of view is a minor problem.

Recent studies of electron backscattering have pointed out important limitations on the resolution and geometry of relief structures that can be fabricated in polymer films [7.26–30]. In mask making, substrate backscattering effects can be reduced somewhat by using materials of low atomic number. The effects of electron backscattering are far more serious in low contrast-high sensitivity polymers than in high contrast-low sensitivity polymers such as polymethyl methacrylate. Although the slow speed of PMMA is objectionable in direct device farbrication, it is usually acceptable in mask making.

The general problem of slow speed in scanning electron beam lithography is serious only if patterns are exposed directly on device substrates. In mask making, the economics are quite different, and one can afford to work with slow speeds, small fields, and hence less distortion.

7.1.6 Projection Electron Lithography

Two types of electron projection instruments have been developed:

1) systems similar to a conventional transmission electron microscope (CEM) in which an electron opaque mask is imaged on a substrate [7.31–34];

2) systems similar to an image tube, in which a pattern on a photoemitter is projected onto a substrate [7.35, 36].

In the past, Type 1 instruments have produced some very high resolution patterns, but over limited areas and with distortions that could not be tolerated in surface wave device applications. Recently, a system has been developed at IBM which significantly reduces distortion [7.32]. Whether such a system would be useful for fabricating surface acoustic wave devices is not known at this time. Problems caused by electron backscattering from the substrate, by electrical charging, and by stray fields can be anticipated as well as difficulties in making the required masks.

Westerberg et al. have developed a Type 1 system in which an electrostatic slit lens images an array of apertures to produce a grating [7.33, 34]. Interdigital electrode surface acoustic wave transducers with fundamental frequencies up to 2.8 GHz have been made with this system, which has the virtues of speed, simplicity, and low cost.

Considerable effort has been spent on developing Type 2 image tube systems [7.35, 36]. The distortion, precision, and orientation of the projected pattern depend upon the uniformity and stability of electric and magnetic fields, and this leads to serious technological and contamination control problems. For example, fringing fields at the edges of a sample cause pattern distortion. It is unlikely that Type 2 systems will be useful in fabricating surface wave devices.

7.1.7 X-Ray Lithography

X-ray lithography [7.17–19, 37–43] is a shadow printing technique, similar in principle to photolithographic contact printing. The wavelengths used range from ~4 to 44Å, and as a result diffraction effects are reduced considerably. Figure 7.6 illustrates the technique. The mask consists of an absorber pattern on a thin membrane which is semitransparent to x-rays. Since the x-rays are not collimated, there is penumbral blurring, as illustrated in the inset. The x-rays travel in straight lines with negligible diffraction, so this is a purely geometric effect given by

$$\delta = s(d/D)$$

where the symbols are defined in Fig. 7.6. The exact choice of these parameters is based on trade-offs among desired pattern resolution, desired mask-sample gap, and exposure time, which is proportional to $(D/d)^2$. *Sullivan* and *McCoy* [7.42] have described an x-ray lithographic apparatus designed specifically for exposing surface acoustic wave devices.

If an x-ray wavelength of 8.34 Å (K line of aluminum source) is used, the x-ray mask is generally made of a membrane of silicon (~3 μm thick) or mylar (~6 μm thick). On this is held the absorber pattern, which is usually about a 4000 Å thickness of gold. For softer x-rays, such as the 13.3 Å copper L radiation, Al_2O_3 films [7.42] and silicon nitride membranes [7.43] are preferred, and the gold absorber need be only about 1000 Å thick.

Absorber patterns for x-ray masks have been produced by scanning electron beam lithography, photolithography, holography, and x-ray lithography, in conjunction with both metal-liftoff and ion-beam etching. As mentioned above, one could make masks by a combination of scanning electron beam lithography and holography, and thereby achieve device patterns (e.g., high-frequency resonators) not realizable by other means. Figure 7.7 illustrates the high resolution capability of x-ray lithography.

Table 7.3 summarizes the advantages and disadvantages of x-ray lithography. Probably the most important aspect of x-ray lithography, from the point of view of surface acoustic wave devices, is that the radiation propagates in straight lines, undeviated by stray fields or contamination. A faithful replica of the mask is thereby assured. X-ray lithography may be an effective means of mass producing surface acoustic wave devices in the future.

7.2 Techniques for Patterning the Substrate

After the polymer film has been exposed and a relief pattern developed, the next step is to transfer this pattern onto the substrate itself. As indicated in the introductory remarks of this chapter, and summarized in the caption of Fig. 7.1, techniques include etching, doping, and depositing. These techniques are discussed and compared in this section.

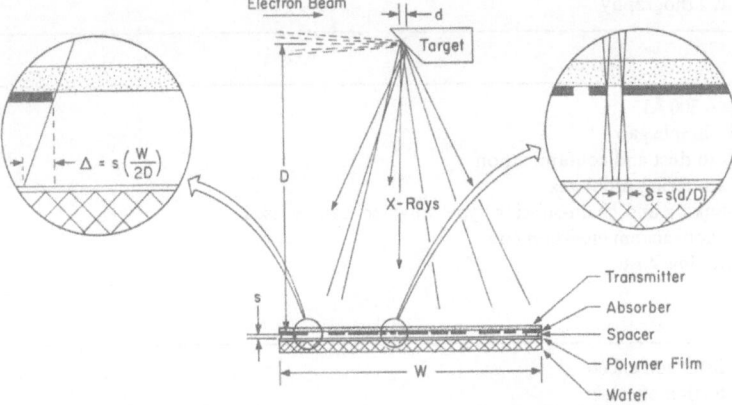

Fig. 7.6. Schematic diagram of a soft x-ray (~ 10 Å) lithography system. The shadow of the mask is recorded in the polymer film. The right inset illustrates the penumbral effect which arises because of the finite angular size of the source. The magnitude of δ can be made arbitrarily small either by reducing the mask-sample gap s or the ratio d/D. The left inset illustrates geometric distortion, which can either be compensated in the mask or reduced by decreasing the ratio s/D

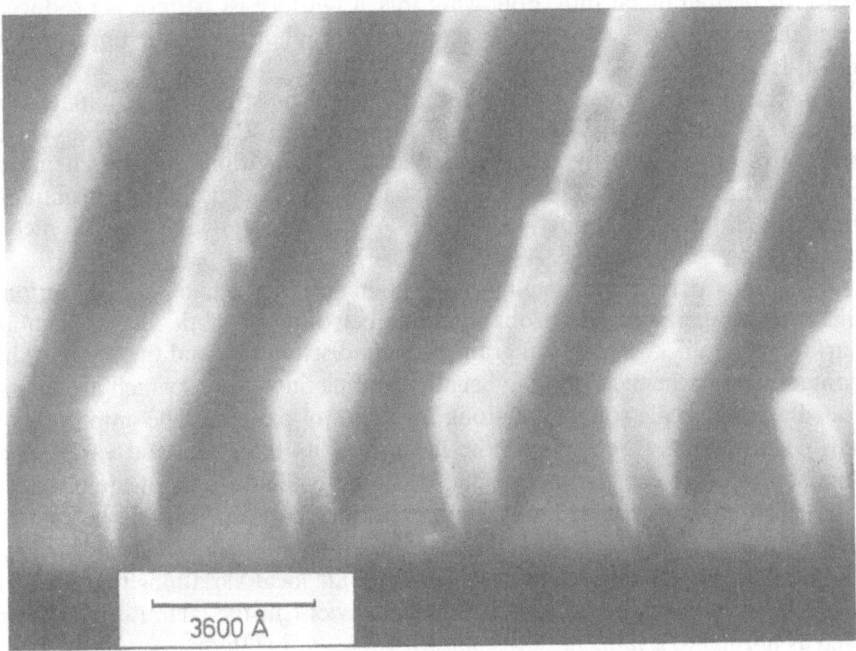

Fig. 7.7. Scanning electron micrograph of the cross section of a grating pattern exposed in PMMA by x-ray lithography using the copper L radiation at 13.3 Å. The spatial period of the grating is 3600 Å, somewhat less than half that in Fig. 7.2. The mask was made by holographic exposure and ion beam etching (micrograph by *P. DeGraff*)

Table 7.3. X-ray lithography

Advantages

1. Resolution (~ 500 Å)
2. Finite mask-sample gap
3. Insensitivity to dust and contamination
4. Absence of scattering problems
5. Both cross-linking and chain-scission type polymers can be used
6. Exposure in nonvacuum environments
7. Simplicity and low cost

Disadvantages

1. Geometric distortion effect
2. Possible distortion of mask

7.2.1 Aqueous Chemical Etching

Aqueous chemical etching has been used for decades to define metal electrode patterns on substrate surfaces, and is still a very widely used technique. To produce metallic patterns for surface wave devices, the substrate is first coated with the desired metal film. Following this, a photoresist pattern is produced over the metal. After appropriate baking of the photoresist, the unprotected metal is etched.

In chemical etching of metal films, an etchant must be chosen that attacks the metal but not the substrate or the photoresist. Some surface wave substrates are chemically reactive and impose limitations on the chemical etchants that can be used. Table 7.4 lists a few etchants useful on chromium, titanium, aluminum, copper, and gold. A more thorough listing of metal etchants can be found in [7.44-46].

The rate of chemical etching is a function of etchant concentration, agitation, temperature, and probably a number of more subtle factors such as surface oxide thickness, crystal grain size and distribution, and the presence of contaminants. Chemical etching usually exhibits fluctuations which are manifested as small-scale roughness along the edges of patterns. The amplitude of this edge roughness depends on the metal and its thickness, and varies as a function of other etching parameters. It is usually less than or of the order of 1 µm. This edge roughness limits the linewidth of patterns that can be etched, and reduces the yield for linewidths near this limit. On surface wave substrates, chemical etching frequently exhibits troublesome local variations that lead to shorts between electrodes. Chemical etchants work in the lateral direction as well as normal to a film, and thus pattern edges shift with time during etching. If the adhesion of the photoresist along an edge is less than perfect, a rapid loss of edge acuity, known as undercutting, occurs.

The main advantages and disadvantages of chemical etching are summarized in Table 7.5. Item 3 refers to the fact that the adhesion of the metal film

Table 7.4. Etchants for various metals

Metal	Etchant
Chromium	a) 165 g of ceric ammonium nitrate, 43 ml of conc. (70 %) $HClO_4$, add water to make 1000 ml b) Bell and Howell HC 300 c) 1 part C25A, 1 part C25B from Film Microelectronics, Inc., Burlington, Mass.[a]
Titanium	a) 1 part HF, 20 parts water
Aluminum	a) 80 ml phosphoric acid, 10 ml water, 5 ml nitric acid b) Techni Strip Au 60 g, add water to make 1000 ml; from Technic Inc., Cranston, Rhode Island[b,c] c) 1 part conc. HCl, 4 parts water d) NaOH in water is an effective etch for aluminum, but is not compatible with AZ1350 type photoresists
Copper	a) 100 g of $FeCl_3$ in 1000 ml of water[b] b) Alki etch, 2 parts A, 1 part B, 1 part water, from Philip Hunt Chemical Corp.[d]
Gold	a) Aqua regia; 1 part conc. HCl and 3 parts conc. HNO_3 b) 10 g iodine, 6 g potassium iodide, 40 ml water[d] c) C35 from Film Microelectronics, Inc., Burlington, Mass.[a,d] d) Techni Strip Au 60 g, add water to make 1000 ml; from Technic Inc., Cranston, Rhode Island[b,c,d]

[a]   Compatible with sputtered ZnO.
[b]   Solution should be filtered to remove particles.
[c]   Compatible with PMMA film (electron or x-ray resist).
[d]   Compatible with $Bi_{12}GeO_{20}$.

can be enhanced by heating of the substrate during deposition, and by use of sputter deposition techniques. Neither heating nor sputter deposition can be readily employed in conjunction with the liftoff process.

In summary, chemical etching is preferred for the fabrication of simple metal patterns on relatively inert substrates such as quartz and $LiNbO_3$ when linewidths are $\gtrsim 5\,\mu m$. The method is particularly suitable if a large number of identical devices are to be made, since this permits optimization of process parameters. Under such circumstances, considerable economy can be realized, and thus the method is compatible with mass production of devices.

7.2.2 Plasma Etching

This technique is a type of chemical etching in which an rf plasma of an etchant gas is used in place of an aqueous etchant solution [7.47, 48]. In the silicon industry the technique is widely used for the etching of Si, SiO_2, and Si_3N_4. In general, undercutting is less marked than with aqueous etching. Recently, plasma etching was shown to be an effective means of producing surface relief gratings for quartz resonators [7.49]. One disadvantage of this approach was an apparant roughening of the surface in the etched areas.

Table 7.5. Chemical etching of metal electrode patterns

Advantages

1. Simplicity and low cost
2. Compatible with mass production
3. Metal deposition methods can be used which produce highly adherent films (e.g., sputtering, heating of substrates)

Disadvantages

1. Difficult to optimize process parameters when a small number of substrates are to be processed
2. Chemical etchants must be compatible with substrates
3. Post-baking can produce dimensional changes
4. Loss of resolution due to ragged edges, lateral etching, and a tendency to undercut
5. Low yield for linewidths near 1 μm and for repetitive type patterns
6. Shorts are the dominant fault

7.2.3 Ion Bombardment Etching

When ions or neutral atoms bombard a solid surface, momentum transfer causes the ejection of atoms from the surface. Thus ion bombardment provides a means of etching any solid material. To etch insulating materials, however, provisions must be made to prevent the buildup of charges that would tend to repel bombarding ions. Charge buildup can be avoided by using either rf or neutralized ion beam techniques. Ion bombardment etching is generally a high yield, high resolution, process and is not subject to the disadvantages of chemical etching listed in Table 7.5.

In rf sputter etching, the substrate is mounted on the rf electrode of a conventional rf sputtering system. Argon gas is admitted into the system to a pressure of about 1×10^{-2} Torr, and the input power is turned on to sustain an rf discharge. A dark space forms around the rf electrode and ions accelerated across this dark space impinge on the substrate, thereby sputtering the unmasked substrate as well as the etching mask itself. A practical advantage of rf sputter etching is that the equipment is rather widely available. It is easily employed to produce high resolution surface electrodes [7.50] or to etch uniform depth surface relief structures. The major disadvantages of rf sputter etching are: 1) material removed sometimes redeposits on the substrate leading to contamination problems [7.51] and somewhat unpredictable etching rates; 2) excessive substrate heating is sometimes observed; and 3) the substrate is part of the rf electrode. The problem of heating can be overcome by adequate heat sinking, and etching in short bursts followed by long cooling times. Alternatively, if metal patterns are used as the etching mask, substrate heating is generally not a problem.

In ion beam etching [7.1, 52–56], the substrate is not part of the ion acceleration system and thus it can be located in a field-free high vacuum region. Etching rates are readily controlled and reproduced. Reactive gases

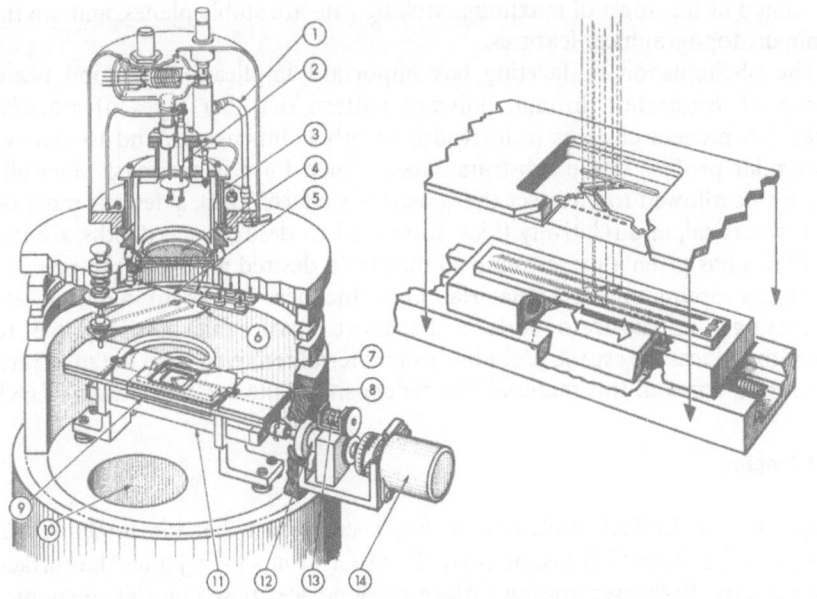

Fig. 7.8. Ion beam etching system utilizing a Kaufmann type source. A substrate (8) with a grating pattern in photoresist is shown mounted on a water-cooled platform (9). This configuration is used to produce a variable depth ("weighted") grating by controlling the amount of time a portion of the grating spends under the chevron aperture (7), which defines the area of etching. Other items indicated are: (1) argon gas inlet, (2) arc filament, (3) magnet, (4) ion beam extraction grids, (5) neutralizing filament, (6) shutter, (10) pumping port, (11) micrometer-slide assembly, (12) rotary-motion high vacuum feedthrough, (13) position indicator, and (14) stepping motor. The beam diameter is 10 cm, and over the central 5 cm diameter the flux is approximately uniform. The accelerating voltage is continuously variable from 200 to 2000 V. The neutralizing filament injects thermal velocity electrons to neutralize the ion beam space charge

such as oxygen can be introduced to alter the relative etching rates of certain materials [7.57], and the energy of bombarding ions can be selected. Probably the most important advantage of ion beam etching relative to rf sputter etching is that it permits one to vary the depth of etching as a function of position on a substrate. In the fabrication of reflective array surface wave devices [7.58–64], this capability is exploited in order to shape a device's amplitude response. Figure 7.8 illustrates an ion beam etching system designed for this application.

The rate at which material is removed from a substrate surface by ion bombardment is a minimum at normal incidence, reaches a maximum between 40 and 60°, and goes to zero at grazing incidence. Because of this, the topography of a surface changes during ion bombardment, and well-defined facets are observed. Theories on the evolution of surface topographies under ion bombardment [7.65, 66] predict that with sufficient etching time, any surface contour will become a horizontal plane, and that prior to the occurrence of this final steady state, facets which are parallel to, perpendicular to, or inclined at the angle of maximum etching rate are stable planes, and are the dominant topographical features.

The phenomenon of faceting has important implications for ion beam etching of substrates through polymer pattern or other types of material masks. To prevent changes in linewidth or other dimensions, and to achieve rectangular profiles in the substrate, facets which form in the mask sidewalls must not be allowed to intersect the substrate surface. Thus, sidewalls must be initially vertical, and relatively thick compared to desired etch depths. *Cantagrel* [7.67] has given a prescription so that for a desired groove depth one can choose an optimum mask material and thickness so as to avoid facets intersecting the substrate surface. Excessive mask thickness relative to linewidth reduces the escape probability of material removed from the substrate. As a result, some of this material can redeposit on the mask sidewalls [7.68].

7.2.4 Doping

On quartz and $LiNbO_3$ substrates it has been shown that ion implantation [7.69] and the thermal diffusion of Ni, Ti, or Cr atoms [7.70] alter the surface wave velocity. Reflective grating surface wave devices based on this phenomenon have been fabricated and tested [7.70, 71]. There does not appear to be any increase in surface wave propagation loss as a result of such doping. On $LiNbO_3$, implantation at dosages in the range 3×10^{15} to 1.5×10^{16} He ions cm^{-2} leads to a velocity decrease, whereas on quartz a velocity increase and density decrease have been observed [7.71]. Diffusion of Ti, Ni, and Cr into $LiNbO_3$ increases the surface wave velocity [7.70]. Increases of 1.4% have been observed and it is speculated that increases as high as 5% are possible [7.70]. Similar comments are made in Section 5.4.1 in connection with in-diffused waveguides.

In the case of ion implantation, a photoresist pattern can serve as the implantation mask. The diffusion of a metal into a substrate is accomplished by first producing, by liftoff or chemical etching, the desired pattern in a thin film of the metal. Following this, the metal is diffused into the substrate surface, typically at a temperature of 1000° C [7.70].

7.2.5 The Liftoff Technique

The liftoff technique, illustrated on the bottom right of Fig. 7.1, consists of depositing a material over a substrate which has on it a polymer film pattern, and then dissolving the polymer. A pattern in the deposited material is left behind on the substrate. It is important that there be little or no continuity between the material deposited on the substrate surface and that deposited on top of the polymer. If such continuity exists, either the "liftoff" will not take place, or relatively violent means such as ultrasonic agitation will have to be used. To prevent continuity, the polymer film sidewalls must be vertical or slightly undercut, and the deposited material must arrive at the substrate at near normal incidence.

In photolithography, vertical and slightly undercut sidewalls can be obtained using intimate contact printing, as illustrated in Fig. 7.2. In combination with electron beam evaporation from a small, distant source, liftoff is 100% reliable with such photoresist profiles, even for metal films up to 2/3 as thick as the photoresist [7.1, 12, 13]. In electron beam lithography, undercut profiles are readily obtained in PMMA. The liftoff technique works reliably, and the requirement of normal incidence deposition can be relaxed somewhat [7.72]. Sidewall profiles are not truly vertical in x-ray lithography because of penumbral exposure. Nevertheless, the sidewalls can be made as steep as desired, and the polymer can be made much thicker than 1 μm. With proper attention to the details, the liftoff technique works well.

A metal pattern can be formed in the interstices of a polymer pattern by electroplating if the substrate is conducting, or coated with a conducting film. High resolution patterns with large thickness-to-linewidth ratios have been obtained in this way [7.41, 73].

One problem area in the liftoff technique is the adhesion of the deposited material to the substrate. Small traces of contamination degrade adhesion and must be removed from those areas of the substrate surface not covered by the polymer pattern. Immersion in an oxygen plasma for a time sufficient to remove about 100 Å of the polymer thickness is the preferred method because it is harmless to most substrates, yet it is specifically effective in removing residual polymer or other organic contamination.

Table 7.6 summarizes the advantages and disadvantages of the liftoff technique.

7.3 Conclusions

This chapter has summarized and updated the findings of two earlier reviews of fabrication techniques for surface acoustic wave devices and added some new material on doping and conformable-photomask lithography.

Of the various techniques available for exposing patterns in polymer films (i.e., optical projection, conventional contact printing, conformable-photomask lithography, holographic recording, scanning electron beam lithography, projection electron lithography, and x-ray lithography), conformable-photomask lithography and x-ray lithography appear to be the ones most compatible with short- and long-term requirements for surface wave devices. The advantages and disadvantages of these two methods are summarized in Tables 7.1 and 7.3, respectively.

Aqueous and plasma etching techniques are most suitable for mass production of moderate resolution devices, whereas liftoff is more suitable when only a few devices are fabricated or when high resolution is required.

Table 7.6. Liftoff

Advantages

1. High resolution and precise dimensional control ($\sim$ few hundred Å)
2. Compatible with chemically reactive substrates, and substrates with rough surfaces
3. Compatible with any deposited metal or combination of metals, and with nonmetals
4. Electrical shorts improbable
5. Processing is the same regardless of whether one or several hundred devices aré handled
6. Photolithographic processing is independent of pattern resolution
7. Yields approaching 100% are possible

Disadvantages

1. Special care required to achieve vertical sidewalls in photoresist (requires intimate contact photolithography)
2. Substrate cannot be heated during deposition
3. Deposited material must arrive at normal incidence to the substrate
4. Substrates must be scrupulously clean to ensure good film adhesion

Chemical doping by diffusion and ion implantation have only recently been applied to surface wave devices. The diffusion technique may offer a cost advantage over ion beam etching in the fabrication of resonators and other reflective array devices.

References

7.1 H.I.Smith: Proc. IEEE **62**, 1361—1687 (1974)
7.2 H.I.Smith: *Surface Wave Devices*, ed. by H. Matthews (John Wiley and Sons, New York 1977) Chap. 4, pp. 165—217
7.3 F.H.Dill: IEEE Trans. ED-**22**, 440—444 (1975)
7.4 F.H.Dill, W.P.Hornberger, P.S.Hauge, J.M.Shaw: IEEE Trans. ED-**22**, 445—452 (1975)
7.5 L.F.Thompson, M.J.Bowden: J. Electrochem. Soc. **120**, 1722—1726 (1973)
7.6 L.F.Thompson, J.P.Ballantyne, E.D.Feit: J. Vac. Sci. Tech. **12**, 1280 (1975)
7.7 M.J.Bowden, L.F.Thompson, J.P.Ballantyne: J. Vac. Sci. Tech. **12**, 1294 (1975)
7.8 H.J.Schuetze, K.E.Hennings: SCP and Solid State Techn. **9**, 31—35 (1966)
7.9 S.Middelhoek: IBM J. Res. Develop. **14**, 117—124 (1970)
7.10 D.T.Bell, D.W.Mellon: Proc. Ultrasonics Symp., ed. by J. deKlerk (Monterey, Calif., Nov. 5—7 1973), IEEE Inc., Cat. No. 73 CHO 807—85U, 486—489
7.11 A.R.Janus, H.M.Gerard, L.Dyal: IEEE Proc. Ultrasonics Symp., ed. by J. deKlerk, Sept. 22—24, 1975, Los Angeles, Calif., IEEE Inc., Cat. No. 75 CHO 994—4SU
7.12 H.I.Smith, F.J.Bachner, N.Efremow: J. Electrochem. Soc. **118**, 821—825 (1971)
7.13 H.I.Smith, N.Efremow, P.L.Kelley: J. Electrochem. Soc. **121**, 1503—1506 (1974)
7.14 J.Melngailis, H.I.Smith, N.Efremow: IEEE Trans. ED-**22**, 496—498 (1975)
7.15 C.V.Shank, R.V.Schmidt: Appl. Phys. Lett. **23**, 154—155 (1973)
7.16 G.Bjorklund, W.W.Hansen: Lab., Stanford University, Stanford, California (private communication)
7.17 S.E.Bernacki, H.I.Smith: Proc. 6th Internat. Conf. on Electron and Ion Beam Sci. and Techn., ed. by R. Bakish, May 13—16, 1974, San Francisco. (The Electrochemical Society, Inc.)

7.18 S. E. Bernacki, H. I. Smith: IEEE Trans. ED-**22**, 421—428 (1975)

7.19 D. Flanders, H. I. Smith: Proc. 7th Internat. Conf. on Electron and Ion Beam Sci. and Techn., May 2—7, 1976, Washington, D.C.

7.20 G. R. Brewer: IEEE Spectrum **8**, 23—37 (1971)

7.21 A. N. Broers, M. Hatzakis: Sci. Amer. **227**, 34—44 (1972)

7.22 F. S. Ozdemir, W. E. Perkins, R. Yim, E. D. Wolf: J. Vac. Sci. Tech. **10**, 1008—1011 (1973)

7.23 E. D. Wolf, F. S. Ozdemir, R. D. Weglein: Proc. Ultrasonic Symp., ed. by J. deKlerk, Nov. 5—7, 1973, Monterey, Calif. (IEEE Inc., New York) pp. 510—516

7.24 O. Cahen, R. Sigelle, J. Trotel: Proc. 5th Internat. Conf. on Electron and Ion Beam Sci. and Techn., ed. by R. Bakish, May 7—11, 1972, Houston, Texas. (The Electrochemical Society, Inc.) pp. 92—101

7.25 D. R. Herriott, P. J. Collier, D. S. Alles, J. W. Stafford: IEEE Trans. ED-**22**, 385—392 (1975)

7.26 R. J. Hawryluk, A. M. Hawryluk, H. I. Smith: J. Appl. Phys. **45**, 2551—2566 (1974)

7.27 R. J. Hawryluk: Ph. D. thesis, Mass. Inst. of Tech., May 1974; also published as Lincoln Laboratory Technical Report, TR511, Nov. 1974

7.28 R. J. Hawryluk, A. Soares, H. I. Smith, A. M. Hawryluk: Proc. 6th Internat. Conf. on Electron and Ion Beam Sci. and Techn., ed. by R. Bakish, May 13—16, 1974, San Francisco. (The Electrochemical Society, Inc.)

7.29 D. F. Kyser, K. Murata: Proc. 6th Internat. Conf. on Electron and Ion Beam Sci. and Techn., ed. by R. Bakish, May 13—16, 1974, San Francisco. (The Electrochemical Society, Inc.)

7.30 T. H. P. Chang: J. Vac. Sci. Tech. **12**, 1271—1275 (1975)

7.31 H. Koops: J. Vac. Sci. Tech. **10**, 909—912 (1973)

7.32 M. B. Heritage, J. Vac. Sci. Tech. **12**, 1135—1140 (1975)

7.33 R. E. Lee, E. R. Westerberg, A. J. Bahr: Proc. Ultrasonics Symp., ed. by J. deKlerk, Nov. 5—7, 1973, Monterey, Calif. (IEEE Inc., New York) pp. 517—521

7.34 L. N. Heynick, E. R. Westerberg, C. C. Hartelius, R. E. Lee: IEEE Trans. ED-**22**, 399—409 (1975)

7.35 B. Fay, D. B. Ostrowsky, A. M. Roy, J. Trotel: Opt. Commun. **9**, 424—426 (1973)

7.36 G. A. Wardly: IEEE Trans. ED-**22**, 414—417 (1975)

7.37 D. L. Spears, H. I. Smith: Elec. Lett. **8**, 102—104 (1972)

7.38 H. I. Smith, D. L. Spears, S. E. Bernacki: J. Vac. Sci. Tech. **10**, 913—917 (1973)

7.39 D. Maydan, G. A. Coquin, J. R. Maldonado, S. Somekh, D. Y. Lou, G. N. Taylor: IEEE Trans. ED-**22**, 429—439 (1975)

7.40 J. S. Greeneich: IEEE Trans. ED-**22**, 434—439 (1975)

7.41 R. Feder, E. Spiller, J. Topalion: J. Vac. Sci. Tech. **12**, 1332—1335 (1975)

7.42 P. A. Sullivan, J. H. McCoy: J. Vac. Sci. Tech. **12**, 1325—1328 (1975)

7.43 R. Feder, E. Spiller, J. Topalion, M. Hatzakis: Proc. 7th Internat. Conf. on Electron and Ion Beam Sci. and Techn., May 2—7, 1976, Washington, D.C.

7.44 L. I. Maissel, R. Glang: *Handbook of Thin Film Technology* (McGraw-Hill, New York 1970)

7.45 R. W. Berry, P. M. Hall, M. T. Harris: *Thin Film Technology* (Van Nostrand Co., Inc., Princeton, N.J., 1968)

7.46 *Handbook of Chemistry and Physics*, ed. by R. C. Weast (CRC Press, Cleveland, Ohio 1974)

7.47 J. R. Hollahan, A. T. Bell: *Techniques and Applications of Plasma Chemistry* (John Wiley & Sons, New York 1974)

7.48 H. Abe, Y. Sonobe, T. Enomoto: Jap. J. Appl. Phys. **12**, 154—155 (1973)

7.49 S. P. Miller, R. E. Stigall, W. R. Shreve: Proc. IEEE Ultrasonics Symp., ed. by J. deKlerk, Sept. 22—24, 1975, IEEE Inc., Cat. No. 75 CHO 994-4SU

7.50 H. I. Smith: Rev. Sci. Instr. **40**, 729—730 (1969)

7.51 L. I. Maissel, C. L. Standley, L. V. Gregor: IBM J. Res. Devel. **16**, 67—70 (1972)

7.52 H. I. Smith, R. C. Williamson, W. T. Brogan: Proc. IEEE Ultrasonics Symp., ed. by J. deKlerk, October 4—7, 1972, Boston, Mass. (IEEE Inc., New York) pp. 198—201

7.53 H. I. Smith, J. Melngailis, R. C. Williamson, W. T. Brogan: Proc. Ultrasonics Symp., ed. by J. deKlerk, Nov. 5—7, 1973, Monterey, Calif., IEEE, New York, Cat. No. 73 CHO 807-8SU, pp. 558—583

7.54 H.L.Garvin, E.Garmire, S.Somekh, H.Stoll, A.Yariv: Appl. Opt. **12**, 455—459 (1973)

7.55 H.L.Garvin: Solid State Tech. 31—36 (Nov. 1973)

7.56 P.Gloersen: J. Vac. Sci. Tech. **12**, 28 (1975)

7.57 M.Cantagrel, M.Marchal: J. Mater. Sci. **8**, 1711—1716 (1973)

7.58 R.C.Williamson, H.I.Smith: Elec. Lett. **8**, 401—402 (1972)

7.59 R.C.Williamson, H.I.Smith: IEEE Trans. MTT-**21**, 195—205 (1973)

7.60 R.C.Williams, V.S.Dolat, H.I.Smith: Proc. Ultrasonics Symp., ed. by J. deKlerk, Nov. 5—7, 1973, Monterey, California, IEEE Inc., Cat. No. 73 CHO 807-8SU, pp. 490—493

7.61 H.Bush, A.R.Martin, R.F.Cobb, E.Young: Proc. Ultrasonics Symp., ed. by J. deKlerk, Nov. 11—14, 1974, Milwaukee, Wisconsin, IEEE Inc., Cat. No. 74 CHO 896-1SU, pp. 494—499

7.62 H.M.Gerard, O.W.Otto, R.D.Weglein: ibid, pp. 197—201

7.63 C.Lardat: ibid, pp. 433—436

7.64 P.C.Myer, R.H.Tancrell, J.H.Matsinger: Proc. Ultrasonics Symp., Nov. 5—7, 1973, Monterey, Calif., IEEE Inc., Cat. No. 73 CHO 807-8SU

7.65 G.Carter, J.S.Colligon, M.J.Nobes: J. Mater. Sci. **8**, 1473—1481 (1973)

7.66 J.P.Ducommun, M.Contagrel, M.Marchal: J. Mater. Sci. **9**, 725—736 (1974)

7.67 M.Cantagrel: IEEE Trans. ED-**22**, 483—486 (1975)

7.68 P.Gloersen: Solid State Tech. (to be published)

7.69 P.Hartemann, M.Morizot: Proc. Ultrasonics Symp., ed. by J. deKlerk, IEEE Cat. No. 74 CHO 896-1SU, pp. 307—310;
P.Hartemann: Appl. Phys. Lett. **27**, 263—265 (1975)

7.70 R.V.Schmidt: Appl. Phys. Lett. **27**, 8—10 (1975)

7.71 P.Hartemann: Proc. Ultrasonics Symp., 1975, IEEE Cat. No. 75 CHO 994-4SU

7.72 M.Hatzakis: J. Electrochem. Soc. **116**, 1033—1037 (1969)

7.73 L.I.Romankiw, S.Krongelb, E.E.Castellani, A.T.Pfeiffer, B.J.Stoeber, J.D.Olsen: IEEE Trans. Mag. **10**, 828 (1974)

Subject Index

Applied Physics

A monthly journal

Board of Editors
S. Amelinckx, Mol. · **V. P. Chebotayev,** Novosibirsk
R. Gomer, Chicago, Ill. · **H. Ibach,** Jülich
V. S. Letokhov, Moskau · **H. K. V. Lotsch,** Heidelberg
H. J. Queisser, Stuttgart · **F. P. Schäfer,** Göttingen
A. Seeger, Stuttgart · **K. Shimoda,** Tokyo
T. Tamir, Brooklyn, N.Y. · **W. T. Welford,** London
H. P. J. Wijn, Eindhoven

Coverage
application-oriented experimental and theoretical physics:

Solid-State Physics   *Quantum Electronics*
Surface Physics   *Laser Spectroscopy*
Chemisorption   *Photophysical Chemistry*
Microwave Acoustics   *Optical Physics*
Electrophysics   *Integrated Optics*

Special Features
rapid publication (3–4 months)
no page charge for **concise** reports
prepublication of titles and abstracts
microfiche edition available as well

Languages
Mostly English

Articles
original reports, and short communications
review and/or tutorial papers

Manuscripts
to Springer-Verlag (Attn. H. Lotsch), P.O. Box 105 280
D-69 Heidelberg 1, F.R. Germany

Place North-American orders with:
Springer-Verlag New York Inc., 175 Fifth Avenue, New York. N.Y. 10010, USA

Springer-Verlag
Berlin Heidelberg New York

Topics in Applied Physics

Founded by H. K. V. Lotsch

Springer-Verlag
Berlin
Heidelberg
New York